TREATISE ON
INVERTEBRATE PALEONTOLOGY

Prepared under Sponsorship of
The Geological Society of America, Inc.

The Paleontological Society   The Society of Economic Paleontologists and Mineralogists
The Palaeontographical Society   The Palaeontological Association

Directed and Edited by
RAYMOND C. MOORE

Revisions and Supplements Directed and Edited by
CURT TEICHERT

Editorial Assistants, LAVON MCCORMICK, ROGER B. WILLIAMS

Part N
VOLUME 3 (OF 3)
MOLLUSCA 6
BIVALVIA

By †L. R. COX, N. D. NEWELL, D. W. BOYD, C. C. BRANSON, RAYMOND CASEY, ANDRÉ CHAVAN, A. H. COOGAN, COLETTE DECHASEAUX, C. A. FLEMING, FRITZ HAAS, L. G. HERTLEIN, E. G. KAUFFMAN, A. MYRA KEEN, AURÈLE LAROCQUE, A. L. MCALESTER, R. C. MOORE, C. P. NUTTALL, B. F. PERKINS, H. S. PURI, L. A. SMITH, T. SOOT-RYEN, H. B. STENZEL, E. R. TRUEMAN, RUTH D. TURNER, and JOHN WEIR

THE GEOLOGICAL SOCIETY OF AMERICA, INC.
and
THE UNIVERSITY OF KANSAS
1971

Library of Congress Catalogue Card
Number: 53-12913

I.S.B.N. 0-8137-3026-0

Text Composed by
THE UNIVERSITY OF KANSAS PRINTING SERVICE
Lawrence, Kansas

Illustrations and Offset Lithography
THE MERIDEN GRAVURE COMPANY
Meriden, Connecticut

Binding
TAPLEY-RUTTER COMPANY
New York City

Published 1971

Distributed by The Geological Society of America, Inc., P.O. Box 1719, Boulder, Colo., 80302, to which all
communications should be addressed.

EDITORIAL PREFACE

Changes in front-page materials given in Volume 1 of Part N (Bivalvia), as well as some other important notations, are brought together here in order that they may receive the notice of *Treatise* users most readily.

ADVISERS.—These now include EDWIN B. ECKEL, D. L. CLARK (The Geological Society of America), J. W. DURHAM, ADOLF SEILACHER (The Paleontological Society), W. M. FURNISH, D. M. RAUP (The Society of Economic Paleontologists and Mineralogists), W. T. DEAN, R. V. MELVILLE (The Palaeontographical Society), J. M. HANCOCK, M. R. HOUSE (The Palaeontological Association).

VOLUMES ALREADY PUBLISHED.—A revised edition of Part V, Graptolithina (xxxii + 163 p., 507 fig.), by O. M. B. BULMAN, directed and edited by CURT TEICHERT, was issued in December, 1970. It is available on orders sent to The Geological Society of America. This applies also to the present volume of Part N, Bivalvia, almost entirely the work of H. B. STENZEL; it is separately priced by The Geological Society of America.

CONTRIBUTING AUTHORS.—Additions not previously listed include: D. H. COLLINS, University of Toronto, Toronto, Canada; J. S. RYLAND, University College of Swansea, Swansea, Wales.

With deep regret the deaths of G DALLAS HANNA in November, 1970, and MARIUS LECOMPTE in August, 1970, must be recorded.

ABBREVIATIONS AND REFERENCES TO LITERATURE.—Additions to abbreviations and serial publications recorded in Volume 1 of Part N are given in the following lists.

Additional Abbreviations

Abh., *Abhandlung(en)*
Acad., Academy
adj., adjective
Afghan., Afghanistan
Alaisk., Alaiskan
Ark., Arkansas
Aug., August

Baluch., Baluchistan
Ber., *Bericht*

cah., *cahiers*
col., column
Colo., Colorado

Conn., Connecticut
Contr., Contribution(s)

Dec., December
dec., decade
Dépt., *Département*
Doc., Document
Duntroon., Duntroonian

expl., explanation

Feb., February
ft., foot, feet

Ga., Georgia
Govt., Government
gr., gram(s)

Hist., History, *Histoire*

Inst., *Institut*, Institute

km., kilometer

livr., livraison

Mass., Massachusetts
Moroc., Morocco
Ms., manuscript
max., maximum
mo., month

Naturhist., *Naturhistorische*
Nov., November

Oct., October
Opin., Opinion

Paleont., Paleontological, *Paleontologicheskikh*
Penin., Peninsula
Prog., Progress

ref., reference
Reg., Region
R. I., Rhode Island
Rupel., Rupelian

Sept., September
Sparnac., Sparnacian
sq., square
subsp., subspecies

Thanet., Thanetian
Torton., Tortonian
Transylv., Transylvania
Turkestan., Turkestanian

unpubl., unpublished
Uzbek., Uzbekistan

vert., vertical
Vierteljahrsschr., *Vierteljahrsschrift*
viz., *videlicet* (namely)

yrs., years

Additional Serial Publications

Acta Biologica Academiae Scientiarum Hungaricae. Budapest.
Akademie der Wissenschaften zu München, Abhandlungen, mathematische-physikalische Klasse, Sitzungsberichte.
[K.] Akademie der Wissenschaften, St. Petersburg [Akademiia Nauk SSSR, Leningrad].
Annales du Muséum. Paris.
Annales des Sciences Géologiques. Paris.
Argentine Republic, Ministerio de Agricultura, Direccion General de Minas y Geologia, Boletín. Buenos Aires.
Beiträge zur Geologischen Karte der Schweiz. Bern.
Geologo-Razvedochnoy Sluzhby tresta "Sredazneft," Trudy (Izdatelstvo Komiteta Nauk Uzbek. SSR). Tashkent.
Imperatorskago Mineralogicheskago Obshchestva, Zapiski. St. Pétersburg.
International Zoological Congress, 15th Session, Comptes Rendus. London.
Japanese Society of Scientific Fisheries, Bulletins [Nihon suisan-gakkai shi]. Tokyo.
Muséum d'Histoire Naturelle du Pays Serbe, Bulletin. Belgrade.
Natuurhistorisch Genootschap in Limburg, Publicatiës. Maastricht.
Natuurhistorisch Maandblad. Maastricht.
Natuurkundige Verhandelingen van de Hollandsche Maatschaappij der Wetenschappen te Haarlem.
New Zealand Institute, Transactions and Proceedings. Wellington.
Polskiego Towarzystwa Geologicznego w Krakówie, Rocznik. Kraków. (Société Géologique de Pologne, Annuaires.)
Prirodnjački Muzej Srpske Zemlje, Glasnik [Muséum d'Histoire Naturelle du Pays Serbe]. Belgrade.
Revue Suisse de Zoologie. Genève.
Schweizerische Geologische Gesellschaft. Basel.
Société Géologique de Pologne, Annuaires. Kraków.
United States Fish Commission, Bulletins. Washington, D. C.

NEW GENERA.—A long-standing policy of the *Treatise on Invertebrate Paleontology* has been the rejection of original publication of new generic taxa, although higher-rank units have been accepted wherever judged to be needful. Mainly this is explained by the adjudged lack of space for adequate diagnosis, descriptions, and illustrations of new genera. Only a very few exceptions have been allowed. The list now is enlarged on the basis of strong pleas submitted by the author and special notice of them given here. They include 1) *Gryphaea (Bilobissa)* STENZEL, n. subgen., p. *N*1099; 2) *Hyotissa* STENZEL, n. gen., p. *N*1107; 3) *Ilmatogyra* STENZEL, n. gen., p. *N*1119; and 4) *Neopycnodonte* STENZEL, n. gen., p. *N*1109.

ERRATA.—Errata and revisions to Volumes 1 and 2 of Part N are included on p. *N*1214.

RAYMOND C. MOORE

PART N
BIVALVIA

By †L. R. Cox, N. D. Newell, D. W. Boyd, C. C. Branson, Raymond Casey, André Chavan, A. H. Coogan, Colette Dechaseaux, C. A. Fleming, Fritz Haas, L. G. Hertlein, E. G. Kauffman, A. Myra Keen, Aurèle LaRocque, A. L. McAlester, R. C. Moore, C. P. Nuttall, B. F. Perkins, H. S. Puri, L. A. Smith, T. Soot-Ryen, H. B. Stenzel, E. R. Trueman, Ruth D. Turner, and John Weir

VOLUME 3

OYSTERS

By H. B. Stenzel

[Louisiana State University, Baton Rouge, Louisiana]

With additions by †L. R. Cox, British Museum (Natural History)

CONTENTS

[Volume 3, p. i-iv, N953-N1224]

INTRODUCTION

The oysters are very successful biologically, if one may measure success by number of individuals living and by territory they occupy. Several species are important as food and as raw materials for high-purity lime used in the chemical and cement industries and have been studied intensively. Fortunately therefore, many biological data are available.

In the geological sciences oysters are useful or important in stratigraphic zonation, in paleoecologic interpretations, in phylogenetic studies, and as the makers of sediments (oyster biostromes and self-sedimentation). Many of the problems concerning oysters in biologic and geologic sciences have defied analysis, notably their evolution, phylogeny, and classification, partly because problems were attacked either with purely neontological or purely paleontological methods. Evidently both sources of information have to be tapped. For this reason the interdisciplinary approach is used here to reach for solutions to these problems.

A major result is the discovery that the oysters, as commonly understood by various authors, are not one monophyletic family but very probably are diphyletic. In other words, the oysters consist of two families: 1) Ostreidae *sensu stricto,* herewith emended to exclude the three subfamilies Gryphaeinae VYALOV, 1936 (*emend.* STENZEL, 1959), Pycnodonteinae STENZEL, 1959, and Exogyrinae VYALOV, 1936, and 2) Gryphaeidae [*nom. transl.* STENZEL herein (*ex* Gryphaeinae VYALOV, 1936, *emend.* STENZEL, 1959)] to comprise the three above-mentioned subfamilies. As far as this chapter is concerned, the vernacular name "oysters" includes two separate families.

Modern summaries of the biology of a few commercially important living species have been published by GALTSOFF (1964), KORRINGA (1952-53), LOOSANOFF (1965), RANSON (1943a), and YONGE (1960). Extensive bibliographies covering living spe-

cies have been assembled by BAUGHMAN (1948), KORRINGA (1952-53), and RANSON (1952). Paleontological works covering larger groups of species are by COQUAND (1869), JOURDY (1924), SCHÄFLE (1929), VYALOV (1948a), WHITE (1884), and WOODS (1913). Living species were described by LAMY (1929-30) and SOWERBY (1870-71) in REEVE.

I wish to acknowledge with gratitude the generous support given before February, 1967, by Shell Development Company, a Division of Shell Oil Company, Houston, Texas, and by Louisiana State University since that time.

Extensive collections and good libraries are prerequisites to the present work. For many years both were not available to me and progress in this work was painfully slow. Were it not for the help and cooperation received from many colleagues at other institutions and in foreign countries the present work hardly could have been completed.

Thanks and appreciation are extended to all these who have given me a helping hand. They are too numerous to list all here in this space. However, I wish to thank specifically: Dr. R. TUCKER ABBOTT formerly of the Academy of Natural Sciences of Philadelphia; Dr. CARLOS C. AGUAYO of the Universidad de Puerto Rico; Dr. VAN REGTEREN ALTENA of Teyler's Museum in Haarlem, Netherlands; Dr. F. BROTZEN of the Swedish Museum of Natural History in Stockholm; Prof. R. M. CARTER of the University of Otago at Dunedin, New Zealand; Prof. W. J. CLENCH of the Museum of Comparative Zoology at Harvard University; the late Dr. L. R. COX of the British Museum (Natural History); Mr. DENNIS CURRY of Northwood in Middlesex, England; Mr. WILLIAM DEMORAN of the Gulf Coast Research Laboratory at Ocean Springs, Miss.; Prof. E. J. DENTON of the

Laboratory at Plymouth, England; Prof. MAXIM K. ELIAS of Oklahoma Research Institute, in Norman, Okla. and Dr. A. H. COOGAN, Kent State University, Kent, Ohio, who translated Russian literature; Dr. C. A. FLEMING and Mrs. A. U. E. SCOTT of the New Zealand Geological Survey; Mme. SUZANNE FRENEIX of the Institut de Paléontologie at Paris; Prof. P. FISCHER formerly of the École des Mines in Paris; Dr. PAUL GALTSOFF of Woods Hole, Mass.; Dr. ERNST GASCHE and Dr. HANS KUGLER of the Naturhistorische Museum at Basel; Dr. L. G. HERTLEIN of the California Academy of Sciences; Dr. E. G. KAUFFMAN and Dr. NORMAN F. SOHL at the U. S. National Museum;

Dr. JIŘÍ KŘÍŽ of the Charles University in Praha; Dr. D. F. MERRIAM of the State Geological Survey of Kansas; Dr. DON McMICHAEL of Australia; Dr. D. P. NAIDIN, Moscow State University, Moscow, USSR; Dr. OLIVER PAGET of the Naturhistorisches Museum in Wien; Dr. GILBERT RANSON of the Musée d'Histoire Naturelle in Paris; Mr. TAKEO SUZUKI of U.C.L.A.; Mlle. JULIETTE VILLATTE of the Laboratoire de Géologie at Toulouse; Mr. H. P. THESEN of the Knysna Oyster Company in South Africa; Dr. J. M. THOMPSON, Department of Zoology, University of Queensland, St. Lucia, Brisbane, Australia; and Dr. BOŘIVOJ ZÀRUBA of the Národní Muzeum in Praha.

ORIENTATION AND AXES

Common normal Bivalvia are capable of moving about on or in a substratum, and while they move forward their sagittal plane of symmetry stands vertical. Direction of movement, hinge axis, mouth-anus axis, and axis through the centers of the two large adductor muscles are parallel or close to parallel. For that reason, it is easy to orient these bivalves anatomically, that is, one can define anterior, posterior, dorsal, and ventral sides readily (Fig. J1).

Not so the oysters. They are permanently immobilized, pleurothetic on the left side, and have no direction of movement. The earliest known fossil oysters show no trace of the anterior adductor' muscle, which probably had disappeared in the ancestors before oysters had evolved, and the axis through the centers of the two adductor muscles cannot be drawn. The two remaining axes form a large angle between them and thereby they present conflicting evi-

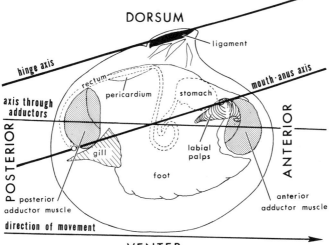

FIG. J1. Principal axes of orientation in common normal isomyarian bivalve as exemplified by *Astarte*, ×1.6 (after Saleuddin, 1965).

The RV has been removed and visceral mass is shown as transparent so that stomach, intestine, and

pericardium are visible. Positions of mouth and anus are shown by circles.

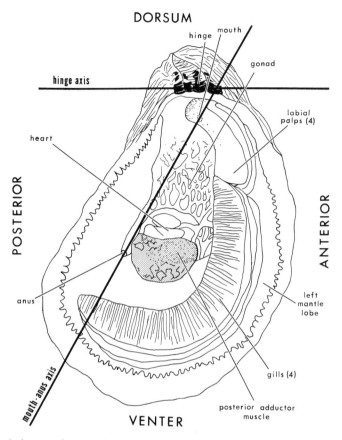

Fig. J2. Principal axes of anatomical orientation in oysters as exemplified by *Crassostrea virginica* (GMELIN, 1791), ×1.6 (Stenzel, n).

The oyster lies on its left side with RV removed so that soft parts are visible, position of mouth indicated, although it is hidden among labial palps; visceral mass shown partly transparent so that heart and gonad are visible.

dence for anatomical orientation of the animal. A choice must be made between these two axes as the means of orienting the mollusk (Fig. J2).

As concerns the animal and its survival, both axes have great functional importance, and one cannot claim that one is more significant than the other. Mouth and anus are part of the indispensable alimentary system, and the hinge is part of the mantle/shell protecting the animal. Rather, it is a question of practicability and general agreement.

If the hinge axis is the guide to anatomical orientation, that is, if it is equated with the anterior-posterior axis, hinge and umbonal region are the dorsum and the valve margins opposite the hinge are the venter of the animal. This is the orientation accepted in all conchological and paleontological literature and by some modern neozoologists, notably GALTSOFF (1964).

Since 1888 several neozoologists (JACKSON, 1888; LEENHARDT, 1926; PELSENEER in NELSON, 1938, p. 7; NELSON, 1938, p. 8; YONGE, 1953) have advocated using the mouth-anus axis instead. This orientation, or modifications of it, has been accepted by many authors since 1938. If one tries to apply the mouth-anus axis precisely, the hinge becomes not quite the anterior and the opposite valve margins become the ventroposterior side of the animal; in other words, many anatomical locations must

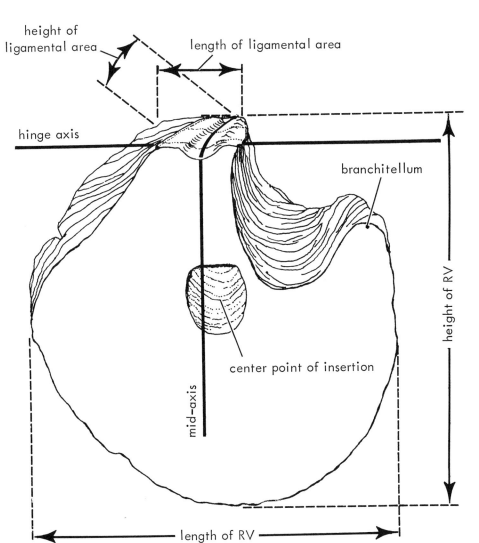

Fig. J3. Height, length, and mid-axis of an oyster as exemplified by RV of *Deltoideum*, ×0.9 (Stenzel, n).

The aragonite pad at insertion of posterior adductor muscle has been leached, so that growth layers of the outer ostracum show through the hole.

have rather long, awkward descriptive terms. In order to simplify them, many neozoologists call the hinge simply the anterior and the opposite valve margins the posterior. The fact that this or other such changes are deemed necessary makes this choice of an axis a poor one.

In fossil oysters, one can locate the mouth-anus axis approximately. The mouth is near the imprint of the Quenstedt muscle (see p. N965), and the anus is at the margin of the posterior adductor muscle imprint (Fig. J2, see J4). However, some imprecision is involved. These anatomical landmarks can be found only if the shells at hand are in excellent condition. On the other hand, the hinge axis can be located with ease and precision on both living and fossil oysters. In short, the hinge axis is the better choice as the definitive guide for locating the anteroposterior axis of the animal.

The terms length, height, and width are

defined as follows. **Length** is the largest dimension obtained by projecting the extremities of the shell onto the hinge axis. **Height** is the largest dimension obtained by projecting the extremities onto the mid-axis of the shell. **Width** is the largest dimension obtained by projecting the outline of the shell onto a line that is at right angles to both the hinge axis and the mid-axis.

The **mid-axis** is a straight line drawn in the commissural plane at right angles to the hinge axis and begins at the mid-point of the ventral margin of the resilifer (Fig. J3).

Length of the ligamental area is the length of its ventral border. **Height of the ligamental area** is measured along the middle of the resilifer following its curves from the umbo to the ventral margin (Fig. J3).

ANATOMY

CONTENTS

The oyster is composed of two · major interconnected subdivisions: 1) **Visceral mass,** containing all those organ systems that must receive protection, namely, the whole digestive, excretory, and reproductive systems and most of the musculature and much of the nervous and circulatory systems. 2) **Gills and mantle/shell.** The latter is composed of a hard portion, the shell, which protects all other organs and a soft portion, the mantle, which carries sense organs and lesser parts of the musculature and of the nervous and circulatory systems. The term mantle/shell, coined by YONGE (1953, p. 443), emphasizes that it is an integral organ unit consisting of both soft and hard parts just as the arm of a vertebrate consists of soft and hard parts (muscles, blood vessels, nerves, hair, fingernails, bones, etc.).

VISCERAL MASS

The visceral mass extends from the hinge to the big adductor muscle, which is part of it. In it are concentrated, in the smallest possible space, organs indispensable to survival of the individual and of the species itself. For protection the visceral mass is as far removed as possible from the open margins of the valves and is tucked behind the adductor muscle, which forms a stout pillar connecting the two valves giving protection to organs lying between it and the hinge (Fig. J4).

A **visceral pouch** enclosing the hairpin loop of the intestine and carrying the urogenital openings wraps around the anterior flank of the adductor muscle.

DIGESTIVE SYSTEM

A detailed description of the digestive system of *Ostrea edulis* LINNÉ, 1758, was given by YONGE (1926).

The **mouth** is at the dorsal end of the groove, and the tips of the gills are at the ventral end of the groove which is flanked by labial palps. There are two pairs of labial palps. Their outline is hatchet-shaped and slightly different in each genus. A short esophagus leads from the mouth to an elaborate stomach. The intestine leaves from the opposite side of the stomach, descends to the visceral pouch, forms a tight hairpin curve in the pouch, returns to form a wide loop around the stomach, and finally leads toward the dorsal flank of the adductor muscle. This terminal part of the intestine is the **rectum** and skirts the dorsal and posterior flanks of the adductor muscle. The anus is at the tip of a papilla at the posteroventral flank of the muscle, where it opens into the cloacal passage of the exhalant mantle chamber (Fig. J4).

In all living oysters, except the Pycnodonteinae, the rectum skirts the dorsal flank of the pericardium. In the Pycnodonteinae, the rectum passes through the pericardium and the ventricle of the heart, and the anal papilla projects more. These relative positions of rectum and ventricle in oysters are of great importance (Fig. J5).

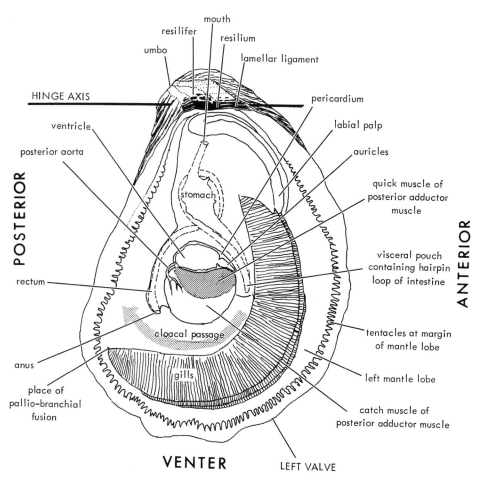

Fɪɢ. J4. Anatomy of an oyster as exemplified by *Crassostrea virginica* (Gᴍᴇʟɪɴ), ×0.8 (Stenzel, n).

This is the common commercial oyster of eastern and southern coasts of North America. The animal rests on its left side; RV and mantle lobe re-moved and visceral mass depicted as partly transparent; shaded arrow indicates exhalant water current in cloacal passage.

EXCRETORY SYSTEM

The **kidneys,** or organs of Bojanus, are two highly contorted tubules encased in a blood-filled sinus, located at the anterior flank of the adductor muscle. The blood coursing through this sinus loses its metabolic waste products to the tubules, which carry the liquid excreta out into the cloacal passage. The tubules discharge by way of a pair of small pores, each located near a genital pore in a short and slitlike uro-genital cleft, which is located on the visceral pouch.

REPRODUCTIVE SYSTEM

The reproductive system is quite simple, because the eggs of oysters are very small and comparatively simple. The reproductive system is composed of two **gonads.** Each is an arborescent system of ciliated ducts ending with a canal at the genital pore.

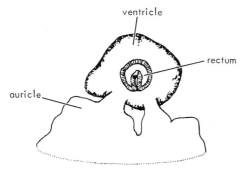

FIG. J5. Features of oyster anatomy. In living Pycnodonteinae rectum passes through both pericardium and ventricle of the heart, as shown in *Hyotissa hyotis* (LINNÉ, 1758) *forma imbricata* (LAMARCK, 1819), ×8 (after Pelseneer, 1911).

Because of this simplicity of structure, the oyster can and does change sex several times in its life. Changeover from male to female requires more vital resources and is slower than the reverse. As one sex phase fades away, the next one starts developing so that the two overlap. For example, during sex change from male to female, gonads carry simultaneously ripe spermatozoa and developing eggs and during the other change gonads carry residual spermatozoa and ripe eggs. In those genera of the Ostreidae that incubate their fertilized eggs and young larvae, one can find at certain times individual oysters that carry simultaneously unfertilized eggs, spermatozoa, and incubating larvae (MENZEL, 1955) (Fig. J6). Young oysters begin their sexually mature life as males. Oysters are protandric alternating hermaphrodites.

Among natural, crowded populations sex change is influenced by food (starvation slows the change), by prevailing temperatures (cold slows it), and by neighboring individuals (old females influence young oysters to remain in the male stage). Old, large oysters are nearly all females.

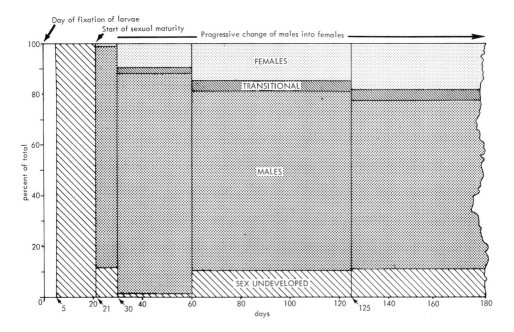

FIG. J6. Sexuality and sex change from male to female in young and small individuals of the incubatory species *Ostrea equestris* SAY, 1834 (=*O. spreta* D'ORBIGNY, 1846) from Aransas Pass, Texas, USA (Stenzel, n).

Five age groups are shown, respectively, 5-21, 22-30, 31-60, 61-125, and over 125 days old, counting from day they became attached to substratum. Transitional stage includes individuals carrying sperm and eggs, sperm and incubating larvae, or sperm, eggs, and larvae. (From data on 439 individuals given by Menzel, 1955.)

MUSCULATURE

The many muscles in the body of the oyster are discussed here in two sets: muscles inserted on the valves at one or both ends and muscles not inserted on the valves. The former set generally leave traces of insertion on the valves and, therefore, are more important in paleontology than the latter. In adult oysters the former set consists of three: 1) The large solitary posterior **adductor muscle,** 2) the pair of small **Quenstedt muscles,** and 3) a group of several anterior **pallial muscles.** The first two are in the visceral mass and are discussed below. The last are in the mantle/shell and are discussed under that heading (Fig. J7).

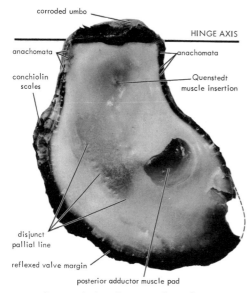

corroded umbo

HINGE AXIS

anachomata

anachomata

conchiolin scales

Quenstedt muscle insertion

disjunct pallial line

reflexed valve margin

posterior adductor muscle pad

FIG. J7. Muscle insertions on valves of oysters as exemplified by RV of *Saccostrea echinata* (Quoy & Gaimard, 1835), living attached to rock surfaces on coast of Queensland, Australia, ×0.73 (Stenzel, n).

ADDUCTOR MUSCLE

Adult oysters are **monomyarian;** they have only one adductor muscle. On the other hand, the full-grown larva, before it affixes itself to its substratum, has an anterior and a posterior adductor muscle like all other normal dimyarian bivalves and their full-grown larvae (see Fig. J38). During metamorphosis from larva to young attached oyster, or spat, the anterior muscle

of the larva shifts position and atrophies quite rapidly, leaving no trace whatsoever. This metamorphosis has been followed many times under the microscope so that there is no question that the solitary adductor muscle of the adult oyster is the posterior one.

Nevertheless, several authors have claimed either that some adult living oysters retain an additional adductor muscle, a vestigial anterior adductor muscle, or that some fossil oysters carry an imprint made by such an additional muscle. All these cases are in error and are based on misinterpretation of the Quenstedt muscles or their imprints on the valves.

JAWORSKI (1913) at first described *Heterostrea* (see Fig. J147) from the Jurassic of Peru as a new taxon of the Ostreidae that was supposedly heteromyarian, that is, had two unequal adductor muscles. Later he corrected himself (JAWORSKI, 1951) by discovering that this Peruvian fossil was incorrectly placed in the oysters and was really a *Myoconcha* J. DE C. SOWERBY, 1824 (see family Permophoridae, p. *N547*). Among the oysters, even among the most ancient ones described so far, there is no case known that would prove that anyone of them had a vestige of the anterior adductor muscle of their dimyarian bivalve ancestors.

The posterior adductor muscle is short and stout; it connects the two valves directly. Contraction of the muscle closes the two valves against the elastic expansion pressure of the ligament and keeps them shut for long periods of time, if need be. The muscle is differentiated over its whole length into two coalescent subdivisions (Fig. J4): 1) the **catch muscle,** white opalescent and opaque in live oysters, composing the ventral or distal part of the muscle and 2) the **quick muscle,** flesh-colored and semitranslucent, composing the dorsal or proximal part of the muscle. The catch muscle is a tonic muscle similar to the tonic muscle fibers in the walls of arteries and intestines of vertebrates. It contracts slowly and can maintain the tension for long periods, even against the pull of outside forces. The quick muscle is phasic; it reacts quickly but does not endure. The two perform together in a division of labor when they contract and close the valves. The relative positions of the two subdivisions in the adductor muscle are the same in all bivalves. The quick muscle is always in proximal position. (See also Fig. J36.)

The relative sizes of the two subdivisions differ from genus to genus. *Ostrea* has a larger catch muscle than *Crassostrea* and *Saccostrea; Hyotissa* has the smallest. The two subdivisions cannot be distinguished in fossils.

Quick action of this muscle is of utmost importance to the survival of the oyster and requires the best possible supply of blood. There are many blood-filled gaps (sinuses) between the fibers of the adductor muscle, and the heart is located next to the muscle so that the artery supplying it is as short as possible (Fig. J4).

Each end of the adductor muscle is inserted on a thin film or pad composed of aragonite, which is deposited by the end of the muscle directly onto the calcitic outer ostracum of the valve. As the animal grows in size the muscle adds more fibers at its ventral flank and simultaneously retreats at its dorsal flank so that the successive aragonite pads grow with delicate growth lines subparallel to the ventral margin of the pad. Thick-shelled individuals have the muscle imprints deeply sunk in the face of the valve. Thin-shelled ones have them flush with that face so that they may be difficult to discern.

The planes of the opposing muscle imprints are tilted so that they converge in distal direction (see Fig. J13). The tilt is greater in more deeply cupped valves and usually the left has greater tilt than the right imprint. In many oysters there is also a tilt toward the posterior side, making the two imprints converge on that side. This tilt depends on the position of the imprint on the valve and on the cuppedness of the individual valve. In many thick-shelled morphs of *Gryphaea, Texigryphaea,* and *Hyotissa* the tilt of the muscle imprint is emphasized by a buttressed ventral border of the imprint which is raised considerably above the adjoining face of the valve (see Fig. J73,*2a,* Fig. J74,*3b,c*).

The adductor muscle and its imprints are on the posterior part of the mantle shell. If one draws the mid-axis on a valve the adductor muscle imprint, or at least the geometric center of it, falls on the posterior side of the mid-axis (Fig. J3). Within this limit there is considerable diversification as to position of the imprint among various genera.

The adductor muscle is more nearly central on the valve in the larger species of *Ostrea s.s.* than in other oysters (see Fig. J23, J109). *Crassostrea* has it closer to the ventral and posterior valve margins (Fig. J8; see Fig. J44). Advanced species of *Flemingostrea* and *Odontogryphaea* have it unusually close to the ventral and posterior valve margins (Fig. J9). *Arctostrea* and *Rastellum* have the imprint very close to the hinge. In the Gryphaeinae and Pycnodonteinae the imprint sits closer to the hinge than to the ventral valve margin (see Fig. J74,*3b*).

The outline of the imprint, which is simultaneously the cross-sectional shape of the muscle, depends very much on the individual shape of the valve (Fig. J8). In other words, the imprints of various individuals of a species are not the same in a strict geometric sense, but their proportions referred to the valve shape are much alike, as is the case with the locations of the imprints on the individual valves.

Outlines of imprints make up two major classes (Fig. J9): the concave class and the convex or orbicular class. 1) The former has concave or nearly rectilinear dorsal margins and imprints are generally longer than high (Fig. J9,*2*). The following outlines belong here: a) crescentic or shaped like crescent moon, ends sharp and concavity deep; b) semilunar or shaped like half moon, ends sharp and concavity shallow (Fig. J7); c) comma-shaped, only one end sharp and concavity deep (see Fig. J133); d) ribbon-shaped, very long and shaped like a curved ribbon (see Fig. J124); e) reniform or shaped like a kidney, ends well rounded and concavity gentle (Fig. J8; see Fig. J23, J109,*1b*).

2) The convex or orbicular class of imprint outlines has convex dorsal margins and is generally as long as it is high. It is characteristic of the Gryphaeidae (Fig. J9,*1*) The following outlines belong here: a) circular, length and height equal (see Fig. J74,*3b;* J76,*1;* J83); b) vertical-oval, height exceeding length (see Fig. J75,*4d*); c) horizontal-oval, length exceeding height (see Fig. J76,*5*); d) oblique-oval, longer

axis placed oblique in valve (see Fig. J9, 1a).

The outlines and positions of the adductor muscle imprints are indicative of the interior anatomical topography of oysters. For this reason they are of utmost importance in the classification of fossil oysters. This was first pointed out and utilized in classification by STENZEL (1959). Some indications of this idea were noted in a special case of a living Australian oyster species by THOMSON (1954, p. 135).

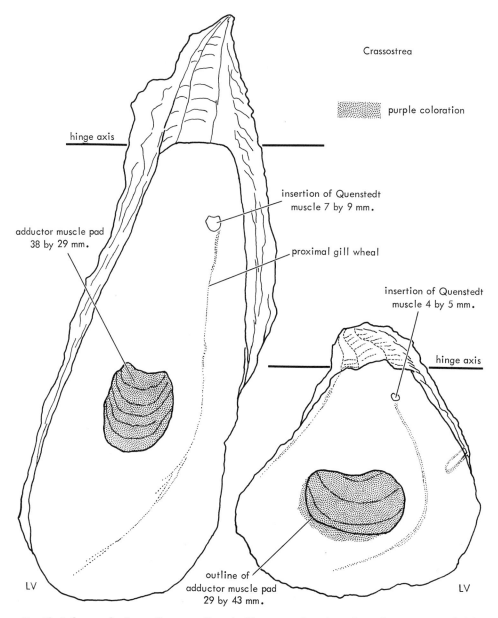

FIG. J8. Influence of valve outline on outline of adductor muscle pad, as shown by *Crassostrea virginica* from coast of New England, USA, ×0.55 (Stenzel, n. Specimens from Branford, Conn., and Providence, R.I., in Museum of Comparative Zoology).

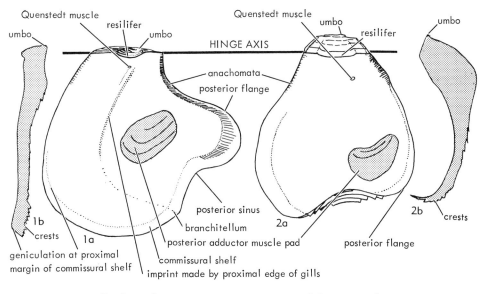

Quenstedt muscle · resilifer · umbo · umbo · Quenstedt muscle · umbo · resilifer · umbo · HINGE AXIS · anachomata · posterior flange · 1b · crests · 1a · geniculation at proximal margin of commissural shelf · commissural shelf · imprint made by proximal edge of gills · posterior sinus · branchitellum · posterior adductor muscle pad · 2a · posterior flange · 2b · crests

Texigryphaea   Odontogryphaea

Fig. J9. Outlines and anatomical positions of posterior adductor muscle pads in RV of oysters (Stenzel, 1959).

In *Texigryphaea roemeri* (Marcou, 1862) (family Gryphaeidae) (*1a,b*) from Grayson Marl (Cenoman.) of Texas outline of muscle pad is convex, orbicular, or oblique-oval. In *Odontogryphaea thir-* *sae* (Gabb, 1861) (*Gryphaea* homeomorph belonging in family Ostreidae) from Nanafalia Formation (Sparnac.) of Alabama (*2a,b*) it is concave or reniform.

In all studies of the adductor muscle imprints one should use only shells that have the aragonite adductor muscle pad and its surroundings perfectly preserved. In many fossil oysters the aragonite pad and the hypostracum are leached, leaving cavities that are either collapsed by rock pressure or are filled with secondary crystalline material or with sediment matrix (see Fig. J24-J26). In many Recent oyster shells the surroundings of the pad are damaged because the calcitic shell cover at the dorsal side of the imprint is very delicate. In all such cases the outline of the muscle imprints may be modified seriously.

QUENSTEDT MUSCLES

Quenstedt muscles are a pair of tiny muscles, each one attached at one end to a valve and ending with the other end among the adoral parts of the gills (Fig. J10). They were so named (Stenzel, 1963b) in honor of their discoverer. Quenstedt (1867, p. 598) noted a tiny muscle imprint between hinge and adductor mus-

cle imprint on shells of *Gryphaea arcuata* Lamarck, 1801, a common fossil of the European Liassic, but he misinterpreted it as the imprint of the vestigial anterior adductor muscle (Fig. J7-J9; see Fig. J23).

These muscles, or their imprints, have been noticed on many oyster species, fossil and living; it seems safe to assume they are present in all oysters. Being inconspicuous they have been forgotten or overlooked by many authors, only to be rediscovered several times. Dall (1880), one of the authors who rediscovered them, was the first to interpret them correctly as probably homologous with the pedal muscles of dimyarian bivalves. The best description is given in Herdman & Boyce (1899, p. 10-12, pl. 2), who dissected *Crassostrea virginica* (Gmelin, 1791 [1790]), a species living on the east and south coasts of North America (Fig. J10).

Each muscle is thickest at its insertion on the valve and thins as it extends inward. It rises at about 50° from its insertion, extends obliquely in ventral direction,

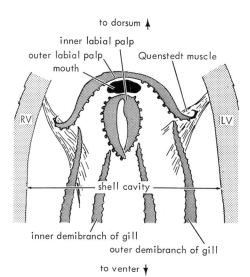

to dorsum ↑
inner labial palp
outer labial palp
mouth
Quenstedt muscle
RV
LV
shell cavity
inner demibranch of gill
outer demibranch of gill
to venter ↓

FIG. J10. Diagram depicting anatomical position of the two Quenstedt muscles in *Crassostrea virginica* (GMELIN), viewed looking in anteroposterior direction (mantle lobes and visceral mass omitted) (after Herdman & Boyce, 1899).

passes close to the bottom of the groove that separates the mantle fold from the outer labial palp, turns forward and splits to spread out mostly among the connective tissues between the inner and outer lamellae of the outer demibranch at the dorsal or adoral end of the gills. The insertion of the muscle on the valve is quite near the position of the animal's mouth. Thus Quenstedt muscle imprints come close to revealing in fossil species where the animal's mouth was located.

Although the two opposing Quenstedt muscles of an oyster converge toward the commissural plane and come close to each other, they do not merge. Each one is firmly attached to its valve and no mantle epithelium separates the muscle fibers from the valve as is also true for the posterior adductor.

Quenstedt muscles are modified anterior pedal muscles *(protractores* or *elevatores pedis),* now modified to perform a totally different function, because adult oysters have no trace of a foot left. Probably they adjust the positions of the dorsal, or adoral, ends of the gills and of the labial palps and aid in the transfer of food particles from gills to mouth. It is inadvisable to call them pedal muscles, because they no longer

function as such and because it is uncertain with which particular pair of the several pairs of pedal muscles in normal bivalves they are homologous.

Many authors have interpreted them as vestiges of the anterior adductor muscle. This interpretation is regarded as erroneous for several reasons. They are not continuous from one valve to the other and they are not rectilinear; the anterior adductor should be continuous and rectilinear. They reach the adoral tip of the gills, that is, the region ventral of mouth and esophagus; the anterior adductor of normal bivalves is wholly on the dorsal side of the mouth and esophagus. During evolution, it would have been a difficult feat for the anterior adductor to cross the alimentary canal from the dorsum to its venter.

CIRCULATORY SYSTEM

Detailed descriptions of the circulatory system were given by LEENHARDT (1926) for *Crassostrea angulata* (LAMARCK, 1819), a living oyster of southwestern Europe, and by AWATI & RAI (1931) for *Saccostrea cuccullata* (VON BORN, 1778), which is a complex superspecies widespread in the Indo-Pacific seas.[1]

The colorless blood of oysters does not contain any respiratory pigments, and the paths taken by the blood are rather uncertain because much of the circulation is through irregular intercommunicating blood sinuses, that is, blood-filled gaps between organs and tissues. Both conditions make for a rather inefficient circulatory system.

A blood-filled, thin-walled sac (**pericardium**) encloses the elongate heart, which has a pair of thin-walled contractile auricles leading into a single larger ventricle sheathed with muscles. The various genera show considerable differences in the configuration of the auricles. The pericardium is on the dorsal flank of the adductor muscle, where this stout muscle protects it from damage by predators and where it is close to organs that need the best possible supply of blood (Fig. J4).

Oxygenation takes place in the gills and in the mantle lobes. Freshly oxygenated

[1] This species was described at first (VON BORN, 1778) as *Ostrea cuccullata,* then the name was changed to *O. cucullata* by VON BORN, 1780. The earlier spelling must stand, although it is objectionable.

blood flows from the gills and mantle lobes to the auricles and passes through valves into the ventricle. Contraction forces the blood from the ventricle into two aortas. The very short posterior aorta leads to the adductor muscle so that this muscle, which must act quickly for the survival of the animal, has a copious supply. The anterior aorta has many branches leading to blood sinuses and also supplies other parts of the animal. Two of the distal branches of the anterior aorta are the large arteries that circle along the margin of each mantle lobe (right and left circumpallial arteries). Venous blood flows back to the kidneys on one side and the gills on the other. Efferent circulation then reaches the auricles.

Circulation of the blood is helped whenever anyone of the muscles contracts, because most of them are surrounded by blood sinuses. Each time one of the muscles contracts some blood is squeezed from a sinus into other blood vessels. Such circulation is mostly oscillatory because there are no valves to direct it. Oscillatory blood movements in the mantle lobes facilitate oxygenation of the blood. In this way the mantle lobes become important to the respiration of the oysters.

NERVOUS SYSTEM

The nervous system of *Crassostrea angulata* was described by LEENHARDT (1926) and that of *Saccostrea cuccullata* by AWATI & RAI (1931).

The central nervous system of an adult oyster consists of loops of nerve strands connecting two pairs of ganglia. The smaller, dorsal pair (cerebropleural ganglia) is between the bases of the labial palps, and each arises from the fusion of two ganglia in the larva. They innervate mouth, stomach, and dorsal parts of gills and mantle lobes. The larger, ventral pair of visceral ganglia is at the adductor muscle and innervates the adductor muscle, nearly all the visceral mass, and most of the gills and mantle lobes. This pair grows to be the chief nerve center of the adult; it receives the nerves coming from the sense organs and sends out nerves to those muscles that must react quickly to outside stimuli, that is, the adductor muscle and the pallial muscles.

It is noteworthy that in the oysters the,

cerebral ganglia are no longer connected to chief sense organs of the animal, because like all the Bivalvia oysters are encased by a protective shell and have lost their heads and cephalic sense organs. It is also noteworthy that, unlike other Bivalvia, the adult oysters have lost all traces of a foot, including the pedal ganglia. This is a consequence of their immobilization.

GILLS AND MANTLE/SHELL

GILLS

The two gills, or branchiae, together form an elaborate strainer structure partitioning the mantle cavity into an inhalant and an exhalant mantle chamber. They are placed so that water cannot flow from one chamber into the other without passing through countless tiny holes, or ostia, of the gills.

The tip of the gills is between the labial palps, a short distance ventral of the mouth. They curve convexly and subparallel to the valve margins and end at the **palliobranchial fusion**, where the end of the gills and the two opposing mantle lobes are united (Fig. J4).

The gills are made of two pairs of demibranchs, one pair at each side of the animal. Each pair of demibranchs is composed of four lamellae joined in form of a W. However, the W is asymmetrical, because the inner two of the four demibranchs are wider, and the left inner demibranch is the widest of all four. In the mid-line the two gills are united by their adjacent edges, and the proximal edge of the outermost lamella of each gill is united with the adjacent mantle lobe. In this fashion the gills separate inhalant from exhalant chamber, leaving the mouth, surrounded by its four labial palps, to open into the inhalant chamber.

The gills carry many microscopic cilia which whip the water driving it through the ostia from inhalant to exhalant chamber. The cilia establish the water current.

MANTLE

The mantle is the soft tissues of the mantle/shell, which completely encloses the adult oyster when it has shut the mantle/shell. On the visceral mass the mantle forms merely a surface membrane, but peripherally to this mass the mantle extends to form two lobes (mantle lobes or pallial lobes). The two mantle lobes reach to the valve margins and are completely free of each other, except at two places: 1) along the hinge, where they join in the mantle isthmus and form a sort of cowl over the mouth and the dorsal ends of the labial palps and 2) at the place of palliobranchial fusion located at the posteroventral edge of the valves.

The mantle lobes of an adult oyster are thick and somewhat solid. They lack the large blood sinuses found in those Bivalvia that have a large active foot and open their valves partly by hydrostatic pressure of the blood in these sinuses.

Peripherally the margins of each mantle lobe carry three parallel small folds. The inner one is the highly mobile pallial curtain, well supplied with muscles and blood sinuses, carrying tentacles at its free edge. The middle one is the tentacular fold and has many tentacles of two sizes, disposed in two rows at its free edge. The outer one is the shell fold and is separated from the tentacular fold by the periostracal groove; lacking tentacles, it lies directly on the surface of the valve. In the periostracal groove, at the base of the outer fold, are the periostracal glands exuding a sheet of conchiolin which is the base layer of the periostracum.

The outer face of each mantle lobe rests loosely on the inner face of the adjoining valve. Along the valve margin this space is closed off from the surrounding sea water by the periostracum, which breaks easily. Firm attachments between shell and soft parts are provided only at those few places where muscles are inserted on the valves, that is, at the insertions of adductor muscle, Quenstedt muscles, and some pallial retractor muscles. The narrow space between mantle lobe and shell wall is called the extrapallial space; it is filled with a mucus-laden liquid secreted by various gland cells in the mantle epithelium. It is in this space that deposition and growth of the shell take place.

MANTLE MUSCULATURE

The mantle lobes have many muscle strands, some arranged in concentric and others in radial patterns. They are well developed at the valve margins where the pseudosiphons form.

The radial muscle strands serve as pallial retractor muscles and extend within the mantle lobes from the vicinity of the visceral mass out to the margins of the mantle lobes, where some of them end in the tentacles on the mantle folds. Each of these muscle strands is surrounded by a pulsating blood sinus. Near the margins of the mantle lobes, in particular, the strands and their surrounding blood sinuses tend to show as radial ridges on the outer face of the mantle lobes. Some of the configuration of the shell at the periphery of the valves may be caused by these ridges (JAWORSKI, 1928).

The pallial line on the inner face of the valves of isomyarian bivalves is a continuous or nearly continuous line, of nearly uniform width, extending from one adductor muscle insertion to the other and is made from the coalescent insertions of many radial pallial retractor muscles. These muscles extend within the mantle lobe from the pallial line radially outward to margins of the lobe. When they contract, they withdraw the margins so that the two valves can be closed tightly without any part of the mantle lobes being caught between the two valves.

The anisomyarian bivalves have a disjunct pallial line. Their mantle retractor muscles are inserted on the valves in bunches of unequal sizes, producing a series of unequal muscle insertions lined up in a crude sequence from one adductor muscle insertion to the other.

In most oyster genera, there is not even a crude sequence of separate muscle insertions, because the retractor muscles begin and end within the mantle lobes. In *Ostrea* and *Crassostrea,* for instance, the mantle retractor muscles that extend to the anterior edge of the mantle lobe are few and thin; their proximal beginnings are near the gills. Those pallial retractor muscles that extend to the ventral edge of the mantle lobes are more numerous and thicker; their proximal beginnings are inserted within the mantle lobes around the ventral flank of the posterior adductor muscle.

Only in one genus (e.g. *Saccostrea*) are the mantle retractor muscles attached directly to the valves and a disjunct pallial line very similar to that of anisomyarian bivalves is developed. As far as this one feature is concerned, this genus is more archaic than other oyster genera, because it still retains the disjunct pallial line of the anisomyarian ancestors of the oysters (Fig. J7; see Fig. J105,*2b, 2d*).

INHALANT MANTLE CHAMBER

The **inhalant** (or incurrent or infra-branchial) **mantle chamber** is simpler in configuration and about four times larger than the other chamber. At the distal margin it is closed at times by the opposing pallial curtains of the mantle lobes along the anterior and ventral valve margins (Fig. J2, J4). On the right and left sides the two mantle lobes enclose it. On the proximal side are the gills, mouth and labial palps. It reaches in a roughly crescentic shape from the mantle cowl and mantle isthmus at the hinge to the palliobranchial fusion.

EXHALANT MANTLE CHAMBER

The **exhalant** (or excurrent or supra-branchial) **mantle chamber** has several special passages and is more complex than the inhalant one.

The gills contain four passages at their adoral end. Each is enclosed by the wall of the visceral mass on one side and by two lamellae of a demibranch on the other. The passages are closed off at the tips of the gills but in the other direction they widen into two larger passages, each of which is enclosed by the wall of the visceral mass and by the four lamellae of a gill. The two branchial passages open into the cloacal passage, which curves around the ventral flank of the adductor muscle and opens into the main part of the exhalant mantle chamber, lying along the posterior valve margin.

The **cloacal passage** receives excretions of the kidneys and genital products from the pair of urogenital clefts on the visceral pouch. The anus projects into the cloacal passage well downstream from the urogenital clefts.

Another **(promyal) passage** was first noted by KELLOGG (1892, p. 396-397). NELSON (1938) was the first to recognize its significance and to elucidate its function. It was named by him the promyal chamber and distinguished by him from the "opisthomyal exhalent chamber," which is here called the cloacal passage of the exhalant chamber. The position taken here is that the oysters, in common with many other

Mollusca, have only two mantle chambers, the exhalant and the inhalant one. All others are merely subdivisions, that is, passages.

The promyal passage is present in one group of living oyster genera and absent in another group (Fig. J11). Unfortunately not all living genera have been investigated in this respect so that information is not as securely founded as might be wished. The genera that have a promyal passage are the nonincubatory genera (e.g., *Crassostrea, Hyotissa, Neopycnodonte, Saccostrea, Striostrea*). Genera that lack a promyal passage are incubatory (e.g., *Alectryonella, Lopha, Ostrea*) (Fig. J11).

The promyal passage separates the right mantle lobe from the visceral mass over the region of the pericardium and between the adductor muscle and the hinge. It connects the branchial passages directly with the main part of the exhalant chamber and bypasses the cloacal passage, allowing an efficient disposal of the waste water current from the gills. There is no counterpart to it on the left side.

SENSE ORGANS

Adult oysters have but few kinds of sense organs: sensory cells in the epithelium of gills and mantle lobes, tentacles at the edges of two of the three mantle folds, and a pair of chemoreceptors in the cloacal passage. The last are asymmetrical; the one on the right side is the larger (AWATI & RAI, 1931, p. 73) and can be seen with the naked eye; the left one is often overlooked (PELSENEER, 1911, p. 27; ELSEY, 1935, p. 152). This heritable anatomical asymmetry indicates that oysters are permanently adapted to a pleurothetic attitude and cannot change from attachment on the left to attachment on the right side.

HARD PARTS

The hard parts of the mantle/shell are the **ligament** and the **shell,** which has two unequal valves. They consist of three organogenic substances: 1) calcite, an allomorph of calcium carbonate crystallizing in the rhombohedral system; 2) aragonite, another allomorph of the same substance, crystallizing in the orthorhombic system, and 3) conchiolin, a complex organic sub-

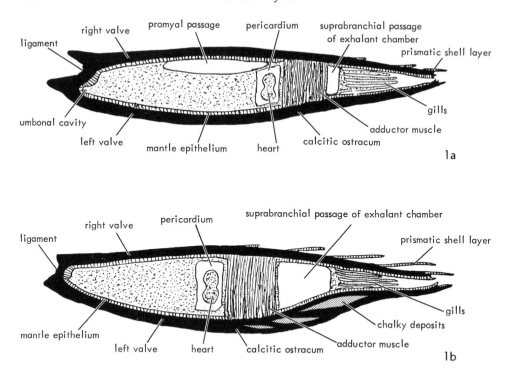

FIG. J11. Anatomical differences between nonincubatory and incubatory oysters, as exemplified by *Crassostrea* and *Ostrea* (from Korringa, 1956).

1a. Crassostrea, a nonincubatory oyster.
1b. Ostrea, an incubatory oyster.

[Diagrammatic sections are approximately along mid-line of shell from umbo at left to ventral margin at right.]

stance containing a mixture of mucopolysaccharides, polypeptides, and scleroprotein fractions. The three organogenic substances are found in different proportions in the various hard parts of the oyster.

Conchiolin is the chief or even exclusive component of the periostracum, the conchiolin scales, lamellae, and fringes of the valves, the ligament, and the delicate internal supports in the gills. Conchiolin is also present in small amounts within the calcareous parts of the valves. A filmy conchiolin network of microscopic thickness envelops each crystal of calcite or aragonite within the valves. As long as this conchiolin film remains untouched by decomposition, it protects the enclosed calcium-carbonate crystals from leaching. The structure of conchiolin matrix and calcium-carbonate enclosures is so tight that permeating water and invading bacteria are kept out and leaching or bacterial decay are retarded.

The amino acids of conchiolin in *Crassostrea angulata* (LAMARCK, 1819), a living species of southwestern Europe, are in percent of total protein: arginine 0.45, histidine 0, lysine 3.55, glycine 15.70, leucine 0.51, tryptophan 0, tyrosine 3.27, valine 0.95, cystine 0, and methionine 1.77. Those of *Ostrea edulis* (LINNÉ, 1758), a species living from Norway to Morocco, are: arginine 2.90, histidine 0.65, lysine 4.30, glycine 15.70, leucine —, tryptophan 0.47, tyrosine 3.05, valine —, cystine 0.98, and methionine 1.62 (ROCHE, RANSON & EYSSERIC-LAFON, 1951).

LIGAMENT

The ligament is strong and flexible within limits. It permits the bivalve to open or shut its two valves. Most Bivalvia have a hinge plate on each valve furnished with interlocking teeth and sockets. So does the oyster larva. However, the adult oyster lacks hinge plates, teeth, and sockets, and no

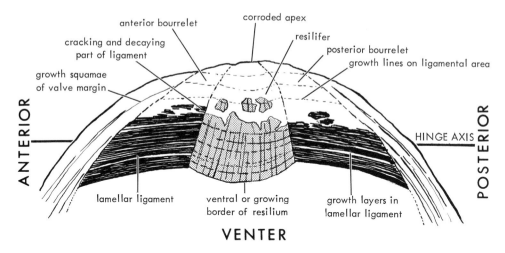

FIG. J12. Ligament and ligamental area on RV of *Ostrea edulis* LINNÉ (Stenzel, n).

vestiges of them are left. Thus the ligament is the sole hinge structure remaining on adult oysters.

The ligament of adult *Ostrea edulis* has been described by BERNARD (1896), RANSON (1940b), and TRUEMAN (1951). It is less modified than that of other genera, for example, *Exogyra,* and serves here as a model (Fig. J12).

It is alivincular and divided into three adjoining parts, lying side by side. 1) The **resilium** (=fibrous ligament of NEWELL, 1937, =inner layer of TRUEMAN, 1951) is in the middle and is fibrous, whitish gray, and semitranslucent. Its fracture pattern is fibrous. Its very fine fibers run parallel to the height of the ligament and about normal to the ventral, or growing, border of the ligament. The fibers are crossed by some light brown growth layers. It is made of naturally tanned protein complexes enclosing very fine white fibers of aragonite (STENZEL, 1962). These white fibers produce the whitish gray overall color. Under compressive stress it is strong, but under tension it is weak. 2) **Lamellar ligament** (of NEWELL, =outer layer of TRUEMAN, =tensilium of OLSSON, 1961, p. 41, 42, 51) composes the two flanking parts of the ligament and is nonfibrous, dark brown, and more translucent. Its fracture pattern is conchoidal. It lacks aragonite and fibrous

structure and is for that reason more translucent. It is made of quinone-tanned protein complexes and is strong under bending stresses.

Some authors have called the resilium the elastic ligament and the lamellar ligament the inelastic one. However, both are elastic and respond well to stresses. Their elastic responses are specialized to different kinds of stresses.

The valves, whenever they open or close, pivot around an axis (**hinge axis,** pivotal axis, cardinal axis) which is parallel to the ventral, or growing, border of the ligament and lies within the ligament itself, about 2 mm. from that border. In fossil oysters, which have lost the ligament through bacterial decay, the ventral border of the ligamental area suffices for locating the hinge axis. During movements of the valves, the resilium is compressed and the lamellar ligament is bent whenever the valves close. Both return to a less strained condition when the adductor muscle relaxes and the valves part. However, even when the muscle is relaxed in a living oyster, the ligament remains under some compression and the valves do not open very wide. Only if the muscle is cut do the valves open really wide.

The ventral border of the resilium juts out a little into the shell cavity; the project-

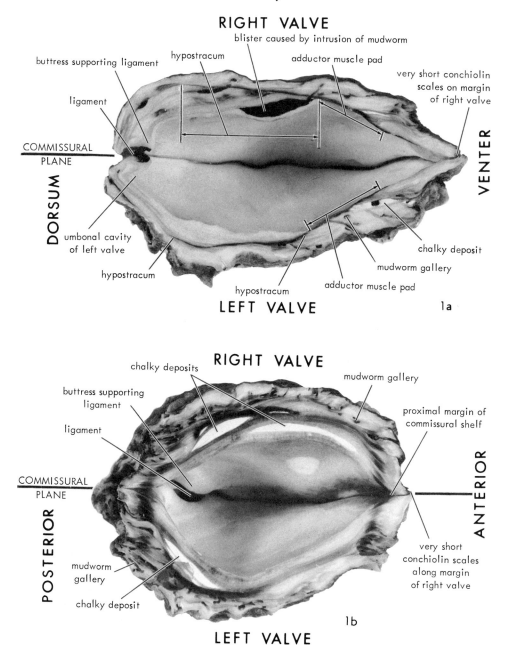

Fig. J13. Sections through shell of *Crassostrea virginica* (Gmelin), specimens from coast of Texas (Stenzel, n).

1a. Dorsoventral section, viewed looking toward anterior edge of shell, ×1.3.
1b. Anteroposterior section, viewed looking toward hinge which is visible at end of shell cavity.

[In this species a purple-colored layer lies immediately beneath the hypostracum so that the latter shows on the photographs as a dark line ending with the adductor muscle pad.]

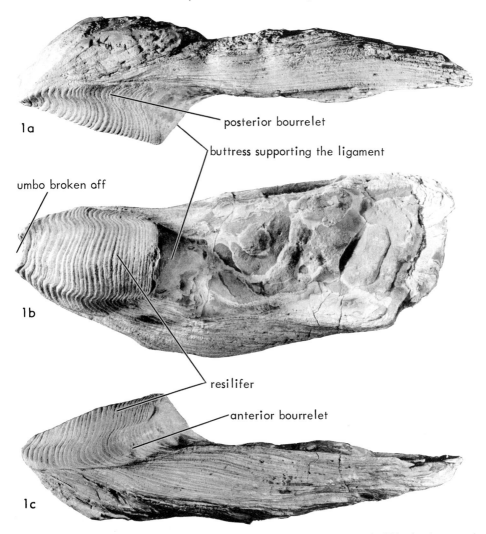

FIG. J14. Umbonal region of RV of *Crassostrea titan* (CONRAD, 1853), Mio., USA(Calif.), showing annual growth layers. [Specimen furnished by courtesy of U. S. GRANT and TAKEO SUSUKI, Univ. California (Los Angeles).]

1a-c. Posterior side, view from left, anterior side, ×0.4 (Stenzel, n). Specimen somewhat weathered and corroded, ventral part missing. Annual growth layers of ligamental area indicate that animal was 33 years old at cessation of growth.

ing part of the resilium of the right valve is strengthened by a supporting **buttress** rising from the right valve (Fig. J13). Through this arrangement the newest and best part of the resilium is placed in the inside of the shell cavity, that is, ventral to the hinge axis. There it becomes subject to compression greater than that of the rest of the ligament. Thus the resilium has improved leverage to open the valves whenever the adductor muscle relaxes. In those oysters which have a particularly great compressive load to overcome when they open their valves, the supporting buttress is outstanding and the ventral border of the resilium is strongly convex. Very good examples are found in the giant elongate oyster shells of certain species of *Crassostrea,* for example, *C. gryphoides* (VON SCHLOTHEIM, 1813) from the Miocene of the Tethyan region, and *C. gigantissima* (FINCH, 1823) from the upper Eocene of the southeastern and southern states of the USA (Fig. J14).

The ligament grows only at its ventral border, the only border in contact with the mantle. The epithelium of the mantle isthmus is replete with gland cells which secrete the conchiolin that makes up the ligament. Therefore, the growth layers of the ligament are parallel to its ventral border (Fig. J14). As the ligament grows, its older, more dorsally situated parts become exposed to sea water, crack, and fall out in pieces. The deterioration is brought about because continuous mechanical employment imposed upon it induces elastic fatigue failure and cracking and because there is invasion of bacteria from the sea water and bacterial decay of the organic substance.

Because the deterioration at its dorsal end cannot be stopped, the ligament must continue to grow at its ventral margin and must shift in ventral direction, even after the soft parts of the oyster have reached ultimate full size and the oyster no more enlarges the size of its shell cavity. Concomitantly the hinge axis shifts also. Growth and progressive ventral shift of the ligament leave a track imprinted on the apical parts of the valves (**ligamental area,** *aire ligamentaire,* or *talon*). The ligamental area has good growth layers parallel to the growing, or ventral, border and may grow to great height, up to 15 cm. or so (Fig. J14; see Fig. J30).

Growth of the ligament is recorded on the ligamental areas of the two valves, except that on many older oysters the umbones, and the ligamental areas with them, are destroyed by a combination of bacterial decay and chemical corrosion (Fig. J14). The ligamental area is divided into three parts, which are growth tracks of the three parts of the ligament. The **resilifer,** in the middle, is the track of the resilium, and its ventral end is the temporarily functioning seat of the resilium during life of the animal. On each side of the resilifer are the growth tracks of the lamellar ligaments, called here the **anterior** and **posterior bourrelets.**

In their simplest forms the resilifer and the two bourrelets are triangles. These shapes are usually greatly modified by individual growth and variability and by generic and subfamilial characteristics. Configuration of the ligamental areas is a feature useful in classification. In all the genera of the Exogyrinae except two, the ligamental area is curved in a consistently regular spiral. The height of the ligamental area of the Exogyrinae is very much greater than its length, and the posterior bourrelet is greatly reduced in length to form a narrow crest (see Fig. J26). The two exceptional genera of the Exogyrinae are like the others in their earlier growth stages, but in old age they cease to have a regular spiral and the posterior bourrelet widens.

The growth and size of the ligament is and stays adjusted to the work it has to perform in lifting the upper valve. The ligament must grow in size, that is, mainly in length of its ventral border, as the load increases. In the first two years growth is rapid; beginning with approximately the third year, growth is less rapid. At about the seventh year, the oyster has commonly reached its full growth and from then on the ligament no longer increases its length but continues to shift in a ventral direction. At that stage the size of the shell cavity and the weight of the soft parts no longer increase with age and anterior and posterior borders of the ligamental area remain parallel (see Fig. J104, J105). This sort of growth gives rise to the high, straplike ligamental areas especially noticeable on the left valves of elongate oysters of great individual age. Examples are *Crassostrea gryphoides* and other species of the same genus. As a rule, the ligament on the right valve does not shift ventrally as much as on the left valve, so that the ligamental area of the right valve is nearly always less high than that of the left. The contrast between the ligamental areas on the left and on the right valve is not large in the Exogyrinae and is very large in the rudistid-like morphs of *Saccostrea* (see Fig. J104, J105).

Multiple Resilifers

More than one resilifer on the ligamental areas of a shell has been found in several unrelated species of oysters. Such individuals, while they were alive, had two or more resilia functioning side by side. Such multiple resilifers are found locally in exceptional individuals among many more normal ones that have only one resilifer,

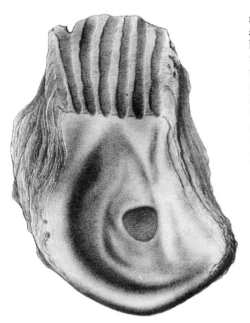

FIG. J15. Multiple resilifers and corroded umbo on RV of *Pernostrea luciensis* (D'ORBIGNY, 1850) (lectoholotype of *P. bachelieri* MUNIER-CHALMAS, 1864), Callov., Sainte-Scolasse-sur-Sarthe, France, ×0.7 (from Munier-Chalmas, 1864).

resilifers. In these shells the ligamental areas appear to be enormously elongated because the umbones are much corroded so that the early parts and the beginnings of the multiple resilifers are amiss (Fig. J15; see Fig. J78). LYCETT (1863, p. 108, pl. 34, fig. 1, 1a) was the first to surmise the true nature of such individuals when he described a right valve bearing nine resilifers as a monstrosity of *Ostrea wiltonensis* LYCETT from the Forest Marble (Bathon.) of Pound Pill near Corsham, Wiltshire, England. It had been found associated with several large and normal specimens of *O. wiltonensis*. Nonetheless, MUNIER-CHALMAS (1864, p. 75) bestowed the name *Pernostrea crossei* onto this extraordinary right valve and proposed the new genus *Pernostrea* for oysters with multiple resilifers. MUNIER-

that is, in exceptional variants of a given species.

One individual of *"Ostrea" wiedeyi* HERTLEIN, 1928, from the Vaqueros Formation (early Miocene) of San Miguel Island, off the coast of southern California, had a simple, single resilium in early growth stages. Rather suddenly, about 20 years before the death of this oyster, it split into two subequal resilia. The two resilifers were small and far apart at first; they lengthened gradually and came close to coalescing before the end (Fig. J16).

A specimen of *Lopha semiarmata* (BÖSE, 1906, pl. 2, fig. 1) from the Cardenas Formation (Maastrichtian) of Cardenas, State of San Luis Potosí, Mexico, has unusually thick shell walls and a ligamental area 8 cm. high. Its resilifer divided three times in its early growth, and near the end two of the four resilifers coalesced. The animal probably lived about five years.

In the Bathonian-Callovian deposits of England and France are a few localities at which some shells of a species have been found that have multiple (up to nine)

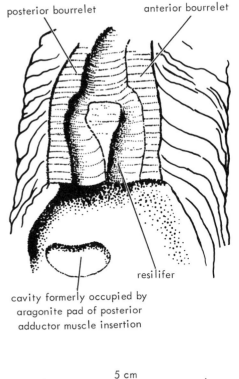

FIG. J16. Multiple resilifers in LV of specimen of *"Ostrea" wiedeyi* HERTLEIN (1928), from Vaqueros Formation (low.Mio.) of San Miguel Island, off coast of southern California, USA (Stenzel, n; specimen from Coll. San Francisco Acad. Nat. Sci.).

CHALMAS described altogether six "species" of the new genus *Pernostrea,* each based on minor differences in shell outline (see Fig. J78). Most were founded on one single valve each. Three of the "species" were from the same locality. MUNIER-CHALMAS probably coined the name *Pernostrea* because the genus *Perna* too has several resilifers and a greatly elongate ligamental area. He may have been influenced by Darwin's work to search for missing links to evolution.

FISCHER (1864, 1865, 1880/87) agreed with MUNIER-CHALMAS as to the biological validity of *Pernostrea* as a genus, but pointed out that *P. bachelieri* MUNIER-CHALMAS, for which MUNIER-CHALMAS had given *"Hab.? (collection de la Sorbonne)"* as the only locality information and for which MUNIER-CHALMAS had failed to give references to an earlier name and description of the species, was nothing else than *Perna bachelieri* D'ORBIGNY (1850, p. 341, no. 212), for which D'ORBIGNY had given the locality Sainte-Scolasse-sur-Sarthe, France. The identity of the two species names and inspection of the type specimen in Paris strongly suggest that MUNIER-CHALMAS used D'ORBIGNY's labeled type specimen for his description.

D'ORBIGNY had described the very same species from two separate localities as *Perna bachelieri* D'ORBIGNY (1850, p. 341, no. 212) and *Ostrea luciensis* D'ORBIGNY (1850, p. 315, no. 341), so that the latter name has page precedence and priority over LYCETT (1863), as accepted by STENZEL (1947, p. 180). Two of the types of *O. luciensis* at the Paleontology Section of the Museum National d'Histoire Naturelle in Paris are very poorly preserved specimens and each has four resilifers. They were described by BOULE (1913, p. 167, fig.) (see Fig. J78,*1*).

The true nature of these isolated cases of multiple resilifers was first proved and clearly stated by MERCIER (1929). The results were reviewed by ARKELL (1933, p. 47-48) and STENZEL (1947, p. 180). Although this Jurassic species is not the only one among the oysters in which arose variants furnished with multiple resilifers, it is remarkable that such variants of this species have been found in at least five places.

To summarize, the genus *Pernostrea* MUNIER-CHALMAS, 1864, as defined by this author and FISCHER is based on variants of a species which have two to nine resilifers. These variants are associated with and rarer than normal variants that have only one resilifer. The name of the species is *P. luciensis* (D'ORBIGNY) [=*Perna bachelieri* D'ORBIGNY, 1850 =*Pernostrea Bachelieri*+ *P. Heberti*+*P. Ferryi*+*P. Fischeri*+*P. Pollati*+*P. Crossei* MUNIER-CHALMAS, 1864].

SHELL

Oyster shells have two unequal valves which fit together without any gape whatsoever. Lack of a gape distinguishes the Gryphaeidae and Ostreidae from the Malleidae, some of which are quite similar and have been confounded with the Ostreidae. All oysters, except two or three species and some individuals of a few other species, are firmly attached by cementation either permanently or during their early postlarval life. They are invariably attached by their left valves. The shell is bilaterally asymmetrical and in many species highly asymmetrical. In general, most of the Gryphaeinae and Exogyrinae and many of the Pycnodonteinae are highly asymmetrical; the Lophinae approach bilateral symmetry the most (see Fig. J132, J139).

In most Bivalvia, the shell wall has four layers. The periostracum is at the outside. The hypostracum and the inner and outer ostracum are inside of the periostracum cover and compose the hard, light-colored shell substance of calcium carbonates encased in a bonding matrix of conchiolin. The **hypostracum** is deposited as a succession of aragonite pads on which the adductor and pallial retractor muscles are inserted. It divides the ostracum into an **inner** and an **outer ostracum.** The latter is between the periostracum and hypostracum; the former between the hypostracum and shell cavity. In most Bivalvia all these layers are composed of the allomorph aragonite. Oysters deviate from this scheme in many ways (Fig. J13; see Fig. J24, J25).

Chemical Composition

The shell is composed of calcium carbonates and minor amounts of conchiolin. The two allomorphs of calcium carbonate, aragonite and calcite, are present, the latter in greater abundance. Adventitious inclusions are found in cavities produced by boring parasites and consist of clay or silt. On heating, such organic components as conchiolin are burned and disappear so that oyster shells are an excellent source of high-purity lime for various industries.

A modern chemical analysis of the shell of freshly killed *Crassostrea virginica,* ground to pass a 100-mesh screen and dried at 110°C. before analysis, is as follows: loss 1000°C. (gravimetric) 43.79 and 43.77; SiO_2 (gravimetric), 1.1, 1.1, and 1.0; R_2O_3 (gravimetric), 1.1 and 1.0; CaO (gravimetric), 54.00 and 53.99; Mg (gravimetric), 0.19 and 0.20; Na (flame photometric), 0.34; K (flame photometric), 0.04; SO_3 (turbidimetric), 0.0; total halides calculated as chloride (turbidimetric), 0.02; metallic constituents (spectrochemical): Al, 0.03; Sb, 0.0; As, 0.; Ba, 0.002; Be, 0.000; Bi, 0.00; B, 0.0; Cd, 0.0; Cr, 0.001; Co, 0.001; Cb, 0.0; Cu, 0.007; Ga, 0.00; Ge, 0.0; Au, 0.0; In, 0.00; Fe, 0.09; Pb, 0.00; Mn, 0.01; Mo, 0.000; Ni, 0.001; Pd, 0.00; Ag, 0.0005; Na, 0.1; Sr, 0.; Ta, 0.00; Sn, 0.00; Ti, 0.00; W, 0.; V, 0.000; Zn, 0.0; Zr, 0.0.

Partial analyses of oyster shells, species unidentified, showed (Chave, 1954, p. 272) 1.30 and 1.80 (spectrographic) $MgCO_3$.

Several modern oyster shells have been analyzed for strontium (Asari, 1950, p. 157; Odum, 1957; Thompson & Chow, 1955, p. 32). The ratio of Sr:Ca (atoms) was found to be 1.01-1.90:1000. Fossil oyster shells (Odum, 1957; Kulp, Turekian, & Boyd, 1952, p. 707; Turekian & Armstrong, 1961, p. 1821) showed 0.81-1.83:1000. These authors were unaware that aragonite as well as calcite are present in oyster shells and that aragonite has a special distribution in them. Strontium is probably concentrated in the aragonite. It is, therefore, quite important to know which particular part of the shell was analyzed and whether the aragonite was leached in the fossil oysters or not. In the absence of any such information the data given by these authors are unreliable.

Periostracum

The periostracum consists exclusively or nearly so of conchiolin; it is very dark, horny in appearance, flexible, and resistant to leaching by sea water, but subject to bacterial decay. As long as it is intact it protects the underlying calcareous shell. It

originates from two sources. The primary source is the periostracal glands located at the base of the outer one of the three folds of the mantle lobe and within the periostracal groove, the groove between the middle and the outer fold. These glands produce the outer layer of the periostracum. This layer in the oysters is a very thin, delicate, soft, hyaline, and elastic sheet. Because it is so fragile it tears easily. It is so delicate and hyaline that it is difficult to find and has been overlooked by many.

As the periostracum is secreted by the glands the sheet folds over to join the edge of the calcareous shell and to close off the extrapallial space. Deposition of the calcareous shell material takes place within the mucus-filled extrapallial space and for that reason the periostracum must come to cling to the outer face of the calcareous shell.

Among the various oysters, *Hyotissa* has a poorly developed periostracum and *Striostrea* has a very strongly developed one (see Fig. J85, J107).

Conchiolin Scales and Lamellae

In many of the oysters, margin and outer face of the right valve carry imbricating conchiolin scales and lamellae. They are thin, flexible, elastic, delicate, olive-green to dark brown, semitranslucent to horny appearing. Each of them is 0.5 to 1 mm. thick at the base and tapers to its free end (Fig. J7).

They are not firmly attached to the valve and split off easily. In addition, they are exposed to bacterial attack and decay so that they do not last long. Even during the life of the animal, the older parts of the shell have already lost them, and fossil oysters are completely devoid of them. The left valve either is devoid of such conchiolin scales and lamellae or has a lesser development than the right valve (see Fig. J107).

They are freshest at the periphery of the right valve and form a fringe around the margin of the calcareous part of the valve. The fringe is flexible and elastic and bends to mold itself tightly onto the inner face of the slightly larger left valve providing a water-tight closure when the shell is shut.

Douvillé (1920) and Astre (1922, p. 163) pointed out that many fossil oyster species have a right valve that is smaller than the left, which extends much beyond the periphery of the other valve (Fig. J17; see Fig. J117,*1a*). They explained this phenomenon by postulating that the right valve of these species had originally a very extensive conchiolin fringe reaching to the periphery of the left valve but that this fringe was lost post-mortem. The discrepancy in the size of the calcareous portions of the two valves is great in *Ostrea* and closely related genera. Examples are *Cubitostrea wemmelensis* (GLIBERT) (1936, p. 60-63, pl. 2, fig. 4a) from the Wemmel Sands (early Barton.) of Belgium and *C. petropolitana* STENZEL & TWINING (STENZEL, KRAUSE, & TWINING, 1957, pl. 11, fig. 10) from the Stone City beds (middle Eocene) of Texas, which have quite small shells but have the right valve 8 to 15 mm. shorter than the left (Fig. J17). These conchiolin scales and lamellae can be called part of the periostracum.

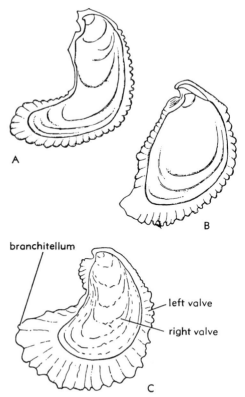

FIG. J17. Size discrepancy of calcareous portion of valves in fossil oysters revealed after decay of conchiolin fringe on RV, as shown in *Cubitostrea,* from Eoc. (Auvers.-Barton.), ×1 (Glibert, 1936).— *A. C. cubitus* (DESHAYES, 1832), Paris basin, France.—*B. C. plicata* (SOLANDER in BRANDER, 1766), Hampshire basin, England.—*C. C. wemmelensis* (GLIBERT, 1936), Sables de Wemmel, Belgium.

Prismatic Shell Layer

The flexible conchiolin scales merge at their bases into rigid calcareous layers of equal thickness. Although the two are continuous, the transition between the distal conchiolinous and the proximal calcareous part of a layer is quite abrupt. The calcareous part is composed of closely packed parallel prisms of calcite in a matrix of conchiolin. The prismatic layers are less opaque and contain more conchiolin than the other calcitic layers of the shell wall. If one dissolves the calcium carbonate with the aid of a chelating agent, the residue is a delicate coherent honeycomb of conchiolin (see Fig. J19, J20, J73, J109, J113).

Under the petrographic microscope (SCHMIDT, 1931) the prismatic layer of *Ostrea edulis* is seen to consist of long prismatic calcite units enveloped by a thin wall of conchiolin (Fig. J18, J19). The calcite units have the shape of parallel, curved, tapering prisms and have irregular polygonal cross sections of various sizes. The tapering curved ends are at an inclination of about 45° to the outer face of the prismatic layer and their taper points toward the outer face. As a prism unit is traced from its tapering outside point toward the interior of the shell cavity, it bends convexly away from the margin of

the right valve and in doing so it achieves gradually a steeper inclination angle to become nearly vertical with reference to the shell wall. Simultaneously the size of the cross section increases, because there is more space available for each prism unit. The prism units evidently did not grow at a uniform rate; for there are darker growth layers that transect uniformly a great many neighboring prisms and are parallel with the inner face of the layer.

Each prism is one crystal of calcite. Most of them have the main optic axis (axis c) approximately normal to the layer. However, the optic axis changes attitude gradually from one end of the prism to the other. There are a few prisms that have a different orientation of the optic axis (Fig. J19, J20).

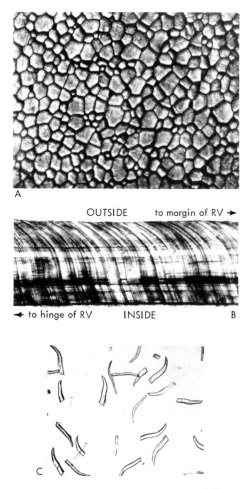

FIG. J18. Prismatic shell layer in oyster shells as exemplified by RV of *Ostrea edulis* LINNÉ (from Schmidt, 1931).—*A.* Horizontal thin section through layer, ×185.—*B.* Vertical thin section through layer, ×185.—*C.* Individual calcite prisms, ×48.

Size and shape of the prism units show considerable differences from genus to genus (RANSON, unpublished data). *Ostrea* has prisms that are more curved than those of *Crassostrea,* which are long, straight, sharp and needle-like. *Flemingostrea* has very long conspicuous prisms (Fig. J20; see Fig. J124). *Hyotissa* does not have any, but *Neopycnodonte* has. As a rule, genera that have well-developed conchiolin scales and lamellae also have well-developed calcite prism layers.

Several oysters that have prominent con-

chiolin lamellae and calcite prism layers also have many fine delicate radial riblets on a surficial layer of the right valve (Fig. J21). This layer is thin and delicate and flakes off readily. In fossil species only a few exceptionally well-preserved specimens retain it on the outer face of the right valve (see Fig. 107,*1c*). Commonly the layer is dark-colored because it is either made entirely of conchiolin or is a prismatic calcite layer rich in conchiolin. Riblets are restricted to this surficial layer, and the immediately underlying, more calcareous and lighter-colored layer shows a faint trace of them at best. Because of its delicate consistency the riblet-bearing layer is better preserved in very young and still fragile oyster shells and dehisces in older individuals. Old individuals may show riblets only on the marginal conchiolin fringes.

The riblet-bearing surficial layer is well developed in young individuals of *Striostrea margaritacea* (LAMARCK, 1819); this living species of East Africa is the type species of *Striostrea* VYALOV, 1936 (see Fig. J107, J108). KORRINGA (1956, p. 34-35) indicated that this layer is well developed on very young individuals, 17 mm. or less high, and makes it possible to distinguish them from young individuals of other, sympatric oyster species. DESHAYES (1860-66, v. 2, p. 106) described the riblet-bearing surficial layer of *Striostrea(?) lamellaris* (DESHAYES) in detail under the redescription of *Ostrea lamellaris* (DESHAYES, 1832) (in DESHAYES, 1824-37, v. 1, p. 372, pl. 54, fig. 3-4). [The type locality of this species is in doubt, and there may be some confusion as to it and other species between the two works of DESHAYES; the original description was based on only the right valve; it is probably from the Bartonian of the Paris Basin.] *Striostrea alabamiensis* (LEA) [=*Ostrea lingua canis* LEA, 1833 (=*O. pincerna* LEA, 1833, p. 91-93, pl. 3, fig. 71-73)] from the Gosport Sand (Auvers.), Claiborne Group, of Claiborne Bluff, Ala., has these riblets which were shown but not explained by HARRIS (1919, p. 8, pl. 1, pl. 2, fig. 1-3, pl. 6, fig. 1, not pl. 3-5, see particularly pl. 6, fig. 1).

Hypostracum

In the Bivalvia the hypostracum is deposited as a succession of thin pads of ara-

prismatic shell layer   ←to ventral margin of valve   prismatic shell layer

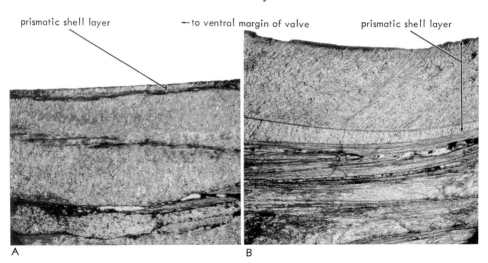

A   B

FIG. J19. Prismatic shell layer in oysters, ×21 (Stenzel, n).

A. Parlodion peel on outside face of RV of *Pycnodonte* sp. from Saratoga Chalk (Maastricht.) near Washington, Hempstead County, Ark. USA. [Prism layer is 0.1 mm. thick.]

B. Thin section on outside face of LV of *Flemingostrea subspatulata* (FORBES, 1845) from Ripley Formation (Maastricht.) of Braggs, Lowndes County, Ala., USA. [Prism layer is 1.35 mm. thick. Specimen by courtesy of ERNEST E. RUSSELL, Miss. State Univ.] [Peel and thin section made under supervision of OTTO MAJEWSKE, Shell Development Co.]

gonite, stacked up one on another, on the last one of which the adductor and various pallial muscles are inserted. The hypostracum of the oysters is modified very much, because they have only one adductor muscle and few, if any, pallial retractor muscles attached to the shell wall. Thus, the hypostracum of most oysters is reduced to only one element, the succession of pads on the last of which the posterior adductor is inserted. It might be better to call it the **adductor myostracum** (OBERLING, 1955a,b, 1964) (Fig. J13, J22).

The pad, or the muscle imprint, can be recognized on fresh shells of modern species by its slightly greater translucency, by its color, which in many species differs from that of the surrounding valve, by its luster, which is mostly brighter and more glistening, differing from that of the rest of the valve, and by its own special growth lines. The growth lines are very fine and subparallel to the ventral margin of the pad (Fig. J7; see Fig. J27, J31, J44).

The last pad and the entire adductor myostracum of the oysters are made of aragonite, whereas the surrounding parts of the valve are made of calcite (STENZEL,

1963b). The fact that the adductor myostracum is made of aragonite has been overlooked generally and statements that oyster shells are made entirely of calcite are erroneous (BØGGILD, 1930, p. 30; RANSON, 1939-41, p. 469; GALTSOFF, 1955, p 118).

As the oyster shell grows in size, the adductor muscle too must grow and the area of the pad must increase. In addition, the muscle must shift position in ventral direction in order to retain its relative location in the shell cavity. Therefore, additional aragonite is deposited on the ventral side of the old pad, new muscle fibers grow onto it, and simultaneously a dorsal part of the pad is vacated by the muscle and covered by freshly deposited calcite (Fig. J23). The first deposit at the dorsal side may be hyaline and exceedingly thin, but additional calcite soon piles up over it. For this reason the growth lines of the pad are parallel only to its ventral border and the hypostracum is buried in calcitic shell material except at the muscle.

If one were able to remove the entire adductor myostracum from a valve, one would see a thin curved sheet of aragonite

shells is prone to selective leaching, because it is made of aragonite and surrounded by calcite. Aragonite is the less stable of the allomorphs of calcium carbonate. Instead of the hypostracum, many fossil oysters have either a cavity (Fig. J26) or a filling composed of fine-grained sediment or a filling of secondary minerals, for instance, crystals of calcite (Fig. J22). In the second case, the hypostracum was leached selectively before the surrounding oozy sediment had a chance to solidify so that it could flow into the cavity (Fig. J24). This is common in chalks and marls. In the last case, the surrounding sediment either had already solidified before the leaching could take place or consisted of a sediment too coarse to penetrate into the cavity or the cavity was so narrow that the sediment could not penetrate. Wherever oyster-bearing sediments do not contain any aragonitic fossils (isomyarian bivalves, gastropods, scaphopods, ammonites, or others), it is most likely that selective leaching of aragonite has taken place, that the oysters have lost the hypostracum, and that the aragonitic fossil remains have disappeared. This is the case in most Mesozoic oyster-bearing deposits. Many authors in describing fossil oysters have either failed to recognize or failed to

Fig. J20. Prismatic shell layers in *Flemingostrea*. Posterior slope of exterior of RV of *Flemingostrea subspatulata* (FORBES, 1845) from Ripley Formation, U. Cret. (Maastricht.), near Braggs, Lowndes County, Ala., showing prismatic shell structures at the broken ends of several successive growth squamae, ×2 (Stenzel, n).

in the shape of a slender and acute triangle with a convexly curved base line. The base line is the ventral border of the last pad and the tip of the triangle is at the prodissoconch. The plane of the triangle is curved to conform with the dorsoventral curvature of the valve. In the Exogyrinae the adductor myostracum is curved in a spiral conforming with the spirality of the valve. If a shell of an exogyrine oyster is cut normal to the commissural plane the hypostracum is seen to spiral in and out of the cut surface and may be cut twice along its path (Fig. J24, J25).

The hypostracum of dead or fossil oyster

Fig. J21. Surficial riblet-bearing layer on exterior of RV of *Pycnodonte vesicularis* (LAMARCK, 1806) from "tuffeau de Maestricht," U.Cret.(Maastricht.), St. Pietersberg south of Maastricht, Netherlands, ×2.25 (Stenzel, n).

calcite replacement of myostracum · to inside face of left valve · calcite replacement of myostracum

↕ to venter

↕ to hinge

A · to outside face of left valve · B

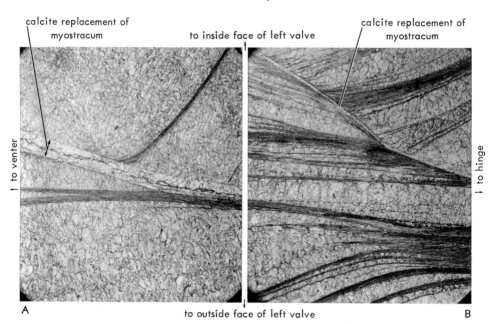

Fig. J22. Shell preservation in *Pycnodonte* sp., ×21.5 (Stenzel, n).

Sections of LV from Saratoga Chalk (Maastricht.) near Washington, Hempstead County, Ark., USA, showing recrystallized calcite replacement of aragonitic myostraca. Sections are cut along tracks of myostraca. The vesicles of the vesicular shell structure are filled completely with secondary calcite. [Thin sections made under supervision of OTTO MAJEWSKE, Shell Development Co.]—*A.* Adductor myostracum.—*B.* Myostracum of Quenstedt muscle.

state that the hypostracum is leached (compare PFANNENSTIEL, 1928, p. 396, figure 8) (see Fig. J74).

Chalky Deposits

Patches of soft, chalky-white, opaque, and porous shell material, so-called **chalky deposits,** are visible on the inner face of the valves in many oysters. They appear amorphous, can be cut and powdered easily by a knife, and consist of microscopic crystals of calcite (MEDCOFF, 1944; KORRINGA, 1951) (Fig. J13; see Fig. J109).

Their sizes vary greatly and their outlines are irregular, although rounded. Many larger patches are apparently confluent smaller patches. Their diameters vary from 1 mm. to several centimeters. Their cross section is flat and thin, tapering to their margins; at their margins many are puffed up above the level of the surroundings. Their distribution is erratic, and patches on the two opposing valves do not match. They are never large enough to

myostracum of Quenstedt muscle, covered by calcite

muscle pad of Quenstedt muscle · white fibers in resilium

myostracum of adductor muscle, covered by calcite · adductor muscle pad

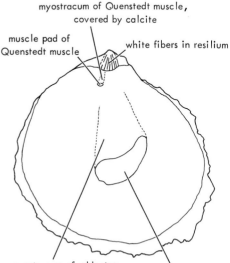

Fig. J23. Aragonite in shell of oysters as exemplified by RV of living species *Ostrea edulis,* ×0.7 (Stenzel, n).

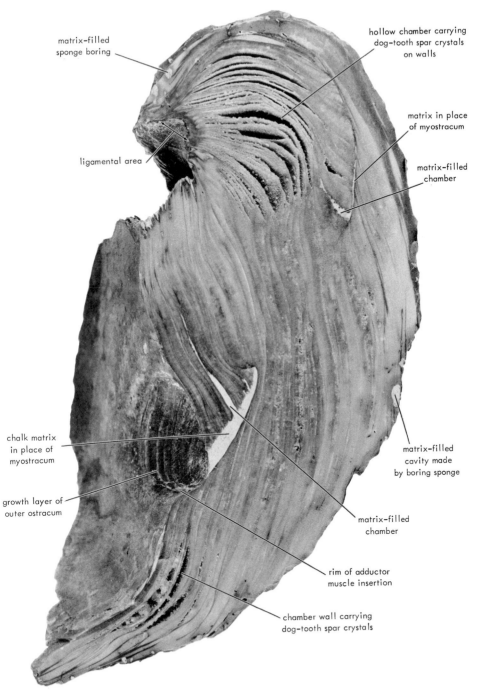

matrix-filled
sponge boring

hollow chamber carrying
dog-tooth spar crystals
on walls

matrix in place
of myostracum

matrix-filled
chamber

ligamental area

chalk matrix
in place of
myostracum

matrix-filled
cavity made
by boring sponge

growth layer of
outer ostracum

matrix-filled
chamber

rim of adductor
muscle insertion

chamber wall carrying
dog-tooth spar crystals

Fig. J24. Shell preservation in *Exogyra*, ×1.2 (Stenzel, n).

Section through LV of *E. (E.) erraticostata* STEPH-ENSON (1914), from Pecan Gap Chalk (Campan.) of Travis County east of Austin, Texas, USA. Section is cut at right angles to commissural plane through ligamental area and ventral valve margin. Chalk matrix was able to penetrate far into valve, because aragonite composing the adductor muscle pad and adductor myostracum had been leached before the surrounding chalk ooze had lithified. In-truded chalk matrix is cut twice because of spiral arrangement.

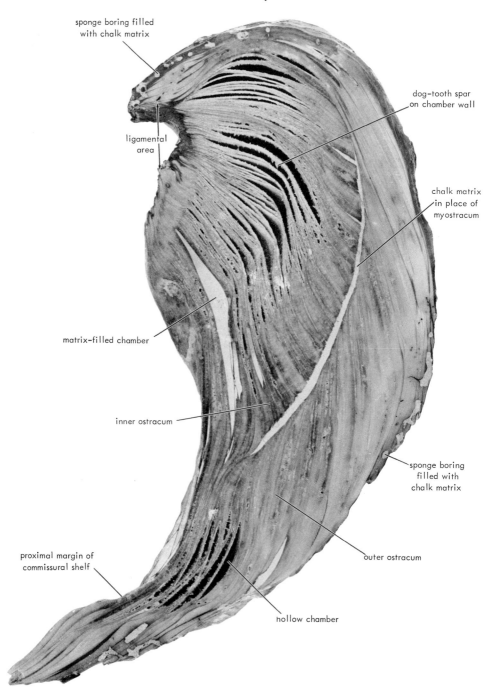

sponge boring filled
with chalk matrix

dog-tooth spar
on chamber wall

ligamental
area

chalk matrix
in place of
myostracum

matrix-filled chamber

inner ostracum

sponge boring
filled with
chalk matrix

proximal margin of
commissural shelf

outer ostracum

hollow chamber

Fig. J25. Shell preservation in *Exogyra,* ×1.2; section through same LV shown in Fig. J24, cut parallel to other section but in plane 2 cm. in front of it. Chalk matrix is intersected in only a single place, because the myostracum spiral is cut near its outside curve (Stenzel, n).

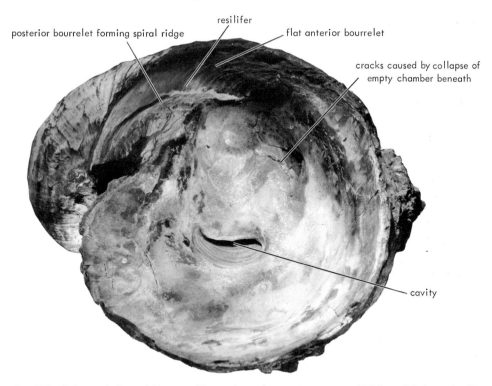

posterior bourrelet forming spiral ridge

resilifer

flat anterior bourrelet

cracks caused by collapse of empty chamber beneath

cavity

Fig. J26. Shell morphology of *Exogyra (Exogyra) erraticostata* STEPHENSON (1914), ×0.7 (Stenzel, n).

Specimen is LV from Brownstown Marl (Santon.) of Lamar County, Tex., showing 3 mm. thick cavity where aragonite of the adductor muscle pad and adductor myostracum has been leached; oblique view up into shell cavity. [Growth lines visible at muscle imprint are growth layers of outer ostracum of valve, beneath. Imprint outline is distorted because of oblique perspective. (Specimen courtesy of Mrs. GERRY KIENZLEN, Arcadia Park, Tex.)]

cover the entire face of a valve and they never are found in the aragonitic parts of the valve. They never are found within the aragonite pad under the adductor muscle, but they can be present at its ventral border where they gradually become covered by the growth of the pad. They form intermittently and become covered by other, more solid shell deposits; new chalky deposits may form again above.

In *Ostrea edulis* chalky deposits form in all possible places on the face of the left valve, but are least common under the visceral mass and most abundant under the exhalant chamber (ORTON & AMIRTHALINGAM, 1927). They are much less frequent and much less extensive on the right valve, but have the same average distribution.

In *Crassostrea virginica* the center of distribution of chalky deposits is under the visceral mass. They are less common under the cloacal passage and in the region surrounding the center. In the right valve they are less common and less extensive. In *Crassostrea cuttackensis* (NEWTON & SMITH, 1912) of Bombay, India, the pattern is similar (DURVE & BAL, 1961).

The mode of formation of chalky deposits is enigmatic, but hypothetical explanations have been proffered. ORTON & AMIRTHALINGAM (1927) assumed that chalky deposits are laid down rapidly in the larger, liquid-filled spaces of the extrapallial space. These larger gaps originated when the visceral mass had shrunk after spawning. Chalky deposits are concentrated in the region of the cloacal passage because there the mantle tends to sag away from the shell wall. However, this explanation would run counter to the fact that the upper, or right, valve has fewer and smaller

Bivalvia—Oysters

chalky deposits than the left. Whatever the correct explanation may be, chalky deposits do not require as much shell material as solid shell layers.

Chambers

Some oysters build their shells in such a way as to leave hollow chambers between solid layers repeatedly. The chambers have various sizes and are thin and lenticular; their tapered margins run into narrow crevices (Fig. J24, J25). During the life of the animal they are filled with liquid, which becomes entombed when the chamber is finished. If the liquid contains organic materials, such as mucus, the occluded liquid putrefies. Most chambers are in the left valve and under the visceral mass, that is, between umbo and insertion of the adductor muscle.

Thin walls enclosing the chambers break easily; accordingly, in fossil oysters the chambers are either crushed in by the load of overlying sediment (Fig. J24, J26) or are filled or partly filled with dog-tooth spar (calcite) or other secondary minerals. They become filled with fine-grained rock matrix if the chamber walls get broken while the surrounding sediment has not yet lithified and is able to invade the chamber.

Chambers occur in the Exogyrinae, all of which lived in euhaline waters and many on the sea bottom beyond shallow nearshore waters. In this subfamily, and perhaps even in all, changes of salinity cannot have been responsible for the formation of chambers within the shell.

Evidence indicates that chambers make appearance preferentially. The more deeply cupped shell shape becomes, the more readily chambers form. Extreme chamber building in the umbonal half of the left valve is a conspicuous and characteristic feature of those ecomorphs of *Saccostrea* that grow to a high conical shape and resemble rudists. Their right valve is simple, flat, operculiform and free of chambers.

Although one would expect chambers to be abundant in the umbonal half of the left valves of *Gryphaea* and its homeomorphs, they are not seen in this polyphyletic group of genera. Rather, the umbonal halves of the left valves in these separate and unrelated genera are filled solidly with shell material (see Fig. J73). The absence of chambers in this place appears to be an adaptation to their particular mode of life.

The origins of chamber construction are explained here by the following hypothesis. In oysters, the calcium carbonate for the shell is not derived from their food and does not become available to the mantle lobes of the animal by way of the digestive system. Rather, calcium ions enter the mantle lobes directly from sea water flowing through the mantle cavity, and travel from inner to outer face of the mantle lobes where they enter the extrapallial space and accumulate to become available for shell deposition. Therefore, growth of the oyster shell and additional deposition of shell layers is independent of the intake of food and of the growth of the fleshy parts of the animal. In some environments favorable to shell formation, shell building can outpace the growth of the internal, fleshy parts of the animal. In that case, the size of the shell cavity may become too large for the fleshy parts it harbors. The fleshy parts, however, must stay connected to the valve margins by the edges of the mantle lobes. In order to do so, the fleshy parts must move up toward the valve margins as these margins grow and must vacate a corresponding space within the shell cavity. The vacating is done preferentially in the far interior of the shell cavity, that is, between the visceral mass and the umbo of the left valve. After the vacating has been accomplished the mantle cover on the visceral mass deposits a shell layer forming the partition that closes off the cavity and a chamber is formed. During all these shifts the mantle isthmus must remain in position resting against the ligament, because this must grow and function continuously.

If this hypothesis is found to be correct, the chambers signify temporary rapid shell growth at the margins of the valve affected. Oysters that show numerous chambers (e.g., *Saccostrea*) live in environments where rapid shell growth is possible and where rapid growth of the valve margins to form deeply cupped shell shapes is an adaptive advantage to the survival of individual oysters. Crowding by competitors might produce such a situation.

Vesicular Shell Structure

Although vesicular shell structure must have been seen by LAMARCK, when he named the fossil oyster from the Chalk

(Campan.) at Meudon near Paris *Ostrea vesicularis* LAMARCK (1806a, p. 160-161; 1806b, p. 266), he did not describe or mention the structure, and over 100 years elapsed before its significance came to light.

The structure consists of shell layers resembling a sort of foam or spongy honeycomb built up as a network of paper-thin partitions enclosing countless small cavities or vesicles (Fig. J27, J28). The vesicles are taller than wide and have irregular polygonal cross sections of varying sizes. Normally the vesicles of museum specimens are empty. It is entirely unknown, however, whether the vesicles in living oysters are filled with gas or liquid.

The vesicular shell layers are separated from each other by normal horizontal lamellae of solid shell material. The lamellae vary from paper-thin to several millimeters thick, and the proportion of vesicular to lamellar layers is variable from species to species and from local environment to local environment in each species.

Extreme in this respect is *Pycnodonte (Pycnodonte) gigantica* (SOLANDER in BRANDER, 1766) from the Barton Beds (Barton.) of Barton Cliff, Hampshire, England. Most of its lamellar layers are paper-thin; some reach 1 mm. thickness. The vesicular layers are 2 to 9 mm. thick and predominate. Because the vesicles are air-filled in this fossil, the shell is extremely light. The other extreme is seen in *P. (Crenostrea) wuellerstorfi* (ZITTEL, 1864) from the Duntroonian (Oligo.) of New Zealand, which has few vesicular layers.

Vesicular shell structure is found only in the Pycnodonteinae and is an important character for identification of this subfamily (STENZEL, 1959, p. 16, 29-30). In most cases it is readily seen either on a cross break of the shell or on accidentally abraded parts of the valves or on their commissural shelf where it was growing. In some Pycnodonteinae of Cretaceous age, notably in *Texigryphaea,* the vesicles are completely filled with secondary crystalline calcite,

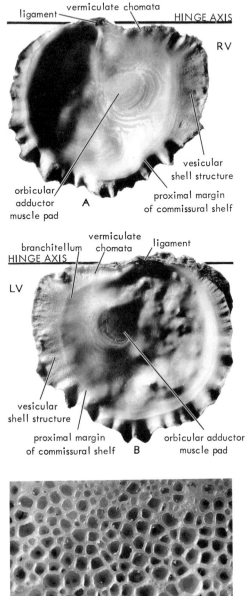

FIG. J27. Vesicular shell structure in *Hyotissa thomasi* (McLEAN, 1941), dredged alive off Key West, Fla. (Stenzel, n). [Specimen by courtesy of R. TUCKER ABBOTT, Philadelphia Acad. Nat. Sci.]— *A-B.* Interior faces of valves, ×0.7.—*C.* Part of commissural shelf, ×10. [Stage of growth is close to end of the construction of vesicles. Lumen of vesicles is partly constricted at the cross walls by hyaline beginnings of a foliated layer.]

to exterior surface

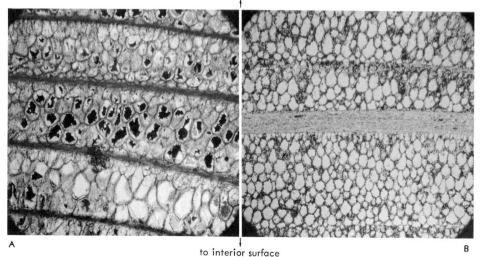

A   to interior surface   B

Fɪɢ. J28. Vesicular shell structure filled with secondary calcite, LV of *Pycnodonte (Phygraea) vesicularis* (Lᴀᴍᴀʀᴄᴋ, 1806), U.Cret., Lüneburg, West Germany, ✕20 (Stenzel, n).

[Alternating thin foliated and thick vesicular layers. Growth of shell wall progressed in direction of interior surface of valve and often began with a layer of small vesicles. Many vesicles have small beginnings and bulbous ends. Several vesicles are partly empty, and secondary crystals have grown from vesicle wall into the lumen.]

A. Thin section under crossed nicols of petrographic microscope valve cut in anteroposterior direction and at right angles to commissural plane.
B. Liquid peel in ordinary light, valve cut at right angles to section shown in *A*, that is, in dorsoventral direction and at right angles to a commissural plane.

making it quite difficult to recognize the structures with the naked eye (Fig. J28). In such cases one needs to study thin sections under the microscope. For that reason Sᴛᴇɴᴢᴇʟ failed to observe it in *Texigryphaea*, although Rᴀɴsᴏɴ (1939-41, p. 64) had already indicated the presence of vesicular shell structure in some species that later were placed in *Texigryphaea* by Sᴛᴇɴᴢᴇʟ (1959, p. 22-29).

Lᴀᴍᴀʀᴄᴋ completely ignored the vesicular shell structure, except that he named one species *Ostrea vesicularis*. He probably never used this structure to recognize the species, for he described the same species from the same type locality under two different generic names and under three different species names (*Ostrea vesicularis* Lᴀᴍᴀʀᴄᴋ, 1806; *O. deltoidea* Lᴀᴍᴀʀᴄᴋ, 1806; and *Podopsis gryphoides* Lᴀᴍᴀʀᴄᴋ, 1819; see Cʟᴇʀᴄ & Fᴀᴠʀᴇ, 1910-18, pl. 14, fig. 46-47; pl. 26, fig. 94-95). He probably did not attribute much importance to this structure. The first to describe the structure was Dᴇғʀᴀɴᴄᴇ (1821, p. 23). Dᴀʟʟ

(1898, p. 676-677) guessed that rapid growth produced vesicular shell structure and asserted that vesicular shell structure rarely attains constancy sufficient to entitle it to systematic significance on even as low as the species level and failed to notice it in several species he described. At first Dᴏᴜᴠɪʟʟᴇ (1907, p. 100, fig. 3-4) seems to have confounded vesicular shell structure with prismatic shell layers and with chalky deposits, naming the three *"couches prismatiques"* or *"structure prismatique."* However, he became later on the first (Dᴏᴜᴠɪʟʟᴇ, 1936b) to see the importance of it as a definitive supraspecific characteristic. Rᴀɴsᴏɴ (1939-41; 1941) followed Dᴏᴜᴠɪʟʟᴇ and greatly expanded our knowledge of this structure firmly establishing its value in classification.

Dᴏᴜᴠɪʟʟᴇ (1936b), Rᴀɴsᴏɴ (1939-41; 1941), and Nᴇsᴛʟᴇʀ (1965) believed that the vesicular shell structure is some form of chalky deposits, that is, that the two are homologous. If that were true, the distributions of the two should be alike. The chalky deposits have their center of distribution under the visceral mass, when they are freshly formed, but the vesicular structure grows chiefly on the commissural shelves.

Although several living species have vesicular shell structure, nothing is known of its origin and growth. The following explanation is entirely hypothetical. All deposits of the calcareous shell wall, including the vesicular structure, grow in the extrapallial space, which is the mucus-filled narrow gap between mantle lobe and calcareous shell wall. The spongy honeycomb must grow in that mucus, and there must be a mechanism by which growth is regulated so that only thin honeycomb-like walls are formed. The most likely process is that the mantle lobes at times release tiny gas bubbles into the extrapallial space. These gas bubbles accumulate in the extrapallial space, because they cannot escape. Gradually they become crowded to form a sort of foam. The gas bubbles crowd each other, but the viscous mucus forms bubble walls that do not rupture easily. By crowding, the bubbles lose their spherical shapes and become irregularly polyhedral, and their cross sections become polygonal. Calcitic shell material crystallizes in the mucus gradually replacing the mucus so that the vesicular calcitic shell structure is a replacement replica of the original foam composed of gas bubbles in mucus. Crystallization and deposition of calcitic shell material begins at the distal flank, at the shell wall, which acts as a seed bed of crystal nuclei, and progresses toward the proximal side.

Fingerprint Shell Structure

In describing a new living species, *Ostrea cumingiana,* Dunker in 1846 (Dunker in Philippi, 1845-47, *Ostrea* p. 82), was the first to describe a unique shell structure, so far known only from that species: "Besides, one notices here and there partly straight, partly swirled sinuous or irregularly dichotomizing somewhat raised lines of brown coloration, which seem to be characteristic of this species . . ."[translated from German].

The fingerprint shell structure is visible in patches on the internal face of the valves. The patches vary in size and are situated on either valve between hinge and adductor

muscle and between adductor muscle and valve margin. They consist of countless narrow curving threads, about 0.3 mm. wide, which are unbranching in some places, confluent in others, and dichotomous in still others. The threads consist of light brown, translucent material, presumably conchiolin. They form a set in a plane conforming to the shell layers. Within each set the threads are equidistant from and parallel with each other in a swirling pattern resembling in size and arrangement the raised lines on a man's fingers. The thread-bearing layers are visible in patches only, because the layers are mostly covered by shell layers free of threads (Fig. J29; see J134, J135).

The fingerprint structure has been seen in only one species, *Alectryonella plicatula* (Gmelin, 1791) (Fig. J29), living in the Red Sea, Indian Ocean, southwestern Pacific, and around southern Japan, that is, in the seas of the tropical and arid climatic belts. Junior synonymous names of this species are: A. *plicatula* (Lamarck, 1819), =A. *cumingiana* (Dunker, 1846), =A. *lactea* (G. B. Sowerby, 1871). It is the type species of *Alectryonella* Sacco, 1897. The origin of this structure is enigmatic.

Interior Topography

Several more or less well-circumscribed features are visible on the inner faces of the valves. Their importance lies in that they are the only means we have to reconstruct parts of the soft anatomy of extinct oysters. The features are (Fig. J7, J8): 1) a pair, right and left, of imprints at the two insertions of the large adductor muscle (see p. N962); 2) a pair of imprints at the insertions of the two Quenstedt muscles (see p. N965); 3) a pair of serial imprints left by the insertions of several pallial retractor muscles (see p. N968); 4) a pair of imprints left by the bulk of the gills (see next paragraph); 5) a pair of commissural shelves (see p. N990); 6) marginal ridgelets and pits, or **chomata** (see p. N990-N994); 7) a buttress supporting the resilifer of the right valve (see p. N974); 8) an umbonal cavity beneath the ligamental area of the left valve (see p. N994); and 9) parasite-induced features (not discussed here).

The gills are somewhat bulky and firmer than the neighboring fleshy parts of the animal, because they are folded over eight-fold and internally stiffened with conchiolin fibers. By their bulk they exert some

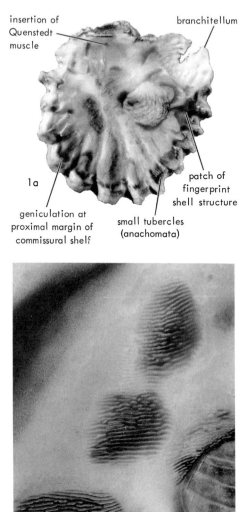

FIG. J29. Fingerprint shell structure on inside face of RV of *Alectryonella plicatula* (GMELIN), living species from Nosi-Bé, Madagascar. [Specimen by courtesy of R. TUCKER ABBOTT, Philadelphia Acad. Nat. Sci.]

1a. RV inside face showing patches of fingerprint shell structure, ×0.7.

1b. Same valve showing fingerprint shell structure in vicinity of adductor muscle pad, ×2.1.

pressure onto the mantle lobes to either side of them and thereby they tend to reduce locally the width of the extrapallial space and the volume of fluid contained therein. Wherever such a pressure is present, the extrapallial space locally cannot furnish enough calcium carbonate and the shell wall there fails to keep up with carbonate deposition elsewhere. The result is a shallow crescentic basin in the form of the gills on the face of the valve. The basin is bounded distally by the inner curb of the commissural shelf and proximally by a faint wheal or curb or by a more pronounced fold, first noted by REIS (1914) and called *Branchialfältchen* or *Hauptkiemenschwelle* (PFANNENSTIEL, 1928). This **proximal gill wheal** (Fig. J8, J9; see Fig. J113) sweeps from the vicinity of Quenstedt muscle in a simple curve past the anterior border of the adductor muscle imprint to the palliobranchial fusion. If the shell has a projecting tip end, or branchitellum, at its posteroventral end, the end of the gill wheal points toward it.

The **commissural shelf** is a flattish band along the periphery of the valve along which opposing mantle lobes touch each other, when the shell is closed and the mantle lobe margins are not withdrawn toward the interior of the shell cavity. A circumferential curb delimits the commissural shelf proximally and encircles the deeper, main part of the shell cavity. The two opposing shelves of a shell and their curbs are better developed in the dorsal than in the ventral half and are the least distinct at the posteroventral part of the shell. Shelves are better developed and have neater curbs in the Exogyrinae, Gryphaeinae, and Pycnodonteinae than in other oysters (Fig. J27; see Fig. J83, J90). In many of the Exogyrinae the commissural shelf of the right valve is reflexed strongly along the anterior valve margin; there the circumferential shelf curb is exaggerated into a sort of internal keel paralleling the right anterior valve margin.

Small ridgelets and pits opposing them are near the hinge on the commissural shelves of many oyster genera. The ridgelets, called here **anachomata,** are on the right and the pits, called **catachomata,** on the left valve. Collectively they are called **chomata.** Although they are mostly in the

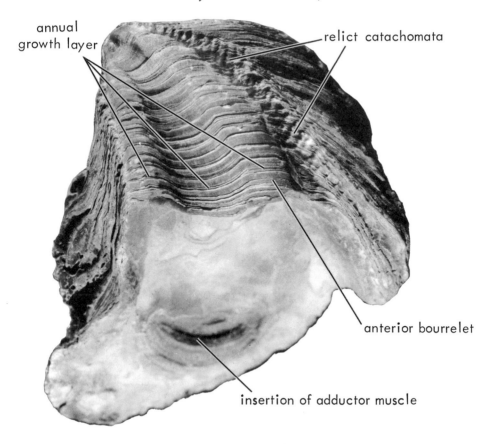

annual growth layer

relict catachomata

anterior bourrelet

insertion of adductor muscle

Fig. J30. Annual growth layers on ligamental area of LV of *Ostrea (Turkostrea) duvali* Gardner (1927), ×1.4 (Stenzel, n).

Specimen is from Caldwell Knob Oyster Bed, Wilcox Group, low.Eoc., of Moss Branch, Bastrop County, Texas. [The oyster is estimated to have been about 15 years old. Pits or catachomata flanking the margin of the anterior bourrelet are relicts of former growth stages. Ligamental area is bent back from plane of valve commissure.]

vicinity of the hinge, they are neither homologous nor analogous to the teeth and sockets on the hinge plates of normal dentate Bivalvia. For that reason, the terms dents or denticles, used by many authors, are inappropriate and downright misleading (Fig. J30, J31; see Fig. J113, J127).

If the anachomata were homologous to the teeth, they would have to have the same or a similar derivation, that is, they would have to grow out of the larval teeth and hinge structures on the prodissoconchs as the teeth and sockets of adult normal Bivalvia do. Repeated studies on individual ontogenies of oysters *(Ostrea edulis* and other species of *Ostrea s.s.)* from the larval to the young adult stage have shown that the chomata are new, postlarval features independent of any hinge features on the prodissoconchs. In normal dentate Bivalvia true teeth and sockets are both found on each valve, but the anachomata of oysters are only on the right and the catachomata only on the left valves. Also, some oysters have chomata all around the peripheries of their valves, including valve margins directly opposite the hinge. True teeth are never found on valve margins opposite the hinges in the Bivalvia.

Functionally also, they do not correspond to teeth and sockets, that is, they are not analogous. True teeth and sockets are inter-

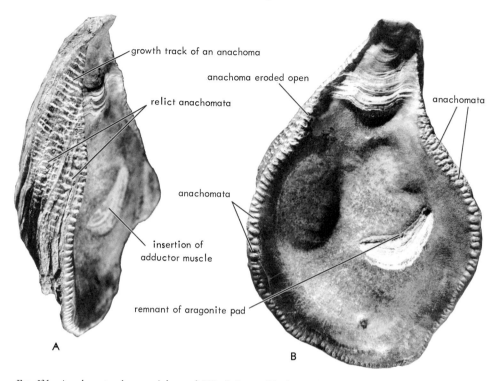

growth track of an anachoma

anachoma eroded open

relict anachomata

anachomata

anachomata

insertion of
adductor muscle

remnant of aragonite pad

A

B

FIG. J31. Anachomata along periphery of RV of *Ostrea (Turkostrea) duvali* GARDNER, ×1 (Stenzel, n).

Specimen is from Caldwell Knob Oyster Bed, Wilcox Group, low.Eoc., of Moss Branch, Bastrop County, Texas.
A. RV viewed from anterior side.
B. Same RV viewed from left side.
[The anterior flank of the valve carries relict anachomata of several preceding growth stages and continuous or nearly continuous growth tracks of many anachomata. Each anachoma is filled with a white powdery chalky deposit. The aragonitic adductor muscle pad has deteriorated to a crumbly mass, which has broken off exposing the growth layers of the supporting calcitic shell wall, except at the dorsal margin of the muscle imprint where the original aragonitic material is still intact.]

locking devices for the continuous guidance of the valves during their movements. Their guidance prevents the valves from closing askew, and in order to accomplish that function in live bivalves they must remain in contact with each other even when the shells are fully opened. Bivalvia that have a large, hydrostatically expandable foot (e.g., Cardiidae) must open their valves rather wide to let the foot move about freely. They have deep sockets and highly projecting teeth that do not lose contact with each other when the animal opens the valves wide. In the oysters, however, chomata lose contact as soon as the animal opens its valves a little. If they exert guidance at all, it can be only minor and restricted to the final stage of shutting the valves. In short, chomata are a novel evolutionary accomplishment of the oysters and their allies that arose during the phylogeny of the superfamily.

The chomata have various characteristic shapes. All Gryphaeinae, *Gryphaeostrea* of the Exogyrinae, and *Crassostrea* of the Ostreinae are entirely devoid of them (see Fig. J101). The genus *Saccostrea* has strongly developed chomata (Fig. J7), high and strong tubercles on the right and deep pits on the left valves along the whole periphery of the valves. In *Ostrea, Cubitostrea, Odontogryphaea,* and other related Ostreinae, the anachomata range in size from simple, low and round tubercles to straight, slightly elongate (up to 3 mm.), unbranching ridgelets.

In *Ostrea* the anachomata are commonly restricted to the vicinity of the hinge. Some species of the genus have only one to five of them, others have a goodly number (10 to 15) in the dorsal half of the right valve. Finally there are species that have great many of them (100 or more), and many individuals of such species have these features encircling the entire or nearly the entire valve. Examples of the latter are *Ostrea (Ostrea) crenulimarginata* GABB, 1860 (p. 398, pl. 68, fig. 40-41), from the Midway Group (Dan.) of the northern coastal plain of the Gulf of Mexico, and *O. (O.) marginidentata* S. V. WOOD, 1861 (p. 27-28, pl. 5, fig. 2) from the lower Brackelsham Beds (Division N of FISHER, Palate Bed, Cuis.) at Brackelsham Bay, Sussex, southern England. These two appropriately named species are closely related, that is, they are members of the same minor phyletic branch in *Ostrea (Ostrea)*. Their chomata are stronger and longer than is usual for *Ostrea s.s.* In contrast, *O. (O.) lutraria* HUTTON, 1873, living off the coasts of New Zealand, has exceedingly few very small, round, tubercle-like anachomata and corresponding catachomata near its hinge (see Fig. J113). In some individuals they are difficult to find and may even disappear with advanced age. They were overlooked by SUTER (1917, p. 86) when he placed this species, under the name *O. angasi* SOWERBY, at the head of his original list of species composing the subgenus *Ostrea (Anodontostrea)*, which he proposed because it differed supposedly from *Ostrea (Ostrea)* by the lack of chomata. The species was later designated the type species of *Anodontostrea* SUTER by FINLAY (1928b, p. 264). Thus *Anodontostrea* differs only quantitatively but not significantly from other species of *Ostrea (Ostrea)* and is best regarded as a junior subjective synonym. *Ostrea s.s.* consists of many extinct and living species that may be arranged in a morphological series on the basis of relative abundance of chomata, and there is no significant gap in this series. Nevertheless, the underlying idea of SUTER's that the presence or absence of chomata is significant in the taxonomy of oysters is sound, even if his observations were incomplete and his examples ill-chosen.

In some species the anachomata project slightly to the outside, beyond the general contour of the right valve. For example,

Odontogryphaea thirsae (GABB, 1861, p. 329-330) from the Nanafalia Formation and other homotaxial formations (Sparnac.) in the northern coastal plain of the Gulf of Mexico and *Ostrea (Turkostrea) duvali* GARDNER (1927, p. 366, fig. 1-4) from the Caldwell Knob Oyster bed (Sparnac.) of the Wilcox Group in central Texas (Fig. J30, J31) have anachomata projecting a fraction of a millimeter beyond the general margin of their right valves so that they are visible from the outside. In contrast, the corresponding left valves have the catachomata a slight distance proximally from the valve margins so that they are not visible except on the inside face of the valves. In these species the very core of each anachoma consists of white, chalky, opaque shell substance and neighboring anachomata are separated by darker, nonchalky, semitranslucent, grayish or brownish, laminated shell substance.

As the right valve grows and shell layer is deposited on shell layer, the anachomata continue to grow in place so that they leave narrow tracks composed of white chalky shell substance. Such tracks may be visible on the outside of the right valve, if the anachomata projected to and beyond the general outline of the valve margins. In that case the track left by an anachoma is a white, narrow, opaque, chalky ridge (Fig. J31). From these ridges it can be deduced that during growth the number of anachomata increases only slightly by dichotomy and intercalation. On the other hand some chomata play out. Because chomata continue to grow in place, the growing ligament seat often bypasses some of them so that the valves have relict chomata on the flanks of their respective ligamental areas (Fig. J30).

Because the chalky cores of the anachomata are less resistant to mechanical and chemical corrosion, they can become corroded or leached in sea water and their tracks can become narrow, sharply defined grooves visible on the outside of the right valves. No such features can be found on the left valves. These features can be used to distinguish left from right valves (see Fig. J128,*1h*).

All Pycnodonteinae have chomata of a special character. In early, primitive genera of this subfamily (e.g., *Texigryphaea*), most species have low, narrow, short (1.5-2.0 mm.), crowded ridgelets and corresponding grooves. On first inspection they are quite similar to those found in many Ostreinae. However, they differ in that they are crowded (about 20 per cm.) and set in series and in that the difference between the right and the left valves is obliterated, that is, the ridgelets cannot be distinguished from the pits because they are so crowded. In addition there is a tendency for some of the ridgelets to join or branch in a simple fashion.

A few species of *Texigryphaea* (e.g., *T. belviderensis*) (HILL & VAUGHAN, 1898, p. 56, pl. 9-10) from the Kiowa Shale (Alb.) of Belvidere, Kiowa County, Kansas, have wider commissural shelves and longer (up to 5 mm.) anachomata, some of which are branched and break up into irregular tubercles. Their ridgelet patterns approach the vermiculate patterns of *Pycnodonte*.

Most of the later, more advanced Pycnodonteinae, with the exception of *P. (Crenostrea)*, have on both their valves wide commissural shelves carrying a vermiculate pattern of ridgelets (see Fig. J83, J84). These are irregular, tortuous (or vermiculate), crowded, and longer (up to 15 mm.); repeatedly they branch away from adjoining ridgelets or coalesce with them and break up into irregular rows of round tubercles.

This pattern has been commented on and figured by many authors. The first to recognize its importance in taxonomy of the oysters was FISCHER DE WALDHEIM (1835). He established the genus *Pycnodonte* chiefly on the basis of this pattern and named it accordingly (πυκνός, Greek adjective, thick, crowded; ὀδους, ὀδόντος, Green noun, masculine gender, the tooth). Later authors to contribute to the evaluation of the vermiculate pattern in taxonomy were DOUVILLÉ (1911, p. 635) and RANSON (1939-41, p. 61; 1941, p. 82). Thanks to their efforts the pattern is recognized now as definitive for a large but well-defined group of species which many authors prefer to regard as a single species-rich and diversified genus and call *Pycnodonta* G. B. SOWERBY (1842). However, *Pycnodonta* is an unjustified emendation of the original validly introduced name *Pycnodonte* and the large diversified group of species breaks easily into several well-defined groups separated by morphological gaps. It seems clear that more than one genus is involved and the large diversified group is really a subfamily, the Pycnodonteinae STENZEL (1959, p. 16, 29-30).

The origin of the chomata is obscure. They are certainly not modified hinge teeth. JAWORSKI (1928) tried to explain them as imprints of radial pallial muscle strands situated at the margins of the mantle lobes. Such strands are indeed present in live oysters and stand out as rounded ridges on the outer face of the mantle lobes. KLINGHARDT (1922, p. 21) tried to explain the vermiculate patterns of the Pycnodonteinae as imprints left by the tentacles that rise from the margins of the mantle lobes. JAWORSKI (1928, p. 345) refuted this assumption by pointing out that tentacles are too small to produce such imprints, are too mobile to stay in place long enough to have much effect, and are held most of the time in a position vertical to the commissural plane of the shell while the oyster is alive. One would expect radial pallial muscle strands and tentacles to produce elongate grooves on both valves. They could not produce ridges on one valve and grooves on the opposite valve. In short, the chomata cannot be explained today.

Only the left valve has an **umbonal cavity.** Several genera of oysters commonly have an open cavity beneath the platform that carries the ligament and ligamental area of the left valve. The cavity changes shape and size during the growth of the oyster; at times it may be entirely obliterated by shell deposits or by the formation of chambers. Although it is so variable in size, its average appearance is distinctive (STENZEL, 1963a).

The deepest umbonal cavities and the most chambers are found in *Saccostrea*, particularly in the high-conical rudist-like ecomorphs of the living complex superspecies *S. cuccullata* (VON BORN, 1778) from the tropical IndoPacific (see Fig. J104-J106). Lesser umbonal cavities are found in *Crassostrea*. No umbonal cavities or few and generally shallow ones are found on *Ostrea*. Very shallow umbonal cavities are present in the Gryphaeinae and Exogyrinae, although their greatly extended left beaks would seem to favor their growth. An important feature of the homeomorphs of *Gryphaea* is that they have only very shallow umbonal cavities and that their extensive spiralling left beaks have no chambers but are filled with solid shell material (see Fig. J73).

Distinction Between Valves

To be able to distinguish the right from the left valve is of practical importance in the study of dead and fossil oyster shells. The following methods are useful for that purpose.

1) If one draws the mid-axis on a valve, the center of the adductor muscle falls to one side of it (see p. N963). The center falls to the left side on the inner face of the left valve and to the right side on the right valve (Fig. J3).

2) The left valve is the larger one. If both valves of the same individual are placed together in their natural positions, margins of the left extend beyond those of the right to a slight or considerable amount, provided the marginal conchiolin scales are discarded (see Fig. J113,2c; J117,1a).

3) The left valve is the more convex. The rule applies if the attachment area is small compared with the size of the whole left valve. If it is large, the opposite may be the case. These conditions vary from one to the other individual and may even reverse during growth of an individual. Individual shells that have grown with a large attachment area on a large flat substratum and have a left valve not much larger than its attachment area have xenomorphic sculpture (see p. N1021) in the form of a smooth convex right valve opposed to the flat attachment area of the left valve (see Fig. J75,1).

Differences in convexities between the two valves are quite large in most Exogyrinae, Gryphaeinae, and Pycnodonteinae, and also in *Saccostrea* and homeomorphs of *Gryphaea*. They are moderate in most Ostreinae and highly variable in *Crassostrea*. Almost equal convexities prevail in the Lophinae and in *Ceratostreon* (Exogyrinae) (see Fig. J92), *Ostreonella* (Ostreinae) (see Fig. J128) and *Hyotissa* (Pycnodonteinae) (see Fig. J85), and in the extremely flat-valved *Deltoideum* and *Platygena* (see Fig. J76, J120).

4) The attachment area is on the left valve in all the oysters. However, to recognize the attachment area one needs to have well-preserved material.

5) The umbonal cavity, if present, is on

Fig. J32. Lack of attachment area on umbo of LV of young *Ilymatogyra arietina* (ROEMER, 1852) from Grayson Clay, Cenoman., near Austin, Travis County, Texas, USA, ×4.8. [The recurved pointed calcitic inner mold of the prodissoconch is seen at tip of the umbo. This specimen is a variant which is costulated in youth.]

the left, and the buttress under the resilifer is on the right valve.

6) Anachomata are present near the hinge of the right valve. The corresponding catachomata are on the left valve. This method is useless in those genera that lack chomata altogether and in most Pycnodonteinae, in which the two cannot be distinguished.

7) The left valve is the heavier. This distinction is quite variable so that it can be used only on a statistical basis.

Attachment

With exceedingly few exceptions, oysters must become attached to a firm substratum at the end of their larval stage. They may remain attached during their entire postlarval life. Some individuals may break loose and continue to live lying loose on the bottom. Some species grow attached to a small fragment of shell and soon grow bigger and heavier than the fragment so that they become virtually free-lying. This is the case in the gryphaeas and their homeomorphs.

The only species discovered so far which grow from their earliest larval stage to old age without ever becoming attached to a substratum are (Fig. J32-J34; see Fig. J64):

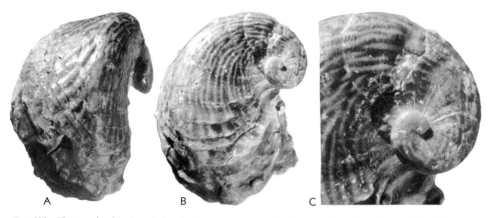

FIG. J33. Traces of coloration, lack of attachment area, and slender polygyral umbo on LV of *Exogyra (Exogyra) tigrina* STEPHENSON (1929) (Stenzel, n).

Specimen is topotype from uppermost bed of Austin Chalk (L.Campan.) of Little Walnut Creek, east of Austin, Travis County, Texas, USA. *A-B.* Exterior views of LV, ×0.85.

C. Tip of LV umbo showing lack of attachment area and semiglobular calcitic internal mold of prodissoconch, ×1.7. [The prodissoconch itself is leached because it consisted of aragonite.]

Ilymatogyra arietina (ROEMER, 1852), from the Grayson Clay (Cenoman.) of Texas and northern Mexico (Fig. J32); *Odontogryphaea thirsae* (GABB, 1861) from the Nanafalia Formation (Sparnac.) of Alabama to Mexico and other species of the genus; and many individuals of *Rhynchostreon suborbiculatum* (LAMARCK, 1801) [=*Ostracites ratisbonensis* VON SCHLOTHEIM, 1813 (=*Gryphaea columba* LAMARCK, 1819, =*Ostrea mermeti* COQUAND, 1862, =*Rhynchostreon chaperi* BAYLE, 1878)] from the Cenomanian and Turonian of Europe, North Africa, and the Near East (see Fig. J97). These oysters are specially adapted, and their shells show no attachment areas at the umbones (see also Fig. J94).

All oysters are attached by their left valves. In no case has it been possible to prove that an oyster had become attached by the right valve. Those cases in which it had been claimed that the right rather than the left valve was attached have been proved to be or are very likely errors of some sort. SAVILLE-KENT (1893, p. 246) claimed that the living Australian species *"Ostrea glomerata"* [COX, 1883, not GOULD, 1850, =*Saxostrea commercialis* (IREDALE & ROUGHLEY, 1933), =*Saccostrea cuccullata* (VON BORN, 1778) subsp. *commercialis* (I. & R., 1933)] . . . "is . . . invariable affixed by its right one" (see Fig. J105). This

claim is no longer supported by modern Australian authors. ROLLIER (1917, p. 587) distinguished the Jurassic genus *Deltoideum* from *Ostrea s.s.* because he surmised it was attached by its right valve. ARKELL (1932-36, p. 149, footnote 1) expressly refuted this claim. The genus has very flat valves and the two valves are most difficult to distinguish. ROLLIER probably mistook the right for the left valve (see Fig. J76).

NELSON (1938, p. 45, footnote 9) noted, among hundreds of normal newly set *Crassostrea virginica* (GMELIN), the living oyster of the North American east coast, one single one attached supposedly by its right valve. He did not illustrate it nor did he offer any proof. Innumerable newly set spat of this species have been seen by many other authors, but no one has duplicated NELSON's observation. Presumably it was erroneous.

Size of the attachment area is quite variable from one to the next individual, even in the same vicinity. It is mostly quite small (diameter of 2 mm.) in *Gryphaea* and its homeomorphs. SWINNERTON (1939, p. xliv and lii; 1940, p. xcviii, fig. 8; 1964, p. 419-420) measured the longest diameter of the attachment area on 658 specimens of *Gryphaea arcuata* LAMARCK (1801) from clay shales in the Zone of *Schlotheimia angulata* (VON SCHLOTHEIM, 1820) in the Granby Limestones, latest Hettangian, Lias-

FIG. J34. Traces of coloration in LV of *Odontogryphaea thirsae* (GABB, 1861) from Nanafalia Formation (Sparnac.) of Alabama, USA, ×1.5 (Stenzel, n).

sic, near Granby, Lincolnshire, England. The average of these 658 diameters, calculated from SWINNERTON's data, is about 2.1 mm. and the average size of the attachment areas is less than 4.5 square mm. Only 2.9 percent of the specimens had diameters in excess of 10 mm. (Fig. J35). Oysters grown onto very small objects or fragments soon become heavier than their substratum. On toppling over, they lift the substratum up from the floor of the sea and become virtually free-lying but continue to grow.

The attachment area has the negative configuration of the substratum. In fossil oysters the configuration of the attachment area is an important clue to the nature of the substratum. In many places, one finds well-preserved fossil oyster shells that grew originally on aragonitic mollusk shells, but these are no longer preserved, because aragonite is subject to selective leaching. The configuration of the attachment area lets one identify the now no-longer-present mollusk shell with some confidence (see Fig. J45).

Some oyster genera also produce clasping shelly processes on the left valve. These processes grow out from the valve margin of the left valve periodically to embrace or enclose the substratum. Most of the Lophinae grow them, also *Gryphaeostrea* of the Exogyrinae. Many of the Lophinae that produce clasping processes grow on live gorgonacean coral stems (see Fig. J47, J129).

Pigmentation

The oysters have many pigmentation patterns. A common pattern is radial streaks of dark colors on the outer faces of the valves. The streaks begin as narrow bands near the umbones and widen concomitantly with growth of the shell. Another common pattern found in tropical oysters is a wide circumferential band of dark color (black, dark brown, or dark purple) along the margins on the inner faces of the valves (see Fig. J85).

Oysters living in warm and tropical regions have darker, more vivid, more varied, and more extensive colorations than those living in cooler climates. This rule applies to all species but is particularly noticeable in those having a large north-south geographic range extending from cool to hot climates. For example, northern populations of *Crassostrea virginica,* living north of Cape Cod along the east coast of North America, have whitish to grayish yellow (5Y 8/4) colors on the inner faces of the valves and have grayish red purple (5RP 4/2) to very dusky red purple (5RP 2/2) muscle imprints. The same species living in the northern part of the Gulf of Mexico has large areas of the inner faces light brown (5YR 6/4) to grayish purple (5P 4/2) and only a few patches remain whitish; the muscle imprints have approximately the same colors.

On the basis of coloration it is possible to distinguish geographic subspecies among living species that have a wide enough range. However, coloration, even traces of it, is rarely found in fossil oysters.

Traces of coloration in fossil oysters are in the form of radial brownish streaks on the exterior of the shell. The brownish color is probably the result of deterioration of purplish. Such brownish streaks are found on *Ilymatogyra arietina* (ROEMER, 1852), from the Grayson Shale (Cenoman.) of Texas and northeastern Mexico; *Exogyra (Exogyra) tigrina* STEPHENSON, 1929 (Fig. J33) and *Exogyra (Exogyra) laeviuscula* ROEMER, 1849, both from the top of the

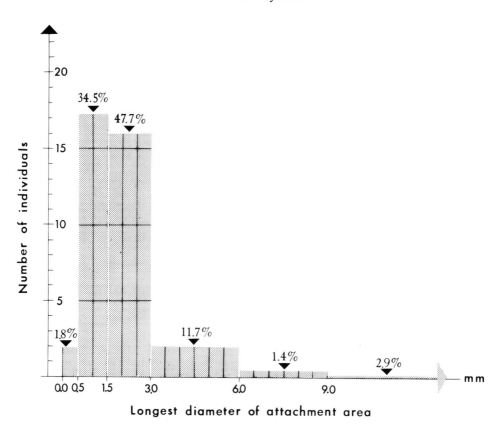

Fɪɢ. J35. Size of LV attachment area of *Gryphaea arcuata* Lᴀᴍᴀʀᴄᴋ (1801) (Stenzel, n; data from Swinnerton, 1964).

Specimens studied come from Granby Is. (latest Hettang., Lias.) south of Granby, Lincolnshire, Eng. Size of attachment area is indicated by length of its longest diameter measured on 658 individuals.

Dessau Chalk (Santon.), Austin Chalk Group, of central Texas; *Odontogryphaea thirsae* (Gᴀʙʙ, 1861) (Fig. J34) from the Nanafalia Formation (Sparnac.) of Alabama to northeastern Mexico; and *Rhynchostreon suborbiculatum* (Lᴀᴍᴀʀᴄᴋ, 1801) from the Cenomanian and Turonian of Europe, North Africa, and the Near East. The last-mentioned species and its color bands has been figured by Gᴏʟᴅꜰᴜꜱꜱ (1826-44, pl. 86, fig. 9c,d), CᴏQᴜᴀɴᴅ (1869, pl. 45, fig. 8-9), and Bᴀʏʟᴇ (1878,

pl. 138, fig. 3) and specimens showing them have been found from County Antrim, Ireland, to the Postelberg in Bohemia, Czechoslovakia. Radial color bands are more common in the Exogyrinae than in other fossil oysters.

Colors in the shell wall of oysters are probably organic metabolic waste products. The kind of food available to the oysters probably determines their coloration. Purple pigments are probably acid-soluble bile pigments.

PHYSIOLOGY

CONTENTS

FUNCTIONS OF ADDUCTOR MUSCLE

The adductor muscle serves to move and close the two valves. It does so against the elastic reaction of the ligament, which is always under compression and pushes the valves apart. So important to the survival of the animal is the adductor muscle and so much work does it perform that it must be supplied with oxygenated blood at all times and in the best possible way. Therefore, the artery supplying it is very short and the heart is right next to the muscle on its more protected, dorsal flank. Thus heart and muscle are functionally tied together. If evolutionary adaptive change induce the muscle to shift position, the heart must follow it inexorably and remain next to it (see p. N1058).

Both quick reactions and long continued contraction are required. These two separate faculties have necessitated specialization of the muscle into two subdivisions side by side.

The tonic subdivision (catch muscle) holds the shell closed against the unceasing push of the ligament and it can do that for long periods of time, if need arises. Oysters can starve or suffocate and die while they hold the shell shut tight. Live oysters can remain shut for 20 to 30 days (MARCEAU, 1936, p. 952).

The phasic subdivision (quick muscle) acts frequently. Oysters never keep their valves in the same positions for long while they are feeding. Throughput of the water current through gills and mantle cavity is regulated by the pallial curtains and the degree to which the muscle lets the valves open. The water current bears suspended food particles and dissolved oxygen for breathing. A series of rapid contractions expels the newly issuing unfertilized eggs from the cloacal passage through the ostia of the gills into the inhalant mantle chamber. Similar movements eject the eggs of the nonincubatory oysters into the surrounding waters. The incubatory oysters eject their half-grown larvae at the end of their incubation period in a similar fashion. Self-cleansing of the mantle cavity depends on the quick muscle and the pallial curtains (see p. N1001).

ALIMENTATION

FOOD

Nannoplankton, that is, various microscopic plants or animals less than 10 microns large, are the food of oysters. Most of it is various one-celled plants or animals, such as algae, bacteria, diatoms, flagellates, and protozoans. Microscopic metazoa such as copepods, free-living nematodes, polychaetes, rotifers, and even tiny fish eggs are also taken. Larvae of oysters or many other sea animals are taken in too.

However, oysters cease to feed while they are ejecting their sex products.

FEEDING

While the oyster is feeding it must open its mantle/shell sufficiently to let in the water current which it establishes itself. Yet, the opening must be kept as small as possible to keep out predators. The extent to which oysters open their mantle/shell while feeding is astoundingly small. The opened gap is 2-3.5 mm. wide at the ventral valve margins of full-grown individuals (shells 80-110 mm. high) of *Crassostrea virginica* (data kindly supplied by Mr. W. J. DEMORAN of the Gulf Coast Research Laboratory at Ocean Springs).

The gills are covered with cilia. Cilia on the gills, particularly those at or near the ostia, beat regularly and thereby set in motion a water current that passes through the ostia from the side of the inhalant chamber to that of the exhalant chamber. No water can go from one chamber to the other without passing through the ostia of the gills. Therefore, the water is strained efficiently and nannoplankton and other particles are collected on the gills. The oyster is a ciliary suspension-feeder.

The gills perform two functions, respiration and food straining. However, respiration is also taken care of elsewhere. The mantle lobes are well supplied with blood vessels and have large epithelial surfaces exposed to the inhalant oxygen-carrying water current. Their help in respiration freed the gills to some extent so that they were able to become efficiently adapted to food-straining.

Individuals of *Crassostrea virginica* pump and strain for brief periods (5-15 minutes)

at the rate of 41 liters per hours at the best and attain a sustained rate of 37.4 liters per hour when temperatures are 24.1 to 24.5°C. (LOOSANOFF, 1958, p. 62; COLLIER, 1959). They cease in cold waters. ALLEN (1962) measured filtration rates of *Ostrea edulis* by using a suspension of the unicellular alga *Phaeodactylum* labeled with radioactive phosphorus. The alga is 4 by 40μ in size and is readily accepted as food by filter-feeding bivalves. They remove the algae from surrounding waters, enabling one to calculate rates of filtration by measuring concentrations at intervals and the time intervals. *Mya arenaria* has a filtration rate of 0.8 milliliters per hour per milligram of dried tissues of all soft parts of the animal; *Ostrea edulis* has a much larger filtration rate than other bivalves and the rate is 6.6 milliliters per hour per milligram of dried tissues.

Nannoplankton strained by the gills becomes enshrouded in mucus and moved by cilia along special pathways over the gills to the adoral tip of the gills, where the food drops onto the labial palps. There unsuitable pieces are shunted aside but food continues to the mouth.

The unsuited particles dribble down to the mantle lobes. Any particles coming in with the inhalant current that are too large to be carried along to the gills have already dropped out and lodged on the mantle lobes. All the rejects are combined, enveloped in mucus, and moved by the cilia on the mantle lobes to discharge areas near the pallial curtains.

During this time the valves are kept quiet for no longer than a few minutes at a time. Repeatedly the adductor muscle contracts quickly and drives the water out from the mantle cavity. At that moment the combined rejects are forcibly ejected as **pseudofeces** from their discharge areas. Thus the self-cleansing mechanism keeps the complicated sieve structure of the gills from clogging.

The pallial curtains are equally mobile. By opening or closing they control the size and location of the expelled current carrying the pseudofeces. Normally the inhalant current is restricted by the pallial curtains and enters through a narrow gap, about 3 cm. long in full-grown oysters. This gap is usually present directly anterior to the adductor muscle or in a slightly more antero-ventral position. This gap is the **inhalant pseudosiphon.**

An **exhalant pseudosiphon** is kept open at a location in direct line with the axis of the cloacal passage. This pseudosiphon is about 1.5 cm. long and gapes about 2 to 3 mm. Oysters that have a promyal passage have a third pseudosiphon opposite the axis of the passage.

DIGESTION

A detailed description of the digestive functions has been given by YONGE (1926; 1960). The entire intestinal tract from esophagus to anus is weakly acid, register-ing average pH concentration from 5.93 to 5.2. The most acid part is the stomach with 5.2-5.6. The mantle cavity itself contains water that is close to neutral (6.8-7.2 pH).

These acid liquids within the digestive system must have some chemical effect on clay minerals accidentally taken in and contained in the excrements (compare AN-DERSON, JONAS, & ODUM, 1958).

SELF-CLEANSING AND SELF-SEDIMENTATION

Problems of sanitation confront oysters, because they are immobilized. Waste mate-rials of two sorts accumulate in their mantle cavities and must be removed. Feces are discharged into the cloacal passage, but pseudofeces accumulate in the exhalant and inhalant mantle chambers, mainly on the pseudofecal discharge area of the latter chamber.

Feces, digested food particles enveloped in mucus, are retained for a short time in the rectum, where they become compacted into long and firm pellets or rods. Then they are expelled by the action of cilia through the anus into the cloacal passage. There they and pseudofeces are picked up by the exhalant water current. They are also moved along by cilia that cover the exposed faces of the mantle lobes. They move on special ciliated pathways that converge at the cloacal discharge area near the cloacal pseudosiphon. Finally they are expelled in a fairly continuous stream while the oyster is feeding.

Pseudofeces are built up from any parti-cles that happen to enter the mantle cavity but are rejected as food or cannot reach the mouth. Most of them are in the inhalant mantle chamber. Balled up with mucus, they are moved by cilia against the inhalant water current toward the pallial curtains, where they accumulate in larger masses at the pseudofecal discharge area near the inhalant pseudosiphon. A sudden contrac-tion of the adductor muscle forces water from the mantle cavity through the pseudo-siphon. This squirt of water carries the pseudofeces with it. Such squirts have been observed to go for a distance of more than one meter.

These powerful cleansing streams are re-peated whenever sufficient pseudofeces have accumulated at the discharge areas in the mantle cavity. In this fashion, the mantle cavity is kept from fouling and the delicate mechanism of the gills as a food strainer is kept from clogging. In addition, the imme-diate surroundings of a live oyster are squirted free and cleaned of encroaching freshly deposited silt, mud, feces, and pseu-dofeces.

The self-cleansing mechanism is a fine adaptation permitting many oysters to live in turbid, sediment-laden waters. Without it some of the genera, notably *Ostrea, Striostrea,* and *Crassostrea,* could not have been able to invade brackish coastal waters produced by mud-laden rivers and to move up into lagoons and river estuaries.

Materials carried in the waters in sus-pension are called **leptopel** and consist of living and dead nannoplankton and of colloidally or otherwise finely divided or-ganic and inorganic detritus. Oysters quite efficiently extract the leptopel from the wa-ters and convert it into feces and pseudo-feces. Both are somewhat compacted and denser than the surrounding waters so that they tend to sink to the bottom in the gen-eral vicinity of the oysters that produce them. Thus oysters bring about **self-sedi-mentation** in their neighborhood and per-form an enormous task of clearing turbid waters and depositing fresh sediments of very fine grain size and very rich in or-ganic matter (LUND, 1957a; 1957b). Oyster beds in brackish, turbid waters, composed of thousands of live individuals, must be effective sediment accumulators, and such beds are local depocenters of rapid sedi-mentation.

Sediments accumulated in this way consist of 1) silt fine enough to be carried in suspension before being strained out by the oysters, 2) clay-size particles of various clay minerals and plant debris, and 3) organic substances, such as oyster excrements, various indigestible organic materials, 4) mucus, which envelops both feces and pseudofeces, and 5) any additional materials brought in separately by waves and currents. Accordingly, the sediments have prevailingly very fine grain sizes. Organic sulfur compounds are present in the organic ingredients of the sediments, and sulfur is correspondingly quite abundant.

Because the sediments are rich in organic components they have large amounts of free energy available for bacteria to engage in decomposition with high rates of oxygen consumption resulting in strongly reducing, anaerobic conditions within the sediments, which can be called a **sapropel.** Hydrogen sulfide (H_2S) and finely divided and dispersed black hydrotroilite ($FeS \cdot nH_2O$) is formed in the sapropel within four days even when aerated sea water passes over it (LUND, 1957a; 1957b). The iron sulfide gradually changes into FeS_2 (pyrite or marcasite). Both the organic components and the finely divided iron sulfide minerals impart dark colors to the sediments, which are mostly stink muds or stink silts with a strong odor of hydrogen sulfide.

The hydrogen sulfide rises in the sediment and reaches the sediment-water interface. There it is oxidized to sulfurous and sulfuric acid. This acid environment at the top of the sediment accounts for the corrosion many oyster shells show. Corrosion is most noticeable around the umbones, which are commonly etched and deeply pitted, because they have been exposed to the acid environment the longest. Nevertheless, waters above an oyster bed are nearly always alkaline (pH=8.3 to 7.7) because of the buffering action of the calcium carbonate of the shells (WELLS, 1961, p. 244).

PROPAGATION

SEXUAL MATURITY

Sexual maturity, that is, ability to reproduce, is reached rather early in the life of the oyster. In most situations it is reached within a year after the larva has settled down and metamorphosed. The length of time required depends on prevailing temperatures and availability of food. Up to a limit, the warmer the waters, the more food is available, the more time the oyster spends feeding, the greater is its rate of pumping a water current, and the earlier the young attached oyster reaches sexual maturity. In tropical and hot temperate climates it takes merely 20 to 30 days. In cool climates it takes longer and in frigid climates it generally takes more than a year.

The size at which an oyster attains sexual maturity depends on the average size of adults of the species. If adults are generally small, maturity is reached at a small size. At Aransas Pass, Texas (MENZEL, 1955, p. 84-85), *Ostrea equestris* SAY, 1834, the small species living on the east coasts of the Americas, reaches maturity 22 to 30 days after fixation of the larvae, when its sizes are 0.5 to 1.0 cm. (Fig. J6). This species is incubatory and it is ready to incubate its own larvae 31 to 60 days after the parental oysters have become attached and are merely 0.8-1.0 cm. in diameter. Species that average a larger adult size, reach maturity at somewhat larger sizes, between 1 and 2 cm.

In any case, sexual maturity is attained while oysters are still very small. This fact is significant in the interpretation of such small fossil oysters as the Jurassic genus *Catinula* and in the evaluation of ideas proposed to explain the disappearance during the Liassic of the stock of *Gryphaea arcuata* LAMARCK, 1801 (see p. *N*1075-*N*1076).

Early sexual maturity is a safety device for the survival of the species. Given a limited span of life of the individual oyster, the earlier it attains maturity the longer does it contribute gonadal products for the survival of the species.

FECUNDITY

The enormous fecundity of oysters is another safety device for the survival of the species. Oysters produce uncountable eggs and spermatozoa. The number of the gonadal products depends on the size of the gonads, which in turn depend on the amount of food available and the age and size of the oyster. As an oyster grows bigger, its fecundity increases correspondingly.

The nonincubatory oysters produce smaller and more numerous eggs than the incubatory oysters. Calculations show that full-grown oysters of the nonincubatory genus *Crassostrea* produce in one spawning season the following numbers of eggs: *C. gigas* (THUNBERG, 1793) 11-92 million (GALTSOFF, 1930a); *C. rhizophorae* (GUILDING, 1828) 99-170 million (MATTOX, 1949); and *C. virginica* (GMELIN, 1791) 15-115 million.

Fecundity of incubatory oysters is more difficult to estimate, because the eggs remain in the mantle cavity for incubation. Probably the number of larvae found incubating in the mantle cavity of an oyster is approximately the same as the number of eggs it had delivered. An individual of *Ostrea edulis* LINNÉ, 1758, 4 years old and about 7.3 cm. in size, can have as many as 2.05 million larvae incubating at one time (WALNE, 1964, p. 303).

Here again prevailing temperatures of the surrounding waters have some influence. Wherever the summer is long and warm, there is more than one spawning period. In tropical regions, oysters spawn intermittently throughout the year (MATTOX, 1949, p. 352). In frigid regions, for instance the Kattegat, north of Denmark, waters may remain too cold for several years in succession and no spawning may take place in those years. In such areas, oysters are close to extinction and may survive only in narrow inlets which warm up better than the waters further out. Such areas are the limits of the geographic range of a given oyster species (see p. *N*1036).

SPAWNING

Sperm or eggs, depending on the sexual stage of the spawning individual, issue from the gonads through the urogenital clefts in the cloacal passage of the exhalant mantle chamber. However, beyond that point their paths are different.

In the male engaged in spawning, cilia on the walls of the urogenital passages move the sperm balls out into the cloacal passage. There the balls break up into spermatozoa which are carried away by the exhalant water current to form a milky stream issuing from the oyster. Discharge may continue for several hours.

In the female, eggs are delivered into the

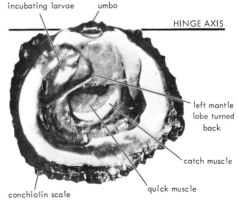

FIG. J36. Incubating oyster, *Ostrea (Ostrea) angassi* SOWERBY, 1871, from South Australia, ×0.6 (Cotton, 1961). Left valve has been removed and left mantle lobe has been turned back to show mass of incubating larvae.

cloacal passage by the same means, but then they are pushed through the ostia of the gills into the inhalant mantle chamber. During this period the oyster ceases to feed and there is no regular water current going from the inhalant to the exhalant mantle chamber. Rather, successive spasmodic and strong contractions of the adductor muscle force water and eggs from the cloacal passage through the ostia into the inhalant mantle chamber.

At this stage of female spawning incubatory genera begin to differ from nonincubatory genera. In the latter, the eggs accumulate in the inhalant mantle chamber near the inhalant pseudosiphon. Shortly afterward some more spasms of contraction of the adductor muscle expel a jet stream of water carrying the eggs through the pseudosiphon out for more than a meter. These forcible expulsions may last from a few minutes to two hours. Fertilization takes place in the surrounding water by haphazard meeting of egg and spermatozoon.

In incubatory oysters, the eggs remain in the inhalant chamber of the mother oyster and drop down onto her gills, labial palps, and mantle lobes. They are fertilized when spermatozoa are drawn into the inhalant mantle chamber by the pumping action of the mother oyster. Thus the chances of fertilization are somewhat better (Fig. J36).

Spawning depends very much on prevailing water temperatures. A certain specific limiting temperature must be attained and maintained for some time before spawning can take place. In effect, this limiting specific temperature sets a limit to the geographic distribution of an oyster species.

SYNCHRONY

One of the mechanisms which oysters have evolved to insure survival of the species effectuates the best possible conditions for fertilization of eggs as they are delivered by the females. This mechanism is mass synchrony or simultaneous discharge of sex products by all mature and ripe males and females in one neighborhood.

In an oyster bed many mature males and females become ready to discharge their individual sex products at about the same time. However, their discharge remains dormant for a time until they are triggered off by a special mechanism. This synchronization insures that as many individuals as possible participate at the same time.

If one male begins to discharge its sperm into water, 6 to 38 minutes later all the ripe males and females in the vicinity start discharging. The result is a maximal density of spermatozoa floating in the water when the eggs are discharged.

The sperm contains a powerful water-soluble stimulant and minute amounts of it suffice to stimulate a neighboring oyster that is drawing in the stimulant material with the inhalant current. Males can be stimulated also by eggs drawn in with the current. In this fashion stimulation can spread rapidly from oyster to oyster over a whole oyster bank, and tidal currents can spread action to neighboring oyster beds.

The stimulant is highly specialized. Sperm from *Ostrea* has no effect on *Crassostrea* species so that genetic isolation is maintained even if the two genera live side by side.

This mass reaction works best on densely settled oyster bottoms, and there must be a minimal population density below which this mass reaction is not attainable. Population density and sexual synchrony are interdependent adaptations in these animals. Because of synchrony the size of an actively and simultaneously interbreeding population of an oyster species tends to be enormously large.

INCUBATION

Several oyster genera incubate their larvae. Most incubatory species have an incubation period lasting 6 to 18 days. However, a species living on the Pacific coast of Chile, *Ostrea (Ostrea) chilensis* Philippi, 1845, which delivers eggs unusually large for oysters, incubates for 5-6 weeks at temperatures of 13-15°C. (Walne, 1963).

While they are incubating, eggs and larvae developing from them rest in the inhalant mantle chamber on the gills, labial palps, and mantle lobes of the incubatory adult (Fig. J36). They receive no nourishment from the adult except whatever they are able to gather from its inhalant water current. They depend on the original food (yolk) contained in the egg. For this reason eggs of incubatory oyster species are larger and richer in yolk. At the end of the incubation period, the adult oyster expels the larvae as a cloud, and the larvae begin their free-swimming larval periods. At the time of expulsion, the larvae already have bivalved shells, which furnish some protection.

Incubation has many advantages. For instance, fertilization of the eggs takes place under protection, in the inhalant mantle chamber of the mother, as spermatozoa are drawn in from the surrounding waters with the inhalant water current of the mother. The eggs have a better chance to become fertilized than would be possible if fertilization had to take place in the surrounding waters, as is the case in the nonincubatory oysters. Because of this aid, a larger percentage of eggs spawned become fertilized so that it is unnecessary for the mother oyster to produce as many eggs. Also, because the eggs are delivered under protection and the larvae remain protected while they incubate, the rate of their survival is much better. Rates of survival are estimated to be about 7 to 80 times as good in the incubatory as compared to the nonincubatory oysters.

These advantages allow the incubatory oysters to reduce the total amount, by weight, of their gonadal products. For this reason, incubatory genera need not have as large gonads in proportion to the entire visceral mass as the nonincubatory ones. As a rule, incubatory oysters have less capacious shell cavities, and their left valves are less deeply cupped. Some of the incubatory genera have flat, shallow left valves mostly devoid of deep umbonal cavities. Deep umbonal cavities prevailing in a genus are an indication that it is nonincubatory.

The only incontrovertible proof of the incubatory habit in a given oyster species is the presence of a brood of larvae within the inhalant mantle chamber of an adult. Such information is difficult to obtain, because incubation takes place only during a few months in a year. Only well-known commercially important oysters living on the coasts of scientifically advanced countries are likely to have been checked for this feature. In addition, it seems well established that nonincubatory genera have a promyal passage.

Incubating larvae have been found in the following genera:

1) *Alectryonella* (Lophinae). The single living species of this genus, *A. plicatula* (GMELIN, 1791) [=*Ostrea plicata* CHEMNITZ, 1785, =*O. plicatula* LAMARCK, 1819, =*O. cumingiana* DUNKER in PHILIPPI, 1846, =*O. lactea* G. B. SOWERBY, 1871] has been seen with incubating larvae (AMEMYIA, 1929; WADA, 1953).

2) *Lopha* (Lophinae). WADA (1953) reported *L. cristagalli* (LINNÉ, 1758) with incubating larvae. According to THOMSON (1954, p. 146-149), *L. cristagalli* is an ecomorph of *L. folium* (LINNÉ) (see p. N1157), and the name is to be written *L. folium* (LINNÉ) ecomorph *cristagalli* (LINNÉ). ROUGHLEY found the *L. folium* (LINNÉ) ecomorph *bresia* (IREDALE) is incubatory (see IREDALE, 1939, p. 397, and THOMSON, 1954, p. 146-149).

3) *Ostrea* (Ostreinae). It is well known that the type species, *O. edulis* LINNÉ, and about 10 other congeners are incubatory.

The following genera are known to be nonincubatory:

1) *Crassostrea* (Ostreinae). The type species, *C. virginica* (GMELIN, 1791), and about 8 other congeners are known to be nonincubatory (CAHN, 1950, p. 12, 17; MATTOX, 1949; PAUL, 1942, p. 9).

2) *Saccostrea* (Ostreinae). The Australian subspecies *S. cucullata* (VON BORN, 1778) subspecies *commercialis* (IREDALE & ROUGHLEY, 1933) is nonincubatory according to ROUGHLEY (1933, p. 318) and so is the typical *S. cucullata* (VON BORN, 1778) according to AWATI & RAI (1931, p. 38) and SOMEREN & WHITEHEAD (1961, p. 14).

3) *Striostrea* (Ostreinae). The type species, *S. margaritacea* (LAMARCK, 1819), is nonincubatory (KORRINGA, 1956).

As to the other living genera, for example, *Hyotissa* and *Neopycnodonte* of the Pycnodonteinae, they have been investigated too little and nothing definite is known of the larval growth periods, although these two are scientifically very important because they are the last survivors of an important phyletic branch of the oysters. One must find anatomical features on them that are somehow linked with their incubatory or nonincubatory habit, as the case may be, in order to have an indication of their mode of larval growth. Such a feature seems to be the promyal passage of the exhalant mantle chamber. As to genera extinct today only features shown on the shell can be used. Here tenuous phylogenetic surmises can lead us to some conclusions.

ONTOGENY

CONTENTS

EGGS

Ripe unfertilized eggs are small and pear-shaped. They turn spherical as they are expelled and fertilized. Nonincubatory genera produce smaller eggs less rich in yolk. Their diameter is 0.046-0.09 mm. Eggs of incubatory oysters have a diameter of 0.09-0.323 mm. The smaller the eggs are the longer does the free-swimming larval period last.

The largest eggs known are produced by *Ostrea (Ostrea) chilensis* PHILIPPI, 1845, living on the Pacific coast of Chile.

LARVAL STAGES

PROTOSTRACUM VELIGER STAGE

It takes fertilized eggs about 30 hours to 18 days to grow into the initial shell-bearing larval stage, depending on abundance of food and prevailing temperatures. This stage has two equal symmetrical equilateral D-shaped hyaline valves connected by a conchiolinous ligament located in the middle of the straight hinge. This is the straight-hinge veliger or protostracum veliger or phylembryo or D-shaped larval stage.

The anterior adductor is formed shortly before the posterior one. At first, the foot is a mere rudiment and a slight concavity between mouth and anus is the beginning of the exhalant mantle chamber.

A disc-shaped ciliated special larval organ, the **velum,** enables the larva to swim. It is located anteroventrally and completely retracts when the larva shuts its shell. In nonincubatory genera the eggs are released into the surrounding waters and grow there into planktonic and free-swimming protostracum larvae. Incubatory species let their fertilized eggs develop into protostracum veligers within the adult's inhalant mantle chamber.

At first, the straight valve margins at the hinge are only slightly thickened and are about half as long as the valves themselves. Later, these margins thicken and lengthen, forming the hinge plates of the prodissoconch, and rudiments of hinge teeth begin to grow as small thickenings. In the larvae of the Gryphaeidae, the thickenings or hinge teeth precursors are lined up in a continuous series all along the hinge, but in the Ostreidae precursors are only at each end of the hinge and a smooth gap separates them. Finally, the protostracum grows 0.095-0.235 mm. long; its height is about 7/8 of its length; its hinge extends for 5/7 of the length of the valves; the umbones are low and have not yet grown to rise above the hinge; the ligament remains in the middle of the hinge (see Fig. J37).

PRODISSOCONCH VELIGER STAGE

This is the stage commonly called umbo veliger larva, because the umbones of the shell are conspicuous. As this stage develops from the preceding protostracum veliger stage, the foot grows larger and more active. Its sole, near its heel-like base, now has a byssus gland. Both adductor muscles are present, and the rudiments of heart, kidneys, and gills make their appearances. A pair of black pigmented spots, or eyes, and a pair of otocysts grow in the wall of the visceral mass near the pedal nerve ganglia. These eyes are probably homologous with the cephalic eyes of other mollusks and appear shortly before the larva is ready to settle down (Fig. J38).

As the prodissoconch grows, its hinge thickens and the rudimentary swellings on the hinge grow into small, well-defined, rectangular, interlocking teeth or rather larval precursors of hinge teeth (Fig. J39). However, the hinge almost quits growing in length so that the valve margins gradually encroach upon and outflank the hinge ends. The valve margins grow progressively more inward, that is, toward the plane of symmetry between the opposing valves, so that the valves become more convexly tumid. The fastest growth is along the anteroventral valve margins, and the valves become quite inequilateral. By these growth tendencies the umbones rise above the level of the hinge and become more opisthogyral and bulgingly globose.

During this larval stage, that is, a considerable time before the larva is ready to settle down and metamorphose into a permanently fixed oyster, the left valve outgrows the right and becomes larger and more convex; its umbo grows more prominent than that of the right valve. This fact attests to the strong genetically fixed roots of the attachment habit of the oysters and

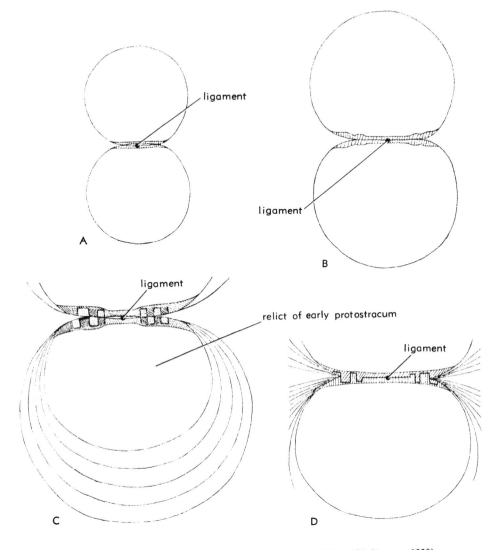

Fig. J37. Growth of larval shell (protostracum) of *Ostrea edulis,* ×170 (Ranson, 1939).

A. After 6 days of growth.
B. After 7 days, showing rudiments of larval hinge teeth and sockets.
C. After 20-25 days, showing well-formed larval precursors of teeth and sockets in arrangement

characteristic of Ostreidae; valves are forced open.
D. Same, hinge in normal life attitude, not forced open, showing growth lines beyond ends of hinge.

explains why they attach themselves exclusively by their left sides.

At the end of this stage, prodissoconchs are 0.25-0.425 mm. long and there are many specific differences in the sizes and configurations of the prodissoconchs among the various species (RANSON, 1960a). There are also generic differences (FORBES, 1967).

BERNARD (1896) extended to the oysters his studies on the ontogenetic development of the hinge structures in the Bivalvia. Although he had available three oyster species, two of them living ones, his study was based only on a species from the Calcaire Grossier (Lutet.) of the Paris Basin, which he identified as *"Ostrea flabellula?"* [=O.

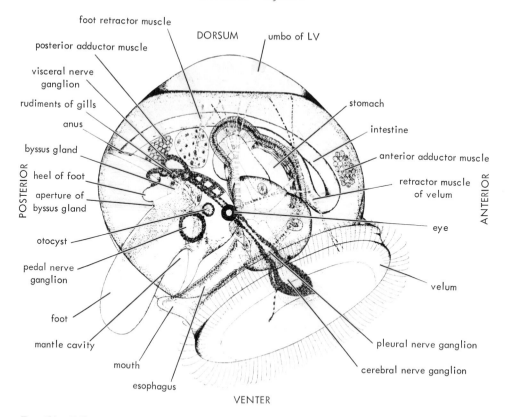

FIG. J38. Full-grown larva of *Ostrea edulis,* at end of prodissoconch veliger stage, ×335 (from Erdmann, 1934).

flabellula LAMARCK, 1806, =*Cubitostrea plicata* (SOLANDER in BRANDER, 1766)].

In so richly fossiliferous a sediment as the Calcaire Grossier, known to contain several sympatric oyster species, it is almost impossible to identify oyster prodissoconchs correctly and to try to relate prodissoconchs and very young attached oysters to fully grown adults of the same species. The young of one species may grow attached to the adults of another, and the more sympatric species are involved the more difficult is the task.

For that reason BERNARD's identification was only tentative. RANSON (1939a; 1941; 1942; 1948b; 1960a, fig. 1-16) demonstrated that the kind of prodissoconch illustrated by BERNARD (1896, p. 447, fig. 15) is found only in the Pycnodonteinae, that is, that BERNARD's identification was erroneous.

The hinge of the prodissoconch of the Pycnodonteinae differs considerably from that of the Ostreidae as was first recognized by RANSON. In the Pycnodonteinae the prodissoconch hinge carries an unbroken series of five interlocking teeth and no smooth gap occurs between them (Fig. J39).

In the Ostreidae, the prodissoconch hinge has two interlocking teeth at each end separated by a fairly long smooth gap. All these teeth are small, rectangular, interlocking, and somewhat higher than long. Each tooth fits into an opposing socket located on the other valve. In the common Ostreidae the anterior set of teeth is weaker than the posterior set. These teeth are larval structures and are the larval precursors of true teeth, homologous with the true teeth of adult eulamellibranchs. However, in oysters the larval tooth precursors never develop into true adult teeth. Strictly speaking, they should be called tooth precursors.

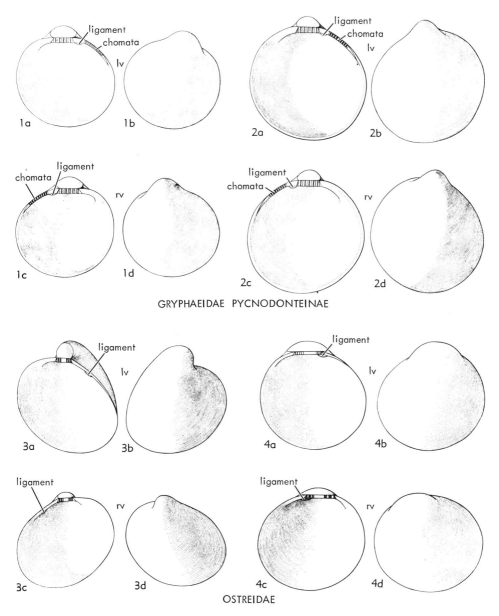

Fig. J39. Full-grown prodissoconchs of oysters, ×76 (Ranson, 1960a). All from living species.

1,2. Gryphaeidae-Pycnodonteinae. — *1a-d. Hyotissa hyotis* (Linné), IndoPacific.—*2a-d. Neopycnodonte cochlear* (Poli), circumglobal.
3,4. Ostreidae.—*3a-d. Crassostrea virginica* (Gme-lin), east and south coasts of North America.—*4a-d. Ostrea edulis* Linné, west and south coasts of Europe, northwest Africa.

The ligament is at first, during the protostracum veliger stage and at the beginning of the prodissoconch veliger stage, in the middle of the hinge so that the two valves open up with the opposing umbones remaining close together (Fig. J37). Grad-

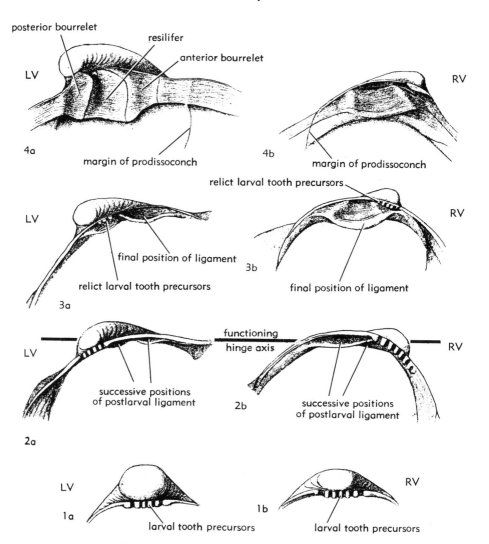

Fɪɢ. J40. Metamorphosis of hinge structures during growth from prodissoconch larval stage to early post-larval stage in oysters. An indeterminate species of pycnodonteine oyster (called *Ostrea flabellula*(?) in Bᴇʀɴᴀʀᴅ) from Calcaire Grossier (mid.Eoc.) of France, ×130 (Bernard, 1896).

1a,b. Prodissoconch showing well-developed larval tooth precursors in pattern characteristic of Pycnodonteinae.

2a,b-4a,b. Successive early postlarval stages showing

obliteration of the larval tooth precursors and growth of ligamental area with progressive shift of positions of the functioning ligament.

ually, as the valves become more inequilat-eral and their umbones more prominent, the ligament migrates along the hinge to-ward the anterodorsal valve margins, leav-ing the umbones and tooth precursors behind in their own places. The shift of the ligament takes place smoothly; at no time is the shell without a well-functioning

ligament. At the end of this shifting, the ligament joins what had been the antero-dorsal margins of the valves and the origi-nal umbones, the hinge, and the tooth precursors come to lie posterior of the ligament (Fig. J40, J41).

Judging by the arrangement of ligament and tooth precursors in the full-grown

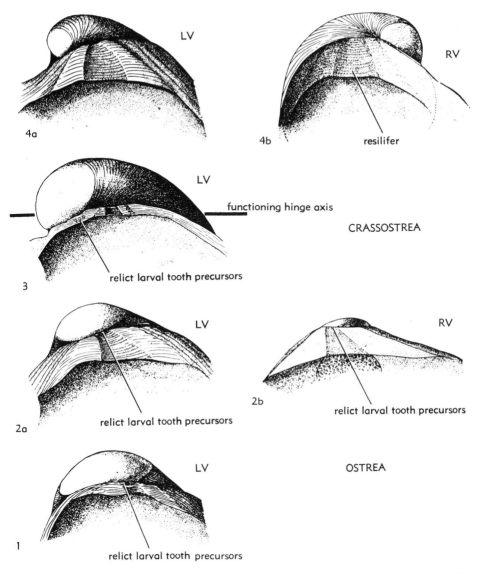

Fig. J41. Metamorphosis of hinge structures during early postlarval growth in oysters (Ranson, 1940).

1-2a,b. Ostrea edulis (Linné, 1758), ×250, three, five, and five days old.
3-4a,b. Crassostrea angulata (Lamarck, 1819),

×250, ×150, and ×250, two, five, and five days old.

prodissoconch, Bernard concluded that in the oysters all the tooth precursors represent only the posterior half of the set normally found in Bivalvia and that the hinge structure of larval oysters would have to be called a demiprovinculum (or half-provinculum), because it differed profoundly from the provinculum of other Bivalvia.

However, Ranson pointed out that the earlier larval stages of the oysters have a provinculum like other Bivalvia in that they have two sets of tooth precursors, one to each flank of the ligament.

Oyster larvae remain planktonic and free-swimming up to 33 days (Erdmann, 1934, p. 7). Incubatory oysters have shorter free-swimming larval periods than nonincubatory ones. Shortest free-swimming periods are recorded (WALNE, 1963; MILLAR & HOLLIS, 1963; HOLLIS, 1963) for *Ostrea (Ostrea) chilensis* PHILIPPI, living on the coast of Chile, and *O. (O.) lutraria* HUTTON, 1873, living on the coasts of New Zealand. They have free-swimming larval periods of a few hours to a few days. These extremes are thought to be special local adaptations of species living in exposed situations among strong currents, where longer periods would allow the larvae to be carried on out to sea into dangerous environments. The adaptations prevent excessive larval mortalities. Durations of larval periods depend also on prevailing temperatures and availability of food for the larvae. Warm and plankton-rich waters accelerate growth.

FIXATION

As the end of the prodissoconch veliger larval stage approaches, the larva becomes ready to attach itself to a substratum and undergoes extensive internal changes and rearrangements of its organs (metamorphosis). Nearly all oyster larvae must become attached to a firm substratum or perish.

Exceptions are few. Rarely do a few larvae of the living species *Ostrea (Ostrea) edulis* and *O. (O.) chilensis* succeed in growing into adults without becoming firmly attached (WALNE, 1963). Only a few extinct species have been discovered that were able to settle on a very soft substratum, such as ooze. All or nearly all their larvae grew into adults lying loose on the sea bottom, and the adults became sexually mature and propagated the species. Such species seen by me are:

1) *Exogyra laeviuscula* ROEMER, 1849, from the top chalk layer of the Dessau Formation (early Campan.) at the top of the Austin Chalk in central Texas.

2) *Ilymatogyra arietina* (ROEMER, 1852), from the Grayson Marl (Cenoman.) and other homotaxial formations of Texas and northeastern Mexico (Fig. J32; see Fig. J94).

3) *Rhynchostreon suborbiculatum* (LAMARCK, 1801) [=*R. columba* (LAMARCK, 1819)], Cenomanian and Turonian of Europe and Syria (see Fig. J97).

These three exogyrine oysters have very slender left beaks noticeably more spiro-

gyral than those of other exogyrine oysters which commonly have fairly large attachment imprints.

4) *Odontogryphaea thirsae* (GABB, 1861) from the Nanafalia Formation and other homotaxial formations (Sparnac.) of Alabama, Louisiana, and northern Mexico (Fig. J34; see Fig. J64).

These exceptional species are divisible into two categories: very specialized Exogyrinae (1-3 of above) and homeomorphs of *Gryphaea* (4). Such species, for some unknown reason, have traces of coloration preserved. They are in the form of brownish radial streaks (see "Pigmentation," p. *N997*, and Fig. J33, J34). Their shells have no visible attachment imprints at the left umbo. They may have incidental attachment imprints at other places caused by crowding of neighboring shells.

The full-grown oyster larva can swim with its velum and crawl about with its foot in search of a suitable location, when it is ready to become affixed to a substratum. Normally the substratum must be firm and clean, free of a film of mud, and free of the slimy gelatinous cover growing algae produce. The larva in search of a suitable location prefers close proximity of young adults of the same species (KNIGHT-JONES, 1951). This preference tends to maintain live oyster reefs and is essential to successful propagation.

Finally, the larva settles down with its umbones held up and its foot and ventral valve margins resting against the substratum. After some irregular movements, the final attitude is assumed: the right valve is on top, the mid-ventral margin of the left valve is pressed flat against the substratum and the commissural plane of the valves is inclined about 30° to it (see Fig. 78, p. *N94*). The byssus gland on the sole of the foot then excretes a drop of liquid which spreads in the narrow space between left valve and substratum. By natural tanning, the sticky liquid turns fairly rapidly into horny, no longer water-soluble conchiolin, firmly cementing the left valve of the larva to the substratum. This is the end of the larval stage and the beginning of rapid metamorphosis of the animal into a young adult.

All oyster larvae, except those of the Exogyrinae, assume approximately the same final attitude with reference to the substratum when they are full-grown and ready to affix themselves. Larvae of the Exogyrinae settle with the posterior margin of the left valve pressed flat onto the substratum so that rapid growth of the free anteroventral margin of the left valve, outgrowing the rest, initiates a spiral growth pattern immediately after fixation. This tilted fixation attitude of larvae of all Exogyrinae at the moment the larvae assume their final attitude just before fixation attests to the monophyletic origin of the subfamily and is one of the justifications for regarding them as a natural taxon and as a subfamily.

Because larvae of normal oysters can settle only where firm and solid substrata are available to them, oysters can inhabit only certain restricted brackish and marine habitats and are unable to occupy large sea-bottom areas that are covered exclusively by soft mud or ooze. Special adaptations are necessary for the oysters to be able to occupy sea bottoms that have only tiny shell fragments scattered on an ooze bottom or are composed of ooze exclusively. The four species listed above were adapted to the last-mentioned, highly specialized situation (compare p. N1071).

METAMORPHOSIS

The shell of the full-grown larva is composed of aragonite (STENZEL, 1964). As soon as the oyster has become attached, it begins depositing new shell material at the valve margins. This new ring is composed of calcite and shows special shell structures, for instance, the prismatic shell layer of the right valve. Henceforth, the major part of the shell deposited by the oyster is made of calcite.

Rapid and radical changes among the organs take place as soon as the larva has become fixed. Organs no longer needed disappear, making room for other needed organs to grow in size and to shift into more favorable positions.

The velum is a bulky locomotor organ in the larva, which is located anteroventrally. By its size it tends to push aside mouth and foot, away from the anterior adductor muscle. Because locomotion is no longer possible after fixation of the larva, the velum

and the foot with its byssus gland atrophy rapidly during metamorphosis. Thus a large space, formerly occupied by the velum, becomes available for other organs to expand or shift into.

Its fixed position prevents the young adult oyster from going after its food; rather, it must gather its food as it passes by. This the oyster does by using the gills as a sieve. As filter-feeding is adopted, the gills must become larger.

However, the mouth must remain near the anterior end of the sieve structure of the gills in order to be in correct position for receiving food from the gills. As the gills grow allometrically, the mouth must shift with them correspondingly. When the velum disappears, the mouth is free to shift forward, toward and against the anterior adductor muscle.

Because the gills have become so long that the mouth lies against the anterior adductor muscle, this muscle must become confined. It can neither shift nor grow in size; its fate is sealed.

In contrast, the posterior adductor muscle becomes unrestricted and can grow in size; it also can shift position to occupy the space vacated by atrophy of the foot and velum and by the shift of the mouth. Thus the posterior adductor muscle gets to occupy a more favorable position as to leverage, effectiveness, and ability to grow in size. Very soon it takes over the work of both adductor muscles and the anterior one disappears. The eyes of the larvae disappear also.

The foot, being a locomotor organ, becomes useless and disappears completely except for some anterior pedal muscles which take on new, as yet poorly understood functions and become the Quenstedt muscles of the adult oyster.

One of the results of the shifts among the organs is that the mouth-anus axis has to rotate counterclockwise during metamorphosis. If the pivotal axis of the hinge is chosen as a fixed reference line for the purpose of measuring the amount of torsion of the mouth-anus axis, that axis in the full-grown larva is at minus 70° and in the adult oyster at plus 47° to the pivotal axis; the total amount of torsion is 117° counterclockwise.

Metamorphosis of hinge structures was first investigated by BERNARD (1896) on fossil material. RANSON (1939a, 1940b) added to these observations, corrected them and gave the metamorphoses of the hinges on two living European oysters, *Crassostrea angulata* (LAMARCK, 1819) and *Ostrea edulis* LINNÉ (Fig. J41).

The structure of the hinge had already undergone considerable change during the growth of the prodissoconch veliger larva. The ligament had shifted from its original position in the provinculum, midway on the hinge, to a position considerably anterior to the larval tooth precursors. In that position the ligament and hinge structure supporting it grow vigorously during and after metamorphosis. The larval tooth precursors have become devoid of function and are left more and more behind, to become obscured by growth of the ligament and its supports. Their obliteration is finished during metamorphosis.

POSTLARVAL STAGE

GROWTH

Rate of growth of the postlarval oyster is influenced by local food supply, prevailing temperatures, and the amount of time the animal is able to keep its shell open and feed.

Most rapid growth is in the first three months after fixation. After that the growth rate declines gradually. Toward the end of a long life (after about eight years), the oyster individual ceases to grow in size as concerns the soft parts of its body so that volume of the shell cavity occupied by the soft parts remains nearly constant, but the shell continues to grow and the whole animal gains weight, except that destruction of the shell by bacterial corrosion and by attacks of shell parasites continues all the time.

Growth rates are smaller in winter than in summer, and growth may cease altogether in severe winters. Consequences of their climatic effect make it possible to estimate the age in years of an individual oyster.

Crassostrea madrasensis (PRESTON, 1916) [better named *C. cuttackensis* (NEWTON &

SMITH, 1912)], living in Madras Harbor on the tropical Indian coast in waters of 26.2-29.6°C., has shell growth rates of 0.27-0.62 mm. a day (PAUL, 1942). This is the fastest growth rate recorded for the shell of an oyster.

LIFE SPAN

Very few oysters have been observed individually for more than one to five years in their natural habitat so that direct observations of the life span are exceedingly scarce. However, there is a record (BJERKAN, 1918) of one old, extra-large individual of *Ostrea edulis* from near Bergen, Norway, which had been watched off and on for 20 years and which must have been 25 or more years old. Its valves weighed 251 gr. (left) and 204 gr. (right valve).

Two methods of estimating the age in years of an individual oyster through inspection of its shell are noted. Because the shell ceases to grow, or nearly so, during the winter, either one large or a group of closely spaced imbrications are produced on the valves in winter. They show better on the left than on the right valve. Thus the age in years should equal the number of such markers found on the left valve.

In a mild winter, however, imbrications may fail to show up altogether. On the other hand, extended or repeated storms and rainy years with long periods of too low salinities may produce additional, non-annual imbrications. This method of estimating the age of an individual oyster is obviously exceedingly difficult or even unreliable (MASSY, 1914) (see Fig. J42).

A better method, first used by POISSON (1946, p. 6), is based on the annual growth layers visible on the ligamental area. The consistency and composition of the shell material deposited in a year changes seasonally. In the colder season, during slow shell growth, the material deposited is darker and richer in organic matter; the shell substance deposited in the warmer months is lighter-colored, less rich in organic matter, and more solid and stronger. These annual layers can be seen when the shell is cut through in dorsoventral direction (BJERKAN, 1918). It is much easier without any sectioning, however, to see the annual layers on the ligamental area. Visi-

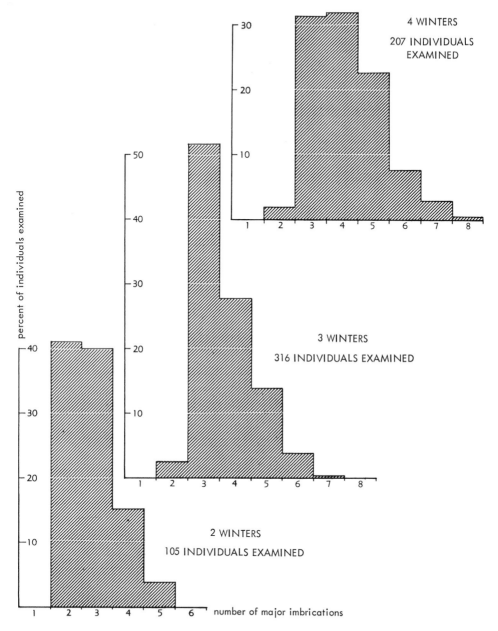

Fɪɢ. J42. Number of major imbrications on LV of *Ostrea edulis,* from Ireland, as related to number of winters survived by the animals, given in percentages of individuals examined (Stenzel, n. From data given by Massy, 1914).

bility of the annual layers on the ligamental area can be improved by a slight amount of buffing or washing with dilute hydro-chloric acid. Even this method used on old and large shells gives an incomplete estimate, because generally the first-grown part of older shells is lost through corrosion during the animal's lifetime (Fig. J14, J30; see Fig. J124).

The oldest oyster seen by me is a specimen of *Crassostrea bourgeoisii* (RÉMOND, 1863) from the Temblor Formation (Mio.) in California (Calif. Acad. Sci., no. 33269), which has a ligamental area 2.0 cm. long and 13.1 cm. high, as measured along the curves of the mid-line. The apex of the ligamental area is missing; the remainder shows 43 annual layers. Counting the annual layers of the ligamental area on the left valve of *Crassostrea gryphoides* (VON SCHLOTHEIM, 1813) from the Miocene of Europe, HARANGHY, BALÁZS, & BURG (1965) determined the life span of this oyster as more than 47 years.

SHAPE, FORM, AND SIZE OF SHELL

CONTENTS

Function of the shell of free-moving, normal Bivalvia, including the distant and unknown ancestors of oysters, is to protect all soft parts of the animal, including the foot. Shape of the shell and its capacity must be adapted to the animal's activities and locomotion and must be adequate to let the foot retract completely when the animal shuts its shell.

Because oysters are sedentary, they have lost the organ of locomotion, the foot. There is no need for their shell to enclose a foot and to be adapted to any kind of locomotion. The absence of these two necessities allows other factors to assume importance and to determine significant features of the shell. There are many such factors and they cause a bewildering profusion of shapes, forms, and sizes.

CONSEQUENCES OF SEDENTARY MODE OF LIFE

Sedentary animals tend to have rounded outlines, that is, cylindrical, conical, or globular shapes, and they tend to develop radially arranged organs. Oysters have the same tendencies, but cannot develop them much.

Ostrea (Ostrea) equestris SAY, 1834, living on the east coast of the Americas, tends to grow in orbicular outline, that is, the height becomes almost equal to the length of its shell, if the individual is not obstructed in its growth by neighboring shells or irregularities of the substratum (GALT-SOFF & MERRILL, 1962, p. 238). Many other oyster species follow this pattern.

Sedentary animals have no weight problems; they do not have to move their bulk and support of it is not difficult in most cases. For that reason, sedentary Bivalvia are commonly larger than mobile ones and have thicker and heavier shells. Many oysters follow this pattern.

INFLUENCE OF GILLS

Influence of the gills on the shapes of shells is proportionately larger in the oysters than in Bivalvia that have a large foot. This is so because oyster gills are proportionately larger and, in addition, their gills do not have to share the mantle/shell cavity with a foot.

The following features indicate influence of the gills: 1) Anterior and ventral valve margins are subparallel to the distal edges of the gills in all oysters. 2) The shells as a whole are noticeably curved to conform with the curved outline of the gills in all oysters. 3) The shell has a more or less conspicuously projecting posteroventral tip end, the branchitellum. It is close to the palliobranchial fusion, and the aboral end of the gills points toward it.

These features are most strongly emphasized in oysters that have a sickle-shaped outline. In them the gills are long and narrow, curved and strap-shaped, and the visceral mass is very small so that the gills determine outline of the shell.

The following genera have sickle-shaped outlines: *Agerostrea* VYALOV, 1936, *Rastellum* FAUJAS-SAINT-FOND [1802?], and *Cu-*

bitostrea SACCO, 1897 (see Fig. J116, J117, J133, J138, J139).

In addition, species of other genera approach or have sickle shapes, although congeneric species do not. Here belongs *"Ostrea"* [probably new genus] *quadriplicata* SHUMARD, 1860, from the Fort Worth Limestone to Mainstreet Limestone, Washita Group (late Alb.) of north-central Texas. Many species of the Exogyrinae related to *Ceratostreon* BAYLE, 1878, or placeable in that genus are narrow and sickle-shaped. An example is *Exogyra* [*sensu latissimo*] *sigmoidea* REUSS, 1844 (v. 2, p. 180), from the basal conglomerate (late Cenoman.-early Turon.) of the vicinity of Praha, Czechoslovakia (ZÁRUBA, 1965).

Agerostrea is related to *Rastellum* and differs from the latter by the smooth median area restricting the plications to the flanks of the shell. It is connected to *Rastellum* by transitional species. *Rastellum* has subequal valves, both ribbed; the zigzag commissure has many long tapering acute-angle points. The muscle pad is comma-shaped and much closer to the hinge than to the branchitellum. *Rastellum*, in turn, is connected to *Lopha* of the Triassic and Jurassic by many transitional species. There seems to be no room for doubt that the three genera have close relationships.

Cubitostrea has a ribbed, but not plicate, left valve and a flat nonribbed right valve; the right valve is considerably smaller than the left. It is closely related to *Ostrea*, from which it arose in Eocene time.

Thus, the sickle-shaped oysters originated from at least two roots. They are a polyphyletic convergent collective group. They are nonestuarine, euhaline, warm-water oysters apparently restricted to very fine-grained sediments, more or less rich in lime.

INFLUENCE OF PSEUDOSIPHONS

INHALANT PSEUDOSIPHON

Much, or even all, of the inhalant water current enters the mantle/shell cavity through the inhalant pseudosiphon, whenever the oyster has its shell open and is feeding (see p. N1000). The part of the gills that is nearest to the inhalant pseudo-

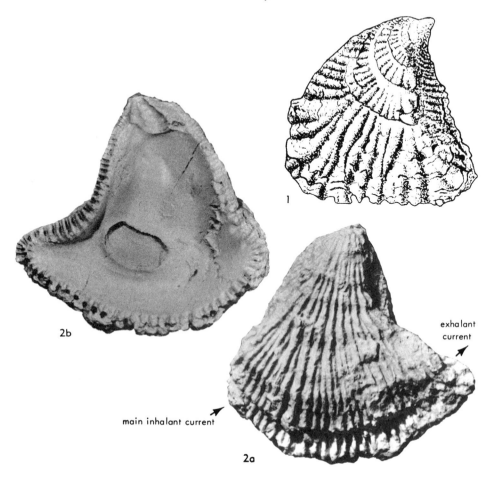

FIG. J43. Triangular shell outlines in oysters.

1. Cubitostrea perplicata (DALL, 1898), Tallahatta Formation, mid.Eoc.(Lutet.), Alabama, USA, exterior of LV, ✕0.7 (Stenzel, 1949).
2. "Lopha" villei (COQUAND, 1862), U.Cret.(Maa-stricht.), Chouf Motta, Algeria; *2a,b,* LV, ext., int., ✕0.7 (Stenzel, n. Specimen by courtesy of C. W. DROOGER, Univ. of Utrecht, Nether-lands).

siphon is in the best position for receiving the current and for collecting food particles. It is to advantage of the animal to provide the best catchment area of the gills at that place. For this reason, the gills of all oysters are widest there and taper thence to both ends of the gills.

How rapidly the gills taper from their widest place depends on the stability, site, and size of the pseudosiphon. If it remains small and opens up at the same place every time the oyster is feeding, the taper of the gills is likely to be rapid and the gills will have the outline of a low elongate triangle

curved in conformity of the shell and the widest place on the gills, the apex of the triangle, will be well defined. On the other hand, if the pseudosiphon is a long slit and opens up variably in a general vicinity, the taper of the gills likely will be much less rapid and there will be no narrowly defined widest place on the gills.

In the first case the shell will have a projecting anteroventral rounded corner. Such shells are more or less clearly trian-gular in outline: *"Lopha" villei* (COQUAND, 1862), Maastrichtian of N.Africa (Fig. J43); *Cubitostrea perplicata* (DALL, 1898) from the upper Tallahatta Formation (Lutet.) of Alabama, USA (see STENZEL, 1949, fig. 3; GEKKER, OSIPOVA, & BELSKAYA, 1962, v. 2, fig. 24-1) (see Fig. J116, J117); *C. prona* (S. V. WOOD, 1861) from the Lattorfian beds of England. Triangular shapes are possibly restricted to oysters that have no promyal passage and in which the convexities of the two valves are subequal.

EXHALANT PSEUDOSIPHON

The exhalant pseudosiphon takes care of most of the exhalant water current of the oysters, and metabolic waste products are expelled through it. These waste products present problems of sanitation; they must be disposed of in such a fashion that they cannot by any chance reenter the mantle cavity of the animal.

The Gryphaeidae are the only oysters in which the exhalant pseudosiphon has great influence on the configuration of the shell. Their left valves are divided by a radial **posterior sulcus** into a main part and a smaller **posterior flange** (see Fig. J74, J83, J84, J87).

In most species the sulcus is a deep groove, in others it is shallow; in some, notably in *Gryphaea arcuata* LAMARCK, 1801, from the Liassic of western Europe, it is evanescent as a groove, but remains indicated by a flexure in the trend of the growth lines. In the strongly plicate genus *Hyotissa* of the Pycnodonteinae the sulcus is difficult to discern from the other sharp plications of the shell. However, it makes a deeper and broader groove on the left valve than the other plications do, and the corresponding radial fold on the right valve is a broader, higher, and more prominently upturned ridge than the other plication ridges in this genus.

At the posteroventral margin of the left valve the sulcus ends with an obtuse, rounded angulation shown by the growth lines of the valves. This angulation tends to become more obscure with age. This angulation is the branchitellum of the Gryphaeidae (see Fig. J74,*3e,f*).

The posterior flange of the left valve has a convex or highly convex cross section differing in shape from species to species. The terminus of the posterior flange at the valve margin is the place at which the exhalant pseudosiphon was located and at which the organic waste products of the oyster were ejected while the animal was alive. The posterior flange is the growth track of the successive positions the exhalant pseudosiphon occupied during the growth of the animal.

LEOPOLD VON BUCH (1835) was the first to attempt an explanation for the posterior flange and sulcus, which he had noticed on *Gryphaea*. According to his explanation the pull of the posterior adductor muscle caused formation of the sulcus. This hypothesis is hardly valid, however, because the shell morphology is determined by the attitude of the shell-secreting edges of the mantle lobes, that is, at the very edges of the valve margins, and these are well removed from the insertion of the adductor muscle. Be it to his credit that he was the first to recognize in print that there is a problem of explaining the feature.

INFLUENCE OF SUNLIGHT

MEDCOF (1949) experimented with *Crassostrea virginica,* growing two sets side by side, one exposed to sunlight and the other shaded. Specimens exposed to sunlight grew (linear growth rate of shell) about 30 percent slower than shaded ones and their shell walls were thinner and had fewer chalky deposits than the shaded set. Those exposed to sunlight had stronger, harder, and denser shells and their shell walls had higher specific gravities.

Oysters growing in the shade had flat or concave right valves mostly; many of those growing in the sunlight had convex right valves. Oysters exposed to sunlight had more chunky compact shells and their left valves were more convex.

The left valves of oysters exposed to sunlight had more prominent ornamentation. The left valves tended to become ribbed or fluted and to have a serrate margin. The right valves had more color streaks (Fig. J44).

Oysters exposed to sunlight had abundant sea lettuce *(Ulva)* growing on them but had fewer mussels *(Mytilus edulis* LINNÉ) and oyster spat attached. Shaded oysters had very little of sea lettuce, but had about three times as many mussels and spat growing on them, on the upper

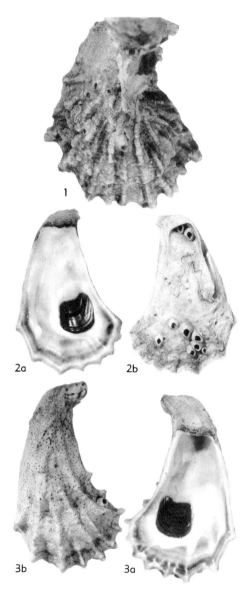

FIG. J44. Shell features of *Crassostrea virginica* (GMELIN) taken alive in Tarpon Bay, Sanibel Island, Florida, USA, ×0.7 (Stenzel, n). [These oysters grew exposed to air at low tide and show the strong ribs characteristic of exposure to sunlight. Specimens courtesy of R. TUCKER ABBOTT, Acad. Nat. Sci. Philadelphia.]—*1.* Outside view of LV.—*2a,b.* Inside and outside views of RV.—*3a,b.* Inside and outside views of LV, same individual.

as well as the lower surfaces. The upper surface of oysters exposed to sunlight was avoided by epibiotic animals.

Exposure to sunlight and different substrata possibly are the causes of extreme polymorphism in a case analyzed for the first time by THOMSON (1954, p. 146-149, pl. 2, fig. 3-4, and pl. 3, fig. 1-4). The case involves the species which he called *Ostrea folium* LINNÉ. He came to the firm conclusion that the names *Ostrea folium* LINNÉ (1758, p. 699), *Mytilus crista galli* LINNÉ (1758, p. 704), and *Ostrea (Pretostrea) bresia* IREDALE (1939, p. 396) all stood for the same polymorphous species, which he chose to call *O. folium,* because it has page priority and because as to shape it is intermediate between the other two (see Fig. J47, J129, J130).

According to THOMSON, individuals of that species which grow in intertidal situations and are attached to a flattish substratum or to each other develop long and strong hyote claspers and well-defined strong plications few in number; they are the ecomorph known as *Lopha cristagalli.* While individuals that remain completely submerged and grow preferentially on gorgonid corals develop a lanceolate outline, many small plications and many small and short claspers embracing the gorgonid stem. These ecomorphs are commonly called *Lopha folium* (LINNÉ) and are intermediate between the other two as concerns plications. *Ostrea bresia* is founded on individuals of orbicular outline furnished with irregular, more rounded plications.

It so happened that *L. cristagalli* became the type species of *Lopha* RÖDING, 1798, *L. folium* the type species of *Dendostrea* SWAINSON, 1835, and *O. bresia* the type species of *Pretostrea* IREDALE, 1939. Thus *Lopha* has clear priority and the other two become its junior subjective synonyms. As to the name for this polymorphous species, THOMSON's choice must stand, because he is the discoverer of these relationships and the first reviser in the sense of Article 24.

The East African mangrove oyster *Saccostrea cuccullata* (VON BORN, 1778) has two distinct ecomorphs: 1) a chunky irregularly shaped oyster, commonly rounded and flat, with fairly thick shell walls and coarsely crenated valve margins, and 2) a spiked form that has thinner, more fragile shells with foliated shell shoots, up to 3 cm. long, extending outward from the shell margins. The spiked form grows at a higher intertidal level and is exposed to air and sunshine for 7.75 to 124 hours during low tides. The other becomes exposed 0 to 20 hours only. So different are the two, that SOMEREN & WHITEHEAD (1961) considered them different species. It is, however, more likely that they are ecomorphs of the same species (see Fig. J54).

INFLUENCE OF SUBSTRATUM

Shell shape is influenced by the substratum in two separate and fundamentally differing ways: 1) Fortuitous nonheritable indivdual effect produced by a local hard substratum resulting in xenomorphism on an individual oyster and 2) the mass effect produced by adaptations of large populations to special widespread facies of the substratum resulting in adaptive, heritable features.

XENOMORPHISM

The outside surface configuration of one valve closely imitating the configuration of the substratum under the other valve is called **xenomorphism** or xenomorphic sculpture. In some individual oysters the umbonal region of the right valve resembles closely the configuration of the hard substratum to which the left valve is or was originally attached. Xenomorphic sculpture is patterned after ammonites, belemnites, snails, trigonias or other bivalves, echinoids, echinoid spines, coral colonies, twigs of sea fans *(Octocorallia)*, or other hard materials, depending on which animal or object the oyster was growing. Xenomorphism is found also in other attached shell-bearing animals, for example, *Anomia* (STENZEL, KRAUSE, & TWINING, 1957), *Myochama, Plicatula,* barnacles, brachiopods.

Xenomorphism in oysters was noted as an unexplainable curiosity by pre-Linnean authors, for instance, D'ARGENVILLE (1742, p. 316, pl. 22, fig. F), and has been described in innumerable cases (JUDD, 1871; SCHÄFLE, 1929; SAINT-SEINE, 1952; STENZEL, KRAUSE, & TWINING, 1957, p. 98-99). The first to give a good description was DEFRANCE (1821, p. 21) and the first to explain it was GRAY (1833, p. 780-784). Good bibliographies are given by SCHÄFLE (1929) and SAINT-SEINE (1952).

The first to apply a definite term to it were STAFF & RECK (1911, p. 166-167). They applied the German adjective allomorph [=allomorphic] to it, contrasting it with automorph. Neither term was defined by them, presumably because both were in current use in Germany in other situations. Although allomorphic has been adopted by several authors, it is inappropriate and is here replaced by xenomorphic.

It is derived from the Greek pronoun ἄλλος, another, that is, a person or thing of the same kind besides the one mentioned before, and the Greek noun μορφη, shape or form. Allomorph (noun) is used correctly in mineralogy and chemistry where it defines two or more different crystalline forms of the same chemical substance. Xenomorphism is derived from the Greek ξενος, strange (adjective) or stranger (noun).

Xenomorphism in oysters remains restricted to the umbonal region of the right valve. The area occupied by it is nearly always devoid of growth squamae or prominent growth lines and has exactly the same size and outline as the attachment area on the left valve of the same individual. In young, small, thin-shelled individuals it may occupy all of the right valve. In old, large, thick-shelled ones it is limited to the right umbonal region, and the rest of the valve has a configuration normal (idiomorphic) for the species and the inside of the valve is smooth (Fig. J45).

If the oyster shell is detached from its substratum, its left valve carries the negative impression of the substratum on its attachment area and the right valve has the corresponding positive xenomorphic configuration, which is a replica of the substratum. However, the replica is always less sharp than the impressions on the left valve. Points or ridges on the right valve are wider and more rounded, the corresponding pits or grooves on the left are sharper and have narrower crests (STAFF & RECK, 1911).

Xenomorphism results in reversed convexities of the two valves in some cases. For example, if the substratum is a flat surface, the attachment area is flat too, but the corresponding xenomorphic area is convex. Thus the usual condition is reversed among the two valves. Were it not so, there would be no room for the soft parts of the animal.

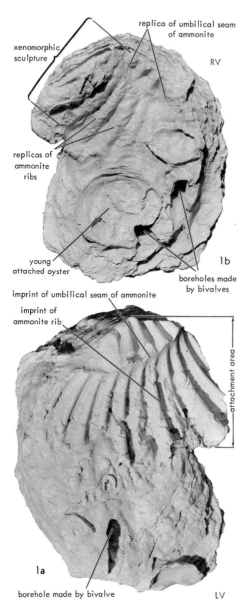

FIG. J45. Xenomorphism of oysters. *Ceratostreon texanum* (ROEMER, 1852), upper part Walnut Formation, L.Cret.(M.Alb.), Coryell County, Texas, USA, showing xenomorphic sculpture patterned after ammonite *Oxytropidoceras,* ×0.6 (Stenzel, n).—*1a.* Attachment area of LV.—*1b.* Xenomorphic sculpture on RV.

Xenomorphism is caused by simple passive molding while the shell is growing at the valve margins. As long as the oyster

is young and its shell is very thin, it must grow closely appressed to its substratum. Were it to grow away from this, it would be too exposed to predators which could crush its shell all too easily. At the very thin growing shell margin, the left valve grows over the substratum and produces a detailed negative impression of it on the attachment area; the right valve margin must leave no gap at the valve commissure so must follow or mold over the left valve margin with all of its irregularities. In this way, the outside surface of the right valve is molded and reproduces the configuration of the substratum (Fig. J46).

As the oyster grows older and bigger, the mantle/shell margins start to turn up and away from the support, passive molding ceases, and the shell attains its specific (idiomorphic or automorphic) configuration. Only after that stage is reached, it becomes possible for growth squamae to diverge from the surface of the shell so that the xenomorphic portion remains free of them. Idiomorphic configuration is commonly accompanied by a thickening of the shell. Additional shell layers are deposited on the inside of the valves, making them thicker, and smoothing out any irregularities on the inside of the valves (Fig. J46).

Obviously xenomorphic sculpture is not heritable. Some oyster species, however, have a heritable predilection to become attached to selected substrata, for example, twigs of Octocorallia. Or in certain areas, there may be only one kind of firm substratum available so that all or most individuals of an oyster species will grow on the same kind of substratum and in this fashion they will all have similar xenomorphic sculptures. For these reasons, xenomorphic sculpture may appear to be a specific characteristic, and many misconceptions based on it are found in the literature.

Oysters growing on the cylindrical twigs of sea fans (Octocorallia) have rather unusual and prominent xenomorphism. It consists of a narrow rounded longitudinal ridge extending along the longest diameter of the right valve. The attachment area is a deep rounded channel, and the left valve has many shelly processes that clasp the twig. The twig of the octocoral, when it is dried out, becomes blackish and longitudinally wrinkled, looking very much like

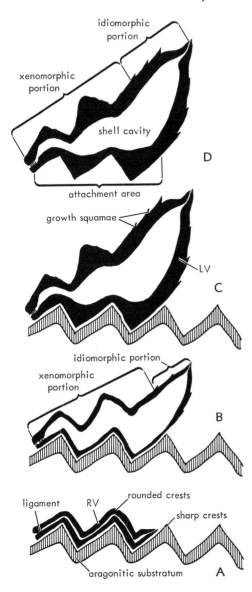

FIG. J46. Origin of xenomorphism on oyster shells (Stenzel, n).

A. Neanic stage of rapid uninterrupted growth showing thin shell walls, narrow shell cavity, lack of growth squamae, and growth closely appressed to substratum, which is here assumed to be a sharply plicated aragonitic shell.

B. Beginning of idiomorphic growth stage showing increasing size of shell cavity, first appearance of growth squamae by intermittent growth, and LV growing away from substratum.

C. Advanced idiomorphic growth stage showing thicker shell walls and more growth squamae, and smoothing out of inside faces of the two

valves through progressive obliteration of their impressed configuration.

D. Freeing of shell from its substratum through disappearance by leaching of aragonitic substratum.

a plant twig or root. It is easy to mistake it for part of a woody plant.

The first such oyster to be noticed was a living tropical species figured by D'ARGENVILLE (1742, p. 316, pl. 22, fig. F), who described it as a *"feuille"* [leaf] and assumed that it had been attached to a *"morceau de bois"* [piece of wood]. Consequently, LINNÉ (1758, p. 699, no. 178) named this species *Ostrea folium*. In many specimens the right valve looks like a lanceolate leaf with the xenomorphic longitudinal ridge as the midrib of the leaf (Fig. J47).

SWAINSON (1835, p. 39) later proposed the genus *Dendostrea* for these "tree oysters" and *Ostrea folium* was made the type species by HERRMANNSEN (1847, v. 1, p. 378).

THOMSON (1954, p. 146-149 and letter dated December 14, 1966) insists that the three: *Lopha cristagalli* (LINNÉ, 1758), which is the type species of *Lopha* (see Fig. J129); *Dendostrea folium* (LINNÉ, 1758), which is the type species of *Dendostrea* (Fig. J47); and *Ostrea (Pretostrea) bresia* IREDALE (1939, p. 396-397, pl. 7, fig. 4), which is the type species of *Pretostrea* (see Fig. J130) really are not three distinct species but actually are ecomorphs of one species. If this is so, *Dendostrea* and *Pretostrea* are junior synonyms of *Lopha* and it becomes clear that shell shape and outline may not be reliable generic features, at least in the Lophinae.

Examples of this sort of xenomorphism in fossil oysters are *"Ostrea" dorsata* DESHAYES (1824-37, v. 1, p. 355, pl. 54, fig. 9-10; pl. 55, fig. 9-11; pl. 14, fig. 1-4) from Bartonian beds of the Paris Basin and *"O." russelli* LANDES (RUSSELL & LANDES, 1940, p. 139-140, pl. 3, fig. 3-6) from the Late Cretaceous Pakowki Formation of Alberta, Canada.

DEFENSE ADAPTATIONS

In order to keep enemy intruders out of the mantle/shell cavity oysters have devel-

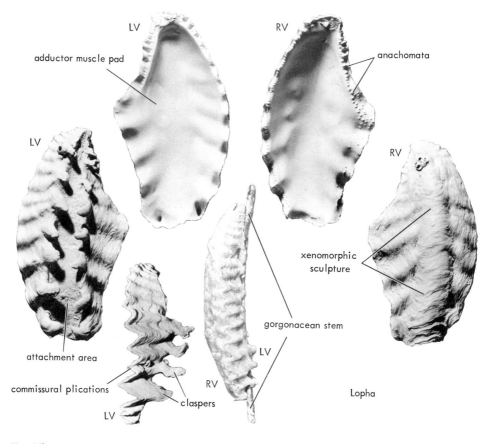

FIG. J47. Ecomorph (2) of *Lopha folium* (LINNÉ, 1758) attached to gorgonacean twig, showing shelly claspers and xenomorphic sculpture, W. Indies, ×0.7 (Stenzel, n. Specimens by courtesy of R. TUCKER ABBOTT, Philadelphia Acad. Nat. Sciences).

oped several defensive features: 1) complete shell closure, 2) commissural shelves, 3) commissural plications, and 4) hyote spines. All these features concern the valve margins or originate at them, where entry might be gained.

SHELL CLOSURE

Oysters can close their shells completely, leaving no gap or chink between the valves. Some of the other bivalves have such gaps; they are reserved for special organs or functions, for example, the byssus, the foot, or the two siphons. When ancestors of the oysters changed their method of attachment from a byssus to cementation of a valve, they gained the advantage of being able to close their valves tightly.

The closure is almost watertight because the thin, flexible conchiolin fringe around the right valve molds itself like a gasket onto the interior surface of the calcareous valve margin of the left. Thus oysters are able to shut out enemies and to survive periods of unfavorable environmental conditions.

For example, the rock oyster *Saccostrea cucullata* (VON BORN, 1778), growing on sea cliffs, is able to survive exposed to air and tropical sun during low tide (MACNAE & KALK, 1958) (see Fig. J50, J51). Many other species are able to survive exposure to air during low tides under less severe conditions. Prolonged periods of unfavorable salinities, either too low or too high, can be survived in the same way.

COMMISSURAL SHELVES

The commissural shelf of a valve is a flattened band along the valve margins. It is better developed on the dorsal than on the ventral half of the valve. It is narrower near the dorsum and widens toward the venter. It is set off from the shell cavity proper by an angulation, which is gentle on the ventral half and strong on the dorsal half of the valve.

The Pycnodonteinae have the best developed shelves (see Fig. J83). In some Exogyrinae (particularly *Exogyra s.s.*), the anterior margin of the right valve has the commissural shelf reflexed at an angle of about 40° from the general plane of the valve (see Fig. J89, J92, J96). This condition makes itself manifest on the outside of the valve by a succession of sharp parallel upstanding concentric crests or growth squamae and on the inside of the valve by a sharply defined commissural shelf delineated proximally by a well-defined smooth ridge.

A commissural shelf makes it more difficult for certain predators to reach the soft parts in the shell cavity. Predators that break pieces of the shell margins in order to gain entry (margin breakers) must break through the entire width of the commissural shelf before they reach the bulk of the soft parts. The commissural shelf is of some use also against small parasites that wait to sneak in while the oyster has its shell open. Quick shutting of the valves may not exclude them, but may catch them before they can finish crossing the shelf.

COMMISSURAL PLICATIONS

Commissural plications take the form of acute-angular zigzags or rounded sinuous undulations of the shell margin at which the opposing valves fit into each other, when the shell is shut. All Lophinae have them, except that some genera (e.g., *Agerostrea*) attain some size before plications show up (see Fig. J133). *Hyotissa* of the Pycnodonteinae commonly has several zigzag plications and the living southwestern Pacific growth form *H. hyotis* (LINNÉ, 1758) *forma sinensis* (GMELIN, 1791) is quite strongly plicate (see Fig. J85). No commissural plications have been found on the Exogyrinae and Gryphaeinae.

The most acute and numerous plications are found on *Rastellum* and in particular on *R. (Arctostrea) carinatum* (LAMARCK, 1806), originally described from Cenomanian beds near Le Mans, Département Sarthe, France, but distributed over most of the globe. The species has a plication prong angle ($2a$) of about 30° and about 3.5 prongs per cm. along the valve margins.

Plications on oyster shells are positioned in such a way that the bisectrix of each prong angle at the valve margin is normal to the generalized commissural plane of the shell. Therefore, as the valves are opened, the width of the slit opened between them remains the same on each side of a given prong.

The plication angles of the various successive prongs on a shell are unequal and form a progressive graded series; the most acute angle of such a series is the one the farthest away from the hinge axis and the least acute is the one nearest. In plicate oysters that are not or little crescentically curved the graded series progresses with continuously decreasing plication angles from the hinge to the region of the branchitellum.

In the highly crescentic and recurved species of *Rastellum (Arctostrea)* and *Agerostrea* the branchitellum reapproaches the hinge axis and the angles of the prongs around the branchitellum are less acute than those halfway between hinge and branchitellum, at the convex outside curve of the shell. Thus in progressing from the hinge to the branchitellum the prong angles first become more acute and then less acute.

Wherever the sizes of the plication angles differ greatly from one plication to the next, the flanks of the plications are curved in order to accommodate the change without interfering with the normal attitude of the bisectrices of the angles. This situation is noticeable in species with few and large prongs around the periphery. In such shells one flank is convex and the other concave; the concave flank always faces the hinge axis.

These rules follow from the geometry of the valves and their rotational movements when the valves open or close and were

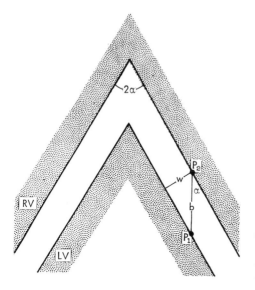

Fig. J48. Geometry of a plication on the margins of an oyster shell (Stenzel, n).

first investigated by Rudwick (1963, p. 138, fig. 3A). Cumings (1903, p. 131-132) was the first to suggest that plicate valve commissures in brachiopods and in oysters are either adaptations for defense against predators trying to enter or reach into the mantle/shell cavity or adaptations to make it impossible for overly large particles to be carried into the mantle/shell cavity. These ideas have been explored further by Herta Schmidt (1937, p. 27-30) and Rudwick (1964).

Two neighboring points (P_1 and P_2), each on an opposite valve, become separated by the distance b, when the plicate oyster opens its valves. Because the commissure is plicate the slit that opens between the valves is narrower than b. The width (w) of the slit is $w = \sin a \times b$. The sin a is always smaller than 1, if the angle a is less than $90°$. Therefore, sin $a \times b$ is always smaller than b, and the width (w) of the slit is smaller, the smaller angle a may be, that is, the more acute the plication angle $(2a)$ (Fig. J48).

HYOTE SPINES

Hyote spines are tubular outgrowths, hollow and cylindrical, that arise periodically from thin edges of the shell margins of oysters. They are open distally, at their tips as well as along their distal flanks. Their tips are rounded ear-shaped openings. They start as accentuations of ridges or costae on the shell. As the shell margins continue to grow, they become closed off at the base and new ones are formed farther on. The older ones are followed by newer ones in radial rows. As long as they are not closed off an extension of the mantle lobe occupies them, lining their shell wall (Fig. J49).

They are best developed on and named after *Hyotissa hyotis* (Linné, 1758), the pycnodonteine oyster living in the Indo-Pacific (see Fig. J85). Other plicate or costate oysters have such spines too, for example, *Exogyra (Exogyra) spinifera* Stephenson, 1941, from the Peedee Formation (Maastricht.) of Robinson's Landing on Cape Fear River, North Carolina.

As long as they are not closed off at the bottom and have a lining of sensitive mantle lobe tissues, they act as sensory organs extending the reach of sensitivity of the mantle edges (Rudwick, 1965). All of them are deterrents to predators making access to live tissues more difficult.

SIZE

Extraordinary shell size and corresponding thickness of the shell wall show up either individually or as a minor generic characteristic.

An individual oyster may reach great size, if by good fortune it manages to survive much longer than its neighbors. Such a case is the individual (listed below on p. N1028) of *Ostrea edulis* from near Bergen, Norway, which probably reached an age of

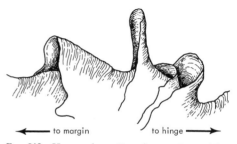

← to margin   to hinge →

Fig. J49. Hyote spines. Four hyote spines arising from radial rib on RV of *Hyotissa hyotis* (Linné, 1758) *forma imbricata* Lamarck, 1819, living at 30 fathoms off southern Japan, ×2 (Stenzel, n).

26 or 27 years, much of it under observation (Bjerkan, 1918). Survival of such large specimens today has become quite unlikely, as human populations have increased and oyster fishing has become intensified. For this reason fossil oysters generally tend to be larger than those of today.

After about eight years the soft parts of an oyster cease to increase in size and the volume of the mantle/shell cavity remains constant, but shell deposition does not cease. The shell increases in weight and size, but the shell cavity does not enlarge. Accordingly, after about eight years the length of the ligament does not increase, but the ligament continues to shift from the dorsum toward the venter and the ligamental area increases in height, if not in length.

Among various oyster genera, *Crassostrea* seems to be capable of growing fastest. Species of this genus tend to reach larger sizes, although there are some that do not grow particularly large. The living northeast Asian *C. gigas* (Thunberg, 1793) grows large rapidly. This species is probably the direct descendant of the giant Miocene species *C. gryphoides* (von Schlotheim, 1813, p. 52). This well-known, widespread, and conspicuous species (Rutsch, 1955) ranged in the Miocene Tethys Sea from Spain, southern France, Switzerland, Swabia, northern Africa, Somalia, and Cilicia in Asia Minor to Japan. Because of its wide geographic range the species has received numerous, supernumerary names. It appears to have crossed the Pacific, for in California there is the giant *C. titan* (Conrad, 1853) from the late Miocene. The group of large crassostreas was discussed by Dollfus (1915) and Dollfus & Dautzenberg (1920, p. 465-469). However, Dollfus erroneously included several species of *Ostrea* and *Pycnodonte* and his claim that *Crassostrea* and *Ostrea*, as well as *Crassostrea* and *Gryphaea*, are connected through transitional species is unacceptable.

MAXIMAL SIZES AND WEIGHTS

The largest oyster shell on record has a left valve 32 by 76 cm. in size (Awati & Rai, 1931, p. 9) and is an individual of *Saccostrea cuccullata* (von Born, 1778); presumably it was collected near Bombay, India. The species is a complex superspecies living widely distributed in the Indo-Pacific.

Many extraordinary sizes reported in the literature are estimates made in the field without benefit of a tape measure and most of them are exaggerated. The following list gives data believed to be trustworthy (H=height; L=length; T=thickness; W=width; Wt=weight).

1) *Aetostreon couloni* (Defrance, 1821) [=*Exogyra sinuata* (Sowerby, 1822)], Lower Greensand (late Apt.), Perthe du Rhone, France: 17.7 by 20.5 cm. [Quenstedt, 1867].

2) *Crassostrea blanpiedi* (Howe, 1937), Chickasawhay Formation (early Miocene), Wayne County, Mississippi, USA: H, 30.5 cm.; L, 15.3 cm.; W, 12.8 cm.; ligamental area, 3.8 by 7.7 cm. [Howe, 1937].

3) *Crassostrea cuttackensis* (Newton & Smith, 1912) [=*C. madrasensis* Preston, 1916], living (Rec.), Pulicat Lake (ca. 70 km. north of Madras), India: H, 43.2 cm.; L, 17.8 cm. [Moses, 1927, p. 552].

4) *Crassostrea gigas* (Thunberg, 1793), living (Rec.), Enoshima Island (near Yokohama), Japan: H, 44.8 cm.; L, 9.5 cm.; H, left ligamental area, 9.5 cm. [Pilsbry, 1890, p. 95].

5) *Crassostrea gryphoides* (von Schlotheim, 1813) [=*Ostrea grandis* Serres, 1843], Miocene marnes argilocalcaires, Tessan (near Béziers), Dépt. Hérault, France: H, 59-60 cm.; L, 7-8 cm.; T (shell wall in center of valve), 4.5-5.0 cm. [Serres, 1843, p. 143-147, pl. 2, fig. 1].

6) *Crassostrea gryphoides* (von Schlotheim, 1813) [=*Ostrea ponderosa* Serres, 1843], Miocene marnes argilocalcaires, Gremian (near Montpelier), Dépt. Hérault, France: Wt, 4.45-4.85 kg.; T (shell wall), 15 cm.+ [Serres, 1843, p. 143, 150-151].

7) *Crassostrea titan* (Conrad, 1853), Neroly Formation (late Miocene), near Bitter Creek, Santa Barbara County, California, USA: H, 44.4 cm.; W,

22.9 cm. (Fig. J14). [EATON, GRANT, & ALLEN, 1941, expl. pl. 4, fig. 3].

8) *Crassostrea virginica* (GMELIN, 1791), subfossil from shell mound (Pleist.), Damariscotta River, southern Maine, USA: H. 35.5 cm.; L, 11 cm. [INGERSOLL, 1881, p. 82, pl. 30, fig. 22].

9) *Crassostrea virginica* (GMELIN, 1791), living (Rec.), Boothbay Harbor, Lincoln County, Maine, USA: distance from hinge to ventral border, 20.6 cm.; H (left ligamental area), 5.5 cm.; H (right ligamental area), 4.5 cm.; L, 9.7 cm.; W (near hinge), 6.5 cm.; Wt (shell), 1175 gm.; Wt (soft parts), 35.8 gm.; Wt (enclosed liquid), 19.2 gm.; Wt (total), 1230 gm. [GALTSOFF, 1964, p. 20-21].

10) *Exogyra (Exogyra) erraticosta* STEPHENSON (1914), Pecan Gap Chalk (Campan.), Austin-Manor road, Travis County, Texas, USA: max. dimension, 20.1 cm.; Wt (both valves and internal mold), 3939.5 gm. [STENZEL, personal collection].

11) *Hyotissa hyotis* (LINNÉ, 1758), living (Rec.), New Caledonia: H, 28.5 cm.; L, 23.4 cm.; W, 19.0 cm. [Mus. Natl. Hist. Nat., Paris, France].

12) *Hyotissa hyotis* (LINNÉ, 1758) [= *Ostrea cristagalli* ROUGHLEY, 1931 (*non* LINNÉ, 1758)], living (Rec.), Great Barrier Reef, Queensland, Australia: H, 24.8 cm.; L, 15.6 cm.;

W, 11.4 cm.; Wt (shell), 1885 gm. [ROUGHLEY, 1931].

13) *Hyotissa hyotis* (LINNÉ, 1758), living (Rec.), Direction Island (Cocos-Keeling Is.), Indian Ocean, depth 10-20 ft.: H, 28 cm.; L, 28 cm.; W, 15 cm.; Wt (right valve), 1673 gm.; Wt (left valve), 2750 gm. [Philadelphia Acad. Nat. Sci.].

14) *Ostrea edulis* LINNÉ (1758), living (Rec.), estimated age 26-27 yrs., near Bergen, Norway: Wt (shell), 1228 gm.; Wt (soft parts), 55 gm.; Wt (enclosed liquid), 67 gm.; Wt (total), 1350 gm. [BJERKAN, 1918].

15) *Ostrea edulis* LINNÉ (1758), living (Rec.), age 10+ yrs., Salcombe Estuary near Salstone, Devonshire, England: H, 17.6 cm.; L, 19.8 cm.; W, 5.9 cm.; Wt (shell), 1038 gm.; Wt (soft parts), 94 gm.; Wt (enclosed liquid), 133 gm.; Wt (total) 1265 gm.; mantle cavity volume, 130 cm³. [ORTON & AMIRTHALINGAM, 1930].

16) *Rastellum (Arctostrea) aguilerae* (BÖSE, 1906) [=*Arctostrea atkinsi* RAYMOND, 1925], Habana Formation (Maastricht.), 1 km. NW of Dos Hermanos on road to Abreus, Santa Clara Prov., Cuba: diameter tip to tip, 23.5 cm.; W (right valve), 7.0 cm. [SOHL & KAUFFMAN, 1964, p. 414-419].

TERMS, CHIEFLY MORPHOLOGICAL, APPLIED TO OYSTERS

Terms regarded most important are in boldface type (as **adductor muscle**); use is not recommended of those printed in italics (as *epidermis*). Some nonmorphological terms have been included for aid to biologists and zoologists who may not be familiar with them. These are enclosed by square brackets. Many terms have general application to Bivalvia (*Treatise*, p. *N*102); additional ones considered by H. B. STENZEL to be especially applicable to oysters are accompanied by an asterisk (*).

aboral*. Pointing away from mouth.

adductor muscle*. Single posterior muscle connecting the 2 valves, tending to close them.

adoral*. Pointing toward mouth.

alate. With wings or auricles.

alivincular. Type of ligament not elongated in longitudinal direction nor necessarily situated entirely posterior to beaks, but located between cardinal areas (where present) of respective valves, with lamellar layer both anterior and posterior to fibrous layer; example, *Ostrea*.

[allochthonous*.] In structural geology, pertaining to rock masses that tectonic forces have transposed to rest on a strange base.

allometric growth*. Growth by unequal rates in different parts of an animal.

[allomorph (noun)*.] Any of 2 or more diverse crystalline forms of the same chemical substance.

[*allomorphic* (adj.)*.] Pertaining to allomorphs. See xenomorphic.

allomorphism. See xenomorphism.

allopatric*. Pertaining to 2 or more species living in different regions.

amphidetic. Extending on both anterior and posterior sides of beaks (applied to ligament or ligamental area); example, *Arca*.

anachomata. *See* chomata.

anisomyarian. With one adductor muscle (anterior) much reduced or absent.

anodont. Lacking hinge teeth. Same as edentulous.

anterior*. Direction parallel to hinge axis more nearly approximating to that in which mouth of animal faces.

anterodorsal margin. Margin of dorsal part of shell in front of beaks.

aorta*. Large tubular blood vessel which carries blood from heart toward other organs.

appressed*. Pertaining to thin foliaceous parts of shell separated from main part of shell by narrow vacant space.

[aragonite*.] Rhombic-holohedral or pseudohexagonal allomorph of calcium carbonate.

arborescent*. Resembling branched tree.

auricle. Earlike extension of dorsal region of shell, commonly separated from body of shell by notch or sinus.

auriculate. With auricles.

[authigenic*.] In mineralogy, pertaining to minerals that have grown in place within sediment.

[autochthonous*.] Having originated in place.

automorphic*. Same as idiomorphic.

beak. Noselike angle, located along or above hinge margin, marking point where growth of shell started.

[beekite.] Concentrically structured silicification center.

bialate. With 2 wings or auricles.

bilobate*. With 2 distinct bulges.

biocoenosis (pl., biocoenoses)*. Natural ecologic unit composed of diverse, but mutually dependent organisms.

biostrome*. Flat extensive biocoenosis composed of sedentary organisms that have hard skeletons or shells and sediment derived from them.

biotope*. Region of uniform environmental conditions and animal populations.

blood sinus* (blood-filled sinus, or blood lacuna). Irregularly shaped blood vessel without special confining walls.

body of shell. In alate or auriculate shells, entire shell with exception of wings or auricles.

Bojanus, organ of*. Kidney of Bivalvia.

bourrelet. Either of two portions of bivalve ligamental area flanking resilifer on its anterior and posterior sides; each comprises growth track and seat of the lamellar ligament. [The posterior bourrelet is flattish in all oysters except the Exogyrinae, in which it is a narrow sharp-crested spiral ridge on the LV and a corresponding groove on the RV.]

branchia* (pl., branchiae). Gill.

branchial passage*. Conduit confined by gills carrying parts of exhalant water stream.

branchitellum (pl., branchitella). Point on posteroventral shell margin of oysters nearest to pallio-

branchial fusion, commonly forming conspicuously projected posteroventral tip on LV, especially in sickle-shaped oysters; aboral end of gills points toward it.

buttress*. Internal shelly projection supporting resilifer or adductor muscle.

buttressed*. Provided with internal shelly projection for support of resilifer or adductor muscle.

byssal gland*. Gland on foot of Bivalvia which secretes byssus.

byssiferous. Possessing a byssus.

byssus. Bundle of hairlike strands by which temporary attachment of bivalve can be made to extraneous objects.

[calcite*.] Rhombohedral-holohedral allomorph of calcium carbonate.

cardinal axis*. *See* hinge axis.

carina. Prominent keel-like ridge.

carinate. With carina or sharp angulation.

cartilage. Old term for internal ligament.

catachomata*. *See* chomata.

catch muscle*. White, opaque, opalescent, tonic portion of adductor muscle.

[chalk*.] Earthy, crumbly limestone.

chalky deposits*. Parts of shell resembling chalk.

chemoreceptor*. Sense organ sensitive to chemical stimuli.

chomata (sing., choma). Collective term for anachomata, which are small tubercles or ridglets on periphery of inner surface of RV, and catachomata, which are pits in LV for reception of anachomata; both generally restricted to vicinity of hinge, but may encircle whole valve.

clasper* (or clasping shelly process). Narrow extension of shell tending to attach to extraneous objects.

cloacal passage*. Passage in exhalant mantle chamber into which excrements and gonadal products are discharged.

commissural plane*. Imaginary plane drawn through valve commissure.

commissural shelf*. Peripheral, shelflike part of shell adjoining commissure.

commissure. Line of junction of 2 valves.

compressed. Relatively flattened.

concentric. With direction coinciding with that of growth lines. (By no means concentric in literal and geometrical sense of the term.)

conchiolin. Material (protein) of which periostracum and organic matrix of calcareous parts of shell are composed.

[conchological*.] Pertaining to conchology (shell science).

[conchology*.] Study of shell shapes.

convexity. Degree of inflation.

costa. Moderately broad and prominent elevation of surface of shell, directed radially or otherwise.

costella. Rather narrow linear elevation of surface of shell.

costellate*. Having costellae.

costule. Same as costella.

crescentic*. Curved like crescent moon.

crossed-lamellar. Type of shell structure composed of primary and secondary lamellae, latter inclined in alternate directions in successive primary lamellae.

demibranch*. One half of gills.

demiprovinculum*. Half-provinculum.

denticle. Small rounded toothlike protuberance on shell. *See* chomata.

dentition. Hinge teeth and sockets, considered collectively.

dichotomize*. To divide by dichotomy.

dichotomous*. Divided by dichotomy.

dichotomy*. Forking of a line or rib into 2 equal parts.

dimyarian. With 2 adductor muscles.

dioecious*. Having male and female reproductive organs in separate individuals.

diphyletic*. Pertaining to group of animals descended from 2 originally diverse ancestral lines.

discordant margins. Margins of closed valves not in exact juxtaposition, but one overlapping other.

disjunct pallial line. Pallial line broken up into separate, mostly unequal muscle insertions.

dissoconch. Postlarval shell.

distal. Pointing or situated away from animal's center.

divaricate. Type of ornament composed of pairs of rather widely divergent costules or other elements.

[dog-tooth spar.] Pointed elongate crystal form of calcite.

dorsal. Pertaining to region of shell where mantle isthmus was situated and valves are connected by ligament (i.e., to region of hinge).

D-shaped larval stage. Stage of larval growth in which valve outline resembles a D and hinge is long and straight.

dysodont. With small weak teeth close to beaks (as in some Mytilacea).

ear. Small extension of dorsal region of shell, commonly separated from body by notch or sinus. Same as auricle.

ecomorph. Infraspecific growth form of species in response to special environment.

ecomorphic. Pertaining to ecomorphs.

edentulous. Lacking hinge teeth.

epidermis. Term used by some authors for periostracum.

equilateral. With parts of shell anterior and posterior to beaks equal in length or almost so.

equivalve. With 2 valves of same shape and size.

euhaline. Pertaining to sea water of normal salinity (around 35 permille).

euryhaline. Capable of living in sea water of a broad range of salinities including brackish waters.

excurrent. See exhalant.

exhalant. Applied to water current within mantle cavity from gills on out and the spaces from which it is departing.

exogyroidal (exogyrate). Shaped like shell of *Exogyra,* that is, with left valve strongly convex and its dorsal part coiled in posterior direction, with right valve flat and spirally coiled.

extrapallial space. Narrow mucus-filled space between mantle lobe and interior face of valve.

[facies.] In stratigraphy, sediment characterized by special mineral assemblage, bedding, and fossil organisms but differing from adjacent contemporaneous deposits.

falciform. Sickle-shaped.

fibrous ligament. Part of ligament characterized by fibrous structure and in which conchiolin is commonly impregnated with calcium carbonate; secreted by epithelium of mantle isthmus and elastic chiefly to compressional stresses.

fingerprint shell structure*. Shell structure, as yet unexplored, resembling thumb prints.

fixation. Process of animals permanently attaching themselves.

fold. Rather broad undulation of surface of shell, directed either radially or commarginally.

foot. Protrusible muscular structure extending from mid-line of body, anteroventrally in more typical bivalves, and used for burrowing or locomotion.

fringe*. Extension of the periostracum conchiolin beyond calcareous part of shell.

[fringe reef.] Crowded oysters growing in band parallel to nearest shore line.

gape. Localized opening remaining between margins of shell when valves are drawn together by adductor muscles.

gashes, radial*. Radial, sharp-edged incisions common on upper valves of some Gryphaeidae.

genital pore*. Opening through which gonadal products issue into cloacal passage.

[glauconite.] Family of soft, green, iron-bearing minerals of diverse compositions, usually in pellet shapes.

globose. Tending toward spherical shape.

growth line. Line on surface of shell, one of usually irregularly arranged series, marking position of margin at some stage of growth.

growth ruga. Irregular wrinkle on surface of shell of similar origin to growth line but corresponding to more pronounced hiatus in growth.

growth squamae*. Scaly extensions of shell arising from shell surface parallel to growth lines.

growth thread. Threadlike elevation of surface of similar origin to growth line.

growth welt. Elongate elevation parallel to growth lines.

gryphaeate. Shaped like shell of *Gryphaea,* that is, with left valve strongly convex and its dorsal part incurved and with right valve flat.

gryphaeiform*. Resembling a *Gryphaea.*

gryph-shaped*. Same as gryphaeiform.

height. Largest dimension obtained by projecting the extremities onto the mid-axis of shell.

hermaphrodite. Animal producing both male and

female gonadal products, not necessarily simultaneously.

heteroclite (adj.)*. Commissural plane that is folded or twisted is heteroclite.

heteromyarian. With one adductor muscle (anterior) much reduced.

hinge. Collective term for structures of dorsal region which function during opening and closing of valves.

hinge axis. Imaginary straight line about which valves rotate.

hinge line. Same as hinge axis.

hinge plate. Shelly internal platform bearing hinge teeth, situated below beak and adjacent parts of dorsal margins, and lying in plane parallel to that of commissure.

hinge tooth. Shelly structure (usually one of a series) adjacent to dorsal margin and received in socket in opposite valve; hinge teeth serve to hold valves in position when closed.

[homeomorph.] Two unrelated species or genera or larger taxa that are superficially similar.

homomyarian. With 2 adductor muscles equal in size or almost so.

[homonym.] Identically same word applied as name for 2 different taxa.

hyote spines. Hollow, tubular and cylindrical shell outgrowths open distally at their tips as well as on their distal flanks, arising periodically from thin edges of shell margin of oysters. [The tip ends are rounded, ear-shaped openings, typically developed on *Hyotissa hyotis* (LINNÉ, 1758).]

[hyperhaline.] Pertaining to waters of higher salinities than normal sea water; above 40‰ salinity.

[hypertely.] Evolution carried beyond the point of optimal adaptation.

hypostracum. Term used in 2 different senses; 1) inner layer of shell wall, secreted by entire epithelium of mantle (original sense); 2) part of shell wall secreted at attachments of adductor muscles and pallial line muscles (later sense: *see* myostracum).

idiomorphic (or automorphic)*. Configuration of valves normal for species and not deformed by crowding or attachment to other objects.

imbricate. Overlapping like tiles or shingles on a roof.

imbrication. Part of shell overlapping like tile on roof.

imprint. Impression on valve left by an organ (either gill or muscle).

incremental line. Same as growth line.

incrustation*. Tight attachment of oysters to rock or other substances.

incubatory*. Pertaining to oysters that incubate their young larvae.

inequilateral. With parts of shell anterior and posterior to beaks differing appreciably in length.

inequivalve. With one valve larger than other.

inflated. Strongly convex.

inhalant. Applied to water current entering mantle

cavity from outside, but before it has passed through gills and to spaces in which it moves.

inner layer of ligament. Same as resilium.

insertion. Place of attachment for a muscle.

interspace. Depression between adjacent costae or other linear surface elevations.

isodont. With small number of symmetrically arranged hinge teeth; examples, *Spondylus*, *Plicatula*.

isomyarian. With 2 adductor muscles equal in size or almost so; same as homomyarian.

isthmus (or mantle isthmus). Dorsal part of mantle connecting the 2 mantle lobes.

keel. Projecting ridge; same as carina.

labial palp*. One of four lappet-shaped organs to either side of mouth.

lacuna (pl., **lacunae**)*. Irregular, blood-filled gaps between various organs in mantle and visceral mass.

lamella. Thin plate.

lamellar ligament. Part of ligament characterized by lamellar structure and containing no calcium carbonate; secreted at mantle edge and elastic to both compressional and tensional stresses.

lamelliform. Like thin elongate plate.

lamina. Thin plate.

left valve. Valve of oyster homologous to valve on left side of mobile Bivalvia.

length. Largest dimension obtained by projecting shell extremities onto hinge axis.

lenticular. Shaped like biconvex lens.

[leptopel*.] Extremely fine material floating in sea water, dead or alive, organic or inorganic.

ligament. Horny elastic structure or structures joining 2 valves of shell dorsally and acting as spring causing them to open when adductor muscles relax.

ligamental area*. Area between umbo and ligament showing growth track of ligament.

longitudinal. Direction parallel to that of cardinal axis.

mantle. Integument that surrounds vital organs of mollusk and secretes shell.

mantle chamber*. One of 2 spaces in mantle cavity between gills and mantle lobes.

mantle fold*. One of 3 small folds at periphery of a mantle lobe.

mantle lobe*. One of 2 flat thin extensions of mantle adjoining the valve.

mantle/shell*. Covering organ system of a bivalve consisting of shell and mantle lobes.

[marcasite.] Iron-sulfide mineral forming rhombic-holohedral crystals; allomorph of pyrite.

[marl.] Earthy, crumbly sedimentary rock about halfway in composition between chalk and clay.

metamorphosis. Process by which larva changes into adult form.

mid-axis*. Imaginary straight line drawn in commissural plane at right angles to hinge axis and beginning at mid-point of ventral margin of resilifer.

[minette iron ore.] Sedimentary oolitic iron ore of Jurassic age in Lorraine and Luxembourg.

monoecious*. Having male and female reproductive organs in the same individuals. Same as hermaphroditic.

monomyarian. With only 1 adductor muscle (posterior).

monophyletic. Pertaining to group of animals descended from only one ancestral line.

morph*. Group of variants of a species united by one or several common characters, but not forming a true population.

mouth-anus axis*. Imaginary straight line drawn through mouth and anus of animal.

muscle imprint. Impression on valve left by a muscle at its place of insertion.

myostracum. Part of shell wall secreted at attachments of adductor muscles.

nacreous. Having a shell structure producing mother-of-pearl luster.

[neontology.] Study of living animals.

[neoteny.] Condition of having immature traits prolonged in later life.

[neozoology.] Same as neontology.

nepionic. Pertaining to early postlarval stage.

nodose. Bearing tubercles or knobs.

nonincubatory*. Pertaining to oysters that do not incubate their larvae.

oblique*. Most extended in direction neither parallel nor perpendicular to hinge axis.

obliquity. Angle between straight dorsal margin and line bisecting umbonal angle (in terminology of some authors); or between dorsal margin and most distant point of ventral margin (in terminology of others).

[olistostrome.] In structural geology, allochthonous layer of disordered rock masses.

operculiform. Shaped like a lid or operculum.

opisthocline*. Sloping in posterior direction from hinge axis (term applied to body of shell).

opisthodetic. Located wholly posterior to beaks (term applied to ligament).

opisthogyral (or opisthogyrate)*. Curved so that beak points in posterior direction (term applied to umbones).

orbicular*. Shaped as an orb; less regular than circular.

orthocline*. Perpendicular to hinge axis or almost so (term applied to body of shell).

orthogyral (or orthogyrate)*. Curved so that beak points at right angles to hinge axis.

ostia*. Tiny holes in walls of gills letting a water current through.

ostracum*. Entire calcareous part of oyster shell.

outer ligament*. Same as lamellar ligament in oysters.

ovate. Shaped like longitudinal section of egg.

pad*. Thin aragonite layer on which adductor muscle is inserted.

pallial. Pertaining to the mantle.

pallial curtain. Innermost of 3 mantle folds at periphery of mantle lobe.

pallial line. Line or narrow band on interior of valve close to margin, marking line of attachment of marginal muscles of mantle.

pallial region. Marginal region of shell interior adjacent to pallial line.

pallial retractor muscles*. Muscles which withdraw peripheral edge of mantle lobe in proximal direction.

palliobranchial fuson.*. Place at which aboral ends of gills and 2 mantle lobes are firmly joined.

[patch reef*.] Reef in outline of a patch.

pedal. Pertaining to foot.

pedal muscles. Muscles activating motions of foot.

[pellet*.] Small, rounded, somewhat elongate body.

pericardium*. Saclike organ enclosing heart.

periostracal glands*. Glands at base of middle mantle fold from which base layer of periostracum issues.

periostracal groove*. Groove housing periostracal glands, between middle and outer mantle fold.

periostracum. Dark, horny, conchiolinic substance covering outside of shell.

phasic muscle*. Flesh-colored, semitranslucent portion of adductor muscle which reacts quickly but does not endure.

pivotal axis*. Axis at ligament around which valves rotate when closing.

pleurothetic*. Resting on its side.

plica. Fold or costa involving entire thickness of wall of shell.

plication. Same as plica.

polychotomous*. Divided into many branches.

[polyhaline*.] Pertaining to brackish water of 16 to 30‰ salinity.

polyphyletic. Pertaining to group of animals derived from diverse ancestral stems.

polytypic. A species encompassing 2 or more geographic subspecies or 2 or more widely divergent ecomorphs.

porcelaneous. With translucent, porcelain-like appearance.

posterior*. Direction parallel to hinge axis more nearly approximating to that in which anus faces and exhalant current issues.

posterior flange*. Flange at posterior of left valve of Gryphaeidae separated from main body of valve by posterior radial groove.

posterior ridge. Ridge passing over or originating near umbo and running diagonally towards posteroventral part of valve.

posterior slope. Sector of surface of valve running posteroventrally from umbo.

posterodorsal margin. Margin of dorsal part of shell posterior to beaks.

primogenitor. Ancestor.

prismatic shell layer*. Layer consisting of many tiny prismatic bodies of calcite.

prodissoconch. Shell secreted by the larva or embryo and preserved at beak of some adult shells.

promyal passage*. Exhalant water passage lying on right side of animal between adductor muscle and mantle isthmus.

prosocline. Sloping (from lower end) in anterior direction (term applied to hinge teeth and, in some genera, to body of shell).

prosogyrate. Curved so that beaks point in anterior direction (term applied to umbones).

protandric*. Pertaining to hermaphrodite animal in which male gonad develops and functions before female one does.

protostracum*. Shell of D-shaped larval stage.

provinculum. Median part of hinge margin of prodissoconch, usually bearing small teeth or crenulations.

proximal*. Pointing or situated near animal's center.

proximal gill wheal*. Low ridge on inner valve surface outlining position of proximal edge of gills.

pseudofeces*. Refuse ejected from mantle cavity that has not passed through intestinal tract.

pseudosiphon*. Outline of 2 opposing mantle edges in form of slit or hole.

pyriform. Resembling shape of pear.

[pyrite.] Iron-sulfide mineral forming cubic-paramorph crystals; allomorph of marcasite.

quadrate. Square, or almost so.

Quenstedt muscles*. Pair of small muscles inserted on valves near mouth of animal.

quick muscle*. Same as phasic muscle.

radial. Direction of growth outward from beak at any point on surface of shell, commonly indicated by direction of costa or other element of ornament.

reniform. Kidney-shaped.

resilifer*. Place of attachment of resilium and its growth track on ligamental area. Term introduced by DALL (1895, v. 3, p. 499).

resilium*. Inner layer of ligament or fibrous ligament. Term introduced by DALL (1895, v. 3, p. 498, footnote).

rib. Moderately broad and prominent elevation of surface of shell, directed radially or otherwise; same as costa.

riblet. Rather narrow linear elevation of surface of shell; same as costella.

right valve*. Valve of oyster homologous to valve on right side of mobile Bivalvia.

sagittal plane*. Anteroposteriorly directed plane of symmetry dividing animal into left and right side.

[sapropel*.] Slimy sediment consisting largely of dead plant and animal debris.

scale*. Thin, flat local projection of outer shell layers.

sculpture. Regular relief pattern present on surface of many shells.

secondary riblet. On shell with riblets of different orders of strength, riblet that appears somewhat later in ontogeny than primary ones and remains weaker than these.

self-cleansing*. Process of removal and ejection of pseudofeces.

self-sedimentation*. Sedimentation produced by self-cleansing process and ejection of feces.

semilunar*. Shaped like half-moon with both ends sharp.

shell fold*. Outer one of 3 mantle folds at periphery of mantle lobe.

sinus. Indentation, embayment.

socket. Recess for reception of hinge tooth of opposite valve.

spatulate. Shaped like a spatula.

speciation. Originating of one or more species by evolution.

spine. Thornlike protuberance of surface of shell.

spirogyral (or **spirogyrate**)***. Curved so that beak is in a distinct spiral.

squamae*. Thin, long, concentric imbrication.

squamose. Bearing scales.

[stenohaline*.] Capable of living in sea water of a narrow range of salinities.

straight-hinge veliger or protostracal veliger*. Same as D-shaped larval stage.

stria. Narrow linear furrow or raised line on surface of shell.

[string reef*.] Crowded oysters in a narrow greatly elongate accumulation.

sulcus. Radial depression of surface of shell.

sulcus, radial posterior*. Groove dividing posterior flange from main body of valve, in left valve of Gryphaeidae.

superspecies*. Monophyletic taxon consisting of several allopatric species too diverse morphologically for inclusion in one single species, but not distinct enough to be raised to status of genus.

surface ornament. Regular relief pattern present on surface of many shells.

sympatric*. Pertaining to 2 or more species occupying the same territory.

taxon (pl., **taxa**). Group of organisms recognized as formal taxonomic unit of any hierarchical level.

tentacular fold*. Middle one of 3 mantle folds at periphery of mantle lobe.

terminal. Forming most anterior or posterior point of valve; term applied to beak.

thickness. Used by some authors to denote the shell measurement here termed inflation, but also commonly applied to the distance between the inner and outer surfaces of wall of shell.

thread. Narrow elevation of surface of shell.

tonic muscle*. White, opalescent, opaque portion of adductor muscle which contracts slowly and can retain tension for long periods of time; same as catch muscle.

transverse. Direction perpendicular to that of cardinal axis in plane of valve margins.

trigonal. Three-cornered.

truncate. With curvature of outline interrupted by straight cut.

tumid. Strongly inflated.

umbo. Region of valve surrounding point of maximum curvature of lontidudinal dorsal profile and extending to beak when not coinciding with it. (Many authors treat beak and umbo as synonymous, but with most shells two distinct terms are needed.)

umbo-veliger*. Last larval stage of oysters.

umbonal cavity*. Part of interior of left valve that lies in umbonal region beneath ligamental area of oysters.

umbonal region*. Part of valve including umbo and its vicinity.

urogenital opening*. Opening through which gonadal products and excretions from organ of Bojanus issue into cloacal passage of exhalant mantle chamber.

valve. One of the calcareous structures (2 in most bivalves) of which shell consists.

veliger*. Velum-bearing larval stage of oyster.

velum*. Large, ciliated, disc-shaped swimming organ of larva.

ventral. Pertaining to or located relatively near to region of shell opposite hinge, where valves open most widely.

ventricle. Heart chamber receiving blood from auricles.

ventricose. Strongly inflated.

vermiculate. Of wiggly outline.

vesicular. Containing small cavities or vesicles.

visceral mass*. Mass of organs from mantle isthmus to adductor muscle.

visceral pouch*. Small pouch-shaped extension of visceral mass on anterior flank of adductor muscle.

width. Largest dimension obtained by projecting shell outline onto a line that is at right angles to hinge axis and mid-axis.

wing. More or less elongate, triangular, distally acute or obtuse, terminal part of dorsal region of shell in Pteriacea, Pectinacea, etc.

xenomorphic (adj.). Pertaining to xenomorphism.

xenomorphism. Special sculpture at the umbonal region of the unattached valve resembling the configuration of the substratum onto which the attached valve is or was originally fixed. Known in the Anomiidae, Gryphaeidae, Ostreidae and other pleurothetic and cemented families. It is on the right valves in oysters and on the left valves in *Anomia* (*see* STENZEL, KRAUSE & TWINING, 1957, p. 98-99). Erroneously called allomorphism by some authors.

DISTRIBUTION

CONTENTS

Today oysters live off the shores of all continents except Antarctica. They exist under various marine climates except the polar ones. Some species have succeeded in colonizing shores of isolated oceanic islands.

Oysters originated in euhaline waters, and to this day the Gryphaeidae are restricted to them. However, among the Ostreidae, groups have evolved that were able to penetrate into brackish estuaries and lagoons. Some species can exist in salinities as low as 10 permille.

DISPERSAL

The best means of dispersal of oyster species are their planktonic larvae. In special cases, however, dispersal is possible even after the oysters have become fixed.

Whenever oysters settle on movable objects or animals they can be carried considerable distances. Driftwood carries oysters. Such oysters can reach sexual maturity, because young oysters become capable of reproduction in a few months when they are 2-3 cm. large and still very thin-shelled. The method of dispersal by driftwood is important wherever mangrove swamps and dense tropical rain forests line the shores.

Sea turtles of the family Cheloniidae, but not *Dermachelys,* often carry attached oysters (W. T. NEILL, letter of May 16, 1964). Jurassic ammonites, dead or alive, carried oysters grown onto their shells (STAFF & RECK, 1911; SEILACHER, 1960) and may have floated considerable distances, before they sank or stranded. How far dead cephalopod shells can float has been shown on *Nautilus* by STENZEL (see *Treatise,* p. K90, Fig. 67).

Although such dispersals are only sporadic and haphazard, they may happen often enough to be quite effective in time, that is, during the geologic life span of a species. Such sporadic dispersals may add 1,000 to 3,000 km. to the geographic spread of a species each time they are successful.

Larvae are the chief means of dispersal, because they are produced in great swarms during every breeding season and because they are planktonic, so that they can be carried far by ocean currents. The larvae remain planktonic up to 33 days (IMAI, HATANAKA, *et al.,* 1950, p. 75) or even 50 days (ERDMANN, 1934, p. 7).

Tidal and oceanic currents distribute and transport the larvae. In some cases great distances are covered. To estimate how far currents can carry oyster larvae one must ascertain the maximal possible velocity of the current that is sustained over the duration of the larval period. Successful long-distance dispersal may be possible only under exceptionally optimal conditions and may happen only once in a century.

Assuming a current velocity of 200-250 cm. per second (=173-216 km. per day), which has been recorded for the Gulf Stream (WORTHINGTON, 1954), and a free-swimming larval period of 6 days, the computed travel distance amounts to 1,000-1,300 km. A short larval period has to be assumed, because only in warm waters do larvae have a good chance to survive and in such waters larval periods are short. Therefore, maximal distance for successful dispersal of an oyster species by its larvae planktonic in an ocean current is about 1,300 km. Evidently, oysters have good dispersal and geologic migration rates.

Euhaline oysters have better chances of survival during dispersal than those which require brackish waters. The latter must find brackish waters at the end of larval transport and such waters are hardly available around isolated oceanic islands.

With such dispersal distances possible, oysters can cross large bodies of water by island hopping. A species established around many islands and over large oceanic areas nevertheless can maintain occasional gene flow from one isolated island to the other so that the species may remain intact and attain an enormous geographic spread during long periods of geological history.

A living example is the euhaline, shallow-water gryphaeid species *Hyotissa hyotis* (LINNÉ, 1758). It lives in warm tropical to subtropical waters from southern Japan in the north to the Persian Gulf and Red Sea in the west, to Inhaca Island, Mozambique, off the east coast of Africa in the southwest, and the Tuamotu Islands and Clipperton Island (HERTLEIN & ALLISON, 1966, p. 139) in the southeastern Pacific Ocean.

Because of these abilities to spread it is not surprising to find the giant Miocene brackish-water oyster *Crassostrea gryphoides* (VON SCHLOTHEIM, 1813) in so many places along the shores of the Miocene Tethys Sea and its bays. No island hopping was required, because continuous coastlines were available from the western outlet of the Tethys Sea at the Atlantic Ocean to the eastern ends on the Indian and Pacific Ocean shores.

The high migration rates of oysters explain also certain former geographic ranges. The euhaline Cretaceous exogyrine genus *Gyrostrea* is well represented during the Turonian and Campanian in central Asia (MIRKAMALOV, 1963) and there is a lone species in southwestern Texas and adjoining Mexico, *G. cartledgei* (BÖSE, 1919) from the Del Rio Clay (early Cenoman.). This species appears to have neither ancestors nor descendants in the latter region; its appearance and disappearance here is best explained by migrations.

Very closely related and approximately contemporaneous species of *Exogyra (Exogyra)* are found in Late Cretaceous deposits of North Africa (*E. overwegi* BEYRICH, 1852) and of the Atlantic and Gulf coastal plains of North America (*E. costata* SAY, 1820, and its closest relatives). These two allopatric groups are morphologically more similar to each other and probably more closely related than they are to other, collocal but not contemporaneous species of *Exogyra*.

The sudden appearance of *Gryphaea arcuata* LAMARCK (1801) in western and central Europe and possibly in other less well-known regions during the early Liassic is probably best explained by migration from afar, as was first suggested by HALLAM (1962, p. 574).

LIMITATIONS TO DISPERSAL

In spite of the excellent means of dispersal available to them, today's shallow-water oyster species are bound by very definite limitations to their dispersal. They are the open-water barriers, climate barriers, and salinity barriers. Similar limitations must have been effective in the geologic past.

OPEN-WATER BARRIERS

Open-water barriers are stretches of open oceanic water not "bridged" by shoals or island chains. Too wide to be crossed in one direction, they are also too long to be outflanked in any way. Their two ends, at the north and south, are closed off by climate barriers so that outflanking becomes impossible.

The Atlantic Ocean today is such a barrier. Both ends abut against polar or at least cold-climate areas too rigorous for shallow-water oysters. All shallow-water oyster species, except one, are restricted to one or the other side of the Atlantic and no species has succeeded in crossing it since the beginning of the Pleistocene or earlier, in spite of the swift and far-reaching Gulf Stream. The only species living on both sides of this ocean is the warm-water *Lopha cristagalli* (LINNÉ, 1758) ecomorph *folium* (LINNÉ, 1758). This tropical and subtropical oyster grows on octocorallian branches and has a worldwide equatorial range. It must have had good means of spreading in the near geologic past, perhaps during the Miocene, but has not evolved since then into a chain of separate, provincial species.

The East-Pacific open-water barrier separates the Hawaiian and Tuamotu Islands on the west and coasts of the Americas and Galápagos Islands in the east (EKMAN, 1934). This barrier limits the species-rich, shallow-water, marine fauna of the Indo-Pacific. Island hopping is possible and has taken place in the past from the shores of Persia, Arabia, and East Africa to this barrier. East of the barrier, on the west coasts of the Americas, the IndoPacific species are replaced by other, in many cases quite similar, separate species. For example, the Indo-Pacific warm-water species *Hyotissa hyotis* (LINNÉ, 1758) is replaced along the shores of the Americas by *H. fisheri* (DALL, 1914), found from Mexico to Ecuador, off the Galápagos Islands, and in the Gulf of California.

At some periods of the geologic past, island hopping may have been much easier than today. Islands existed where only seamounts or submarine ridges now are left.

CLIMATE BARRIERS

Temperature tolerances are different in the various oyster species. The extreme limit of specific temperature tolerance of a given species is its climate barrier. Such barriers are of two different kinds (HUTCHINS, 1947): 1) summer temperatures, meaning the critical warm temperatures that must be reached before reproduction can take place and ontogeny be completed, 2) winter temperatures, meaning the critical low temperatures below which mature individuals of the species die.

These two critical temperatures become evident particularly where they inhibit the poleward dispersal of a given species. Dispersal poleward is limited by whichever of the two factors is the more restrictive.

SUMMER TEMPERATURES

Wherever temperatures during the summer are too low to induce spawning or stay warm enough for too short a time span for the larval period to reach a successful finish, the species cannot propagate itself and becomes extinct. In marginal regions, the oyster species may be able to propagate successfully only once in ten years. Once the period of time between successful years becomes longer than the median life span of the species in that region, however, the species must disappear. Consequently, the poleward limit of a species fluctuates backward and forward with secular climatic changes.

The northwestern European oyster *Ostrea edulis* requires summer temperatures of only 15°C. for success in propagation. It breeds in the chilly Kattegat once in about 10 years and survives there only with difficulty. Farther north, in northern Norway, it survives only in a few isolated, especially favorable localities. These places have a hothouse effect. They are in narrow waters exposed to the full sun and protected by hills or mountains so that water temperature can rise while cold winds or water currents cannot reach them. The oysters originally reached these scattered localities during a former climatic optimum; they are relics.

Because of the Gulf Stream and the low summer temperature requirements of *Ostrea edulis* this species reaches the Arctic Circle in Norway. Its most northerly populations are recorded from Troena Island, just south of the Arctic Circle, and from near Rodoyosen on Tjotta Island, off the Norway coast at 65° 50′ N. The species is not known from Iceland, Spitsbergen, and the Faeroes Islands, but lives off the Shetland and Hebrides Islands. Because its summer temperature requirements are extremely low for oysters, the species is the only one that can survive on the coasts of Europe north of France. *Crassostrea angulata* (LAMARCK, 1819), the other European species, propagates successfully on the coasts of southwestern Europe and France, but only rarely when it is transplanted to England.

Similarly, in the northeast Pacific, *Ostrea conchaphila lurida* CARPENTER has the lowest summer-temperature requirements of any oyster and it is the only oyster able to survive north of San Diego, California. It is reported from as far north as Sitka, Alaska.

Some species of *Ostrea s.s.* appear to have in common their ability to spawn and to complete their larval periods at lower summer temperatures than species of other genera. Species of *Crassostrea* require higher summer temperatures for successful propagation. For example, *C. virginica* requires about 20°C. before it will begin to spawn. Because *Ostrea s.s.* is incubatory and *Crassostrea* is not, and because larvae of incubatory oysters finish their larval periods in shorter time, it is likely that incubation in *Ostrea s.s.* is used as an adaptation that makes it feasible for some species to spread poleward into regions not accessible to other oysters.

WINTER TEMPERATURES

Most oysters are quite sensitive to winter air temperatures, because they live in shallow waters. Those living between tide levels are exposed directly to air temperatures at low tides.

Some species of *Crassostrea* can tolerate repeated freezing and long exposure to freezing. At the Canadian and New England end of its long geographic range, *C. virginica* becomes exposed in the winter to frigid air or to very low water temperatures during low tides. It can withstand freezing near-solid for 4 to 6 weeks (NELSON, 1938, p. 55; KANWISHER, 1955). Although as much as 54 percent of their body fluids become ice at −15°C., the animals survive. Ice masses form between muscle fibers within the adductor muscle. The fibers are pushed aside by ice crystals and clumped into separate bundles. As ice forms the remainder of the body fluids become richer in NaCl and other electrolytes, and this raises their freezing temperature. Only euryhaline oysters can stand such changes of internal salinity (KANWISHER, 1959).

Crassostrea virginica ranges on North America's east coast farther north than any other species. It reaches the Baie des Chaleurs of the Gulf of St. Lawrence, Canada. *Ostrea equestris* SAY, 1834, ranges poleward only to near the entrance to Chesapeake Bay (37° 37′ N, 74° 19′ W, 60 fathoms; GALTSOFF & MERRILL, 1962, p. 241). It is the second one as regards poleward range on this coast.

Similar conditions prevail on the northeast coasts of Asia. There, *Crassostrea gigas* (THUNBERG) ranges farther north than any other oyster, including *Ostrea denselamellosa* LISCHKE, 1869, living off Japan and Korea. The former reaches poleward to near the town of DeKastri in Khabarovskiy Kray (Khabarovsk Territory), USSR, and the south end of the narrows Poliv Nevelskogo at 52°N, between the island Sakhalin and the Asian mainland (SKARLATO, 1960, p. 125, fig. 61).

It is evident that poleward limits of the ranges of *Crassostrea virginica* and similar species of this genus are not fixed by their inability as adults to survive extreme winter temperatures. Rather, at their northern range limits, summer temperatures do not suffice for successful propagation.

In contrast, freezing kills the adults of any species of *Ostrea.* The freezing point of seawater, −1.7° C., is the lowest temperature they can tolerate. They die if they become exposed to cold winter air during low tide. *Ostrea edulis* is intertidal only in the frost-free, southern part of its range, but must live at several meters depth in Norway. *Ostrea equestris* lives between tides or just below low tide level at many places in the Gulf of Mexico and off the Carolinas. However, at the north end of its range, off Maryland, it must live at 60 fathoms. Poleward limits of the ranges of *Ostrea* species are the result of the inability of the adults to survive freezing.

SUMMARY

Some species of *Crassostrea* tolerate extreme cold in winter and require quite warm summer temperatures. In other words, they are adapted to a so-called continental climate offshore. In the northern hemisphere such climates are found on the eastern coasts of continents. In contrast, some species of *Ostrea s.s.* require mild winters and can tolerate cool summer temperatures, that is, they are adapted to a more equable climatic regimen. Such climates are found on the western coasts of continents in the northern hemisphere where warm ocean currents tend to equalize the seasons. For these reasons, *Crassostrea* is the better adapted to penetrate poleward on one side of a continent and *Ostrea s.s.* on the other side. In the southern hemisphere, conditions are reversed. In southern Argentina, *Ostrea puelchana* D'ORBIGNY, 1842 (Paléontologie vol., p. 162), reaches poleward to the Golfo San Matías and is the most southerly oyster species found there.

These observations apply only to some species of the two genera. Other species of both genera do not live outside the tropical and subtropical climatic belts. The warmtemperate and cold-temperate climatic belts have only two shallow-water oyster genera, namely *Crassostrea* and *Ostrea;* in addition, the deep-water genus *Neopycnodonte* lives in those belts. Many additional genera are found living offshore in the arid, subtropical, and tropical climatic belts. Very little is known about their geographic distributions and climatic barriers, however. Only one of them, *Hyotissa,* is discussed below (see p. *N*1040).

SALINITY TOLERANCES

Salinity tolerances and optima can be established by two methods: 1) the more exact method of measuring salinities in experiments or in the vicinity of live animals in their natural environments, and 2) the less exact method of inferring salinities from their biocoenoses using general knowledge of their natural environments. As might be expected, the first method has been applied to only a few commercially important species in scientifically advanced countries. Although inexact, the second method is the more important because it can be applied neatly to both living and extinct species.

Crassostrea seems to be the most euryhaline oyster genus. In *C. virginica* neither

Fig. J50. Incrustations of the rock oyster *Saccostrea cuccullata* (von Born, 1778) on eolian rocks facing open Indian Ocean near Port Hedland, Western Australia. Photographs taken at low tide by H. V. Howe.—*A*. Wave-cut cliff.—*B*. Wave-cut platform.

eggs nor functioning spermatozoa develop unless salinity exceeds 7.5‰. Larvae and young adults grow best at 17.5‰ and tolerate 10 to 40‰. Although nearly all oyster reefs of this species grow in brackish waters, some reefs occur in hyperhaline environments near the south end of the Laguna Madre in southern Texas. In general, the species seems not well adapted to euhaline and hyperhaline salinities, however, and at the southern limits of its range, where a hot climate provides elevated salinities in evaporative lagoons, it is replaced by *C. rhizophorae* (Guilding, 1828), the mangrove oyster of Central America and the Caribbean Islands.

In Laguna Rincón near Boqueron, southwestern Puerto Rico, *Crassostrea rhizophorae* is very numerous and reproduces prolifically although salinities are above 35‰ through 11 months of the year. There, salinities are mostly 38‰. The larvae tolerate even 40.5‰ (Mattox, 1949, p. 348).

Ostrea is polyhaline to euhaline and less euryhaline than *Crassostrea*. *Ostrea edulis,* for example, lives in the basin of the Oosterschelde, an abandoned tributary channel of the Schelde River in the Rhine-Maas-Schelde delta in the southwestern Netherlands,where salinities average 27.5‰ (Korringa, 1941, p. 24-32) and drop to 24 or rise to 31‰. However, *O. equestris* Say at Aransas Pass, Texas, lives in salinities that range from 28.3 to 38.4‰. During the drouth years 1950-57 it thrived in Mesquite Bay north of Aransas Pass, Texas, in salinities of 34.6 to 45.3‰ (Hoese, 1960, p. 331). In most situations, *Ostrea* prefers brackish waters of higher salinities than *Crassostrea*.

Saccostrea is restricted to tropical and subtropical waters of normal salinities; measurements of 35.1-35.45‰ are recorded for it at one place (Macnae & Kalk, 1958, p. 3). It is a rock oyster, growing on hard substrata and on mangrove trees. On the east side of Inhaca Island at the entrance to the Bay of Lourenço Marques, or Delagoa Bay, Mozambique, Portuguese East Africa, *S. cuccullata* (von Born, 1778) grows in profusion on sea cliffs exposed to the open Indian Ocean; there too, the waters must be euhaline.

Striostrea appears to live in euhaline

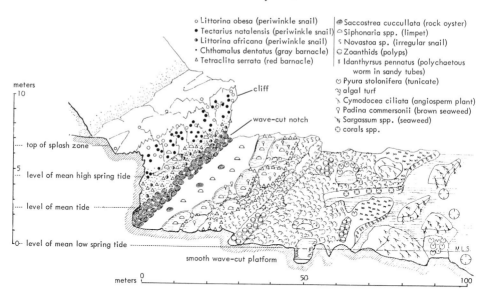

o Littorina obesa (periwinkle snail)
• Tectarius natalensis (periwinkle snail)
ə Littorina africana (periwinkle snail)
· Chthamalus dentatus (gray barnacle)
△ Tetraclita serrata (red barnacle)

⊘ Saccostrea cuccullata (rock oyster)
⌒ Siphonaria spp. (limpet)
ς Novastoa sp. (irregular snail)
☉ Zoanthids (polyps)
ſ Idanthyrsus pennatus (polychaetous worm in sandy tubes)
◡ Pyura stolonifera (tunicate)
⌐y algal turf
↖ Cymodocea ciliata (angiosperm plant)
♀ Padina commersonii (brown seaweed)
↘ Sargassum spp. (seaweed)
◔ corals spp.

cliff
wave-cut notch

meters
10

---- top of splash zone

5
--- level of mean high spring tide

--- level of mean tide

0-- level of mean low spring tide

smooth wave-cut platform

M.L.S.

meters 0 ———————————— 50 ———————————— 100

Fig. J51. Intertidal rock-incrusting oyster *Saccostrea cuccullata* (Born, 1778) on sea cliffs at north end of Inhaca Island, Mozambique, fronting on subtropical Indian ocean and fully exposed to strong wave action, as seen at low water of spring tides, vertical exaggeration, ×2.4 (after Macnae & Kalk, 1958).

waters. Salinities of 33.2-35.0‰ are recorded for *S. margaritacea* (Lamarck, 1819) at one place (Korringa, 1956, pl. 10). It thrives at such salinities at the mouths of the Knysna River and Svartvlei Lagoon alongside the Indian Ocean, Republic of South Africa.

The genus *Hyotissa* is strictly euhaline. Exceedingly few salinity measurements have been made, however. On the west side of Inhaca Island it lives near reef corals, and a salinity of 35.45‰ has been recorded there (Macnae & Kalk, 1958, p. 3, 129). Among the Cocos-Keeling Islands in the Indian Ocean, *Hyotissa*, growing to enormous size, has been found living at depths of 0.6 to 6.0 meters on fine lime sand among coral heads and slabs on the reefs (Virginia Orr, personal communication, September, 1963). On Australian coral reefs, it grows submerged just below low spring tide level (Thomson, 1954, p. 162). Considering its common association with colonial corals there is no doubt that it is a euhaline, warm-water genus. The extinct species of the genus in Tertiary deposits and their fossil associates fully confirm these conclusions. Fossil remains of this genus, so easily identified, are quite valu-

able indicators of warm euhaline environments.

The other living genus of the Pycnodonteinae, *Neopycnodonte,* has a nearly worldwide distribution. It lives in deeper waters and mostly far removed from land. At all places where it has been dredged from the sea bottom it must have been living in oceanic euhaline waters.

Therefore, all living Pycnodonteinae appear restricted to oceanic euhaline waters. This conclusion applies also to the great number of extinct species of the family Gryphaeidae. All of them are found in sediments and in company of faunas that indicate open euhaline seas. Not a single case is known where the association of fossils and sediments indicates brackish water environments for the Gryphaeidae.

SALINITY BARRIERS

Limits of salinity tolerances of a given species serve as salinity barriers to its geographic spread. There are two kinds of barriers: 1) elevated salinities, and 2) lower salinities. The former may be produced by evaporation of sea water along arid coasts or they may consist of merely a lack of brackish water lagoons and estuaries along the coast. Lowered salinities are produced by the influx of fresh waters from rivers.

ELEVATED-SALINITY BARRIERS

Highly elevated salinities arise wherever high evaporation overbalances influx of fresh river water and precipitation of rainwater. In shallow coastal regions and in bays or lagoons with restricted inlets, water currents are often inadequate to fully replenish the evaporated sea water. Then aridity barriers arise. Such areas lie in desert climates and may form unbridgeable barriers to the spread of shallow-water oyster species not adapted to elevated salinities.

The best example is *Ostrea edulis*. Its southernmost known occurrence (LECOINTRE, 1952, p. 39) is off Cape Rhir (or Ghir), 30° 40′ N, near Agadir on the Atlantic coast of Morocco, where it was dredged from a depth of 80 m. This place marks the southernmost place of its geographic range, because here the arid coast on the west flank of the Sahara is highly evaporative and lacks fresh-water streams. South of this aridity barrier many different oyster species and genera thrive on the tropical West African coast.

Another kind of elevated-salinity barrier seems to be responsible for the southern limit to the geographic range of *Crassostrea virginica*. It is not so much an aridity barrier as it is a lack of extensive brackish waters.

South of Yucatan *Crassostrea virginica* is replaced by *C. rhizophorae*. South of the Yucatan Peninsula no large rivers drain from the Central American land bridge into the Caribbean Sea, so that local brackish water environments are rare and occupy only small areas. Because of this lack of suitable brackish habitats the larvae of this species perhaps have difficulties in spreading southward from North America to South America.

On the other hand, *Ostrea equestris*, being adapted to higher salinities, has no such difficulties. It has been able to spread along the east coasts of the Americas from Maryland to the Golfo de San Matías, Argentina.

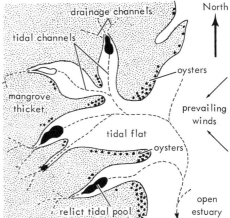

FIG. J52. Diagrammatic map of Mida Creek estuary on tropical coast of Kenya, East Africa, showing distribution of mangrove oyster *Saccostrea cuccullata* (BORN) attached to stilts of mangrove bushes (*Rhizophora mucronata*) (Someren & Whitehead).

INCRUSTATIONS AND REEFS

Incrustations are oysters attached to rocks, cliffs, or mangrove stems and root stilts. These oysters apparently do not accumulate to form thick layers. Only euhaline intertidal species make incrustations.

Reefs are natural accumulations of oyster shells, dead or alive, that rise above the general level of the substratum they are built on. Only coastal brackish-water species form reefs.

ROCK INCRUSTATIONS

Intertidal rock-incrusting oysters live in an extremely severe environment, because low tides expose the animals to air and direct sunlight leading to desiccation, interrupted feeding, and excessive temperatures. During low tides the shells must remain tightly closed. In cold-temperate climates only *Crassostrea* is able to stand this environment with its freezing temperature (see p. *N*1037).

In tropical and subtropical climates only two genera are known to be living in this sort of environment, namely *Saccostrea* and *Striostrea*. The former, represented by the living superspecies *S. cuccullata* (VON BORN, 1778), preferentially grows on rocks and cliffs and is called the rock oyster (Fig.

barnacles

FIG. J53. A mangrove oyster *Crassostrea virginica* (GMELIN, 1791) growing in crotch of mangrove stilt, near Comalcalco, State of Tabasco, Mexico, ×1.4 (Stenzel, n. Specimen donated by J. D. STOEN).

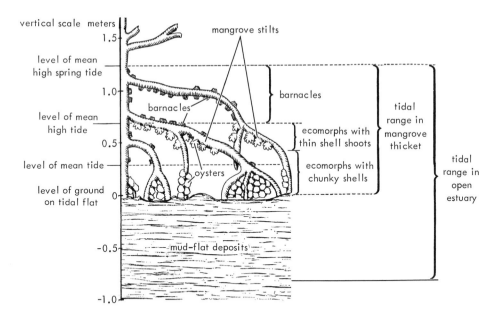

Fig. J54. Vertical zonation of mangrove oyster *Saccostrea cucullata* (BORN) incrusting mangrove stilts in Mida Creek estuary on tropical coast of Kenya, East Africa (Someren & Whitehead).

J50). It grows intertidally just above the upper limit of seaweed cover and below the splash belt occupied by littorinid snails and certain barnacles. Where they are exposed to heavy wave action the oysters occupy a belt 1 meter across, just above the wave-cut rock platform at the base of the cliff (Fig. J50). Where wave action is less strong, the oysters are not limited to such a narrow belt. They grow at various levels below the splash belt occupied by littorinid snails down to mean low spring-tide level (Fig. J51). These incrustations do not differ greatly from those on mangrove where wave action is even weaker.

Exposure to air, sunshine, and sea water bleaches and corrodes the outside of the shells, even while the animals are alive; strong wave action breaks off all protruding delicate growth imbrications and shell-shoots or prevents them from forming. Corrosion and corrasion are very active and dead shells disappear quickly.

MANGROVE OYSTERS

Mangrove trees and bushes grow in much exposed places, and the plants are sensitive to freezes so that they are restricted to frost-free, tropical and subtropical climates, although their distribution is circumglobal. They grow on shallow, intertidal, protected mudflats wherever wave action is minimal. The tangle of stems and stilt roots slows down waves and currents. Therefore, mangrove-covered shores are the sites of deposition of muds. The muds are very dark and rich in organic matter derived from animals and plants. The muds produce hydrogen sulfide gas.

Salinity in the mangrove swamps tends to be 36-38‰ in those parts where oysters grow. Most mangrove swamps have little river water entering them and have high evaporation rates so that salinity can rise to 40-42‰ during dry seasons. Water temperatures are around 31-34° C. in the hot season and 20-30° C. during the cooler, rainy season.

Oysters incrust the mangrove stems and stilt roots only in a narrow band up to 4 m. wide at the edges of the swamps facing open water of tidal channels or of the center of the lagoon (SOMEREN & WHITEHEAD, 1961, p. 9). Edges of swamps facing prevailing winds have more oyster incrustations than the more protected edges. Oys-

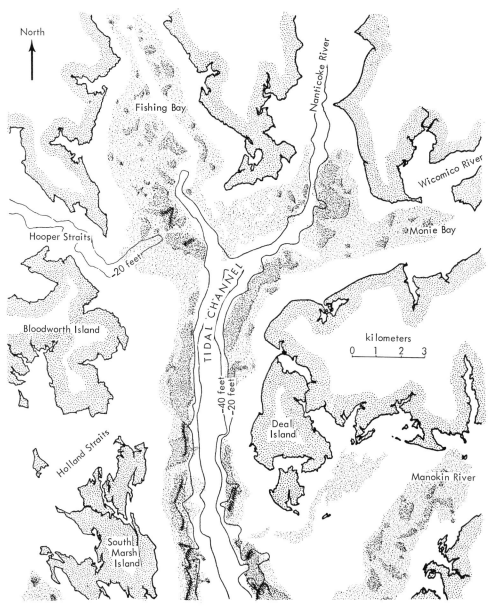

North

Nanticoke River

Fishing Bay

Wicomico River

Monie Bay

Hooper Straits

-20 feet

TIDAL CHANNEL

Bloodworth Island

kilometers
0 1 2 3

-40 feet
-20 feet

Deal Island

Manokin River

Holland Straits

South Marsh Island

scattered oyster growth   medium oyster growth   dense oyster growth

Fig. J55. Fringe reefs of *Crassostrea virginica* on the shoulders at margins of tidal channels in Tangier Sound on east side of Chesapeake Bay, Maryland, USA. Surveyed in September, 1878. From data given by WINSLOW (Winslow, 1882).

ters are absent in the centers of the swamps (Fig. J52; see Fig. J54).

Mud is the reason for these distributions of oyster incrustations. Wherever wave ac-

tion is feeble, mud settles out on the mangrove, and oyster larvae avoid settling on mud-covered substrata. For the same reason oysters tend to colonize the undersurfaces of inclined mangrove stems rather than their top surfaces. The crotch on the underside of tripod-like mangrove stilts is commonly free of mud and is the favorite place for oysters to grow on (Fig. J53).

Incrustations are restricted to between tides. Below low tide level the oysters remain covered by water continuously and therefore, remain exposed to such predators as crabs and fish continuously. These predators are so numerous that they eliminate all young and thin-shelled oysters. Above average high-tide level oysters become exposed to air and sunshine too long to survive; only barnacles can survive there. Periodic exposure of the intertidal oysters to air and sunshine has a strong influence on their growth habits. Oysters do not grow on mud bottoms except where such firm and not mud-covered substrata as gastropod shells are available.

Many different oyster genera have invaded the mangrove biotope: *Crassostrea* through *C. rhizophorae* (GUILDING, 1828) in the Caribbean and West Indies; *Lopha* through *L. folium* ecomorph *cristagalli* (LINNÉ, 1758) in the region from the Indian Ocean to southwestern Japan; and *Saccostrea* through *S. cuccullata* (VON BORN, 1778) and its subspecies in tropical West Africa and East Africa to Honshu, Japan. The situation is indicative of multiple, noncontemporaneous, separate invasions, each by a different genus, and the last invasion probably was by *Crassostrea*.

The various mangrove oysters have several ecomorphic features in common. They tend to be thin-shelled and fragile. They tend to produce thin fragile scalelike shell imbrications or shell shoots and delicate protruding frills (Fig. J53, J54). There is a tendency to clasping shell extensions, which become auxiliary holdfasts. Many of the mangrove oysters have xenomorphic sculpture.

The delicate and thin-shelled features are caused by scarcity of free calcium ions and by abundance of planktonic food in these waters. Shell growth must be rapid, but material to build shell walls is scarce. In addition, the oysters must shut their valves during low tide for some time. During this period their blood tends to become more acid and must be buffered. This is done

FIG. J56. Dagger Reef in San Antonio Bay, Texas, a string reef built by *Crassostrea virginica*. Photograph by courtesy of R. M. NORRIS, Univ. of California at Santa Barbara, Calif.

by some of the calcium carbonate of the shell. In other words, during low tide the shell wall not only quits increasing in thickness but must lose thickness. Because wave action is minimal in mangrove swamps, delicate and thin-shelled features remain undamaged and are not greatly disadvantageous to the animal. In fact, they may be advantageous in keeping enemies at a distance.

REEFS

On the basis of their configurations and the independence of the configurations from the nearest shore line, reefs are classified as fringe, string, and patch reefs. As to their configuration, fringe reefs are the least and patch reefs the most independent from the nearest shore line.

FRINGE REEFS

Fringe reefs are adjacent to the shore. They are common features along the finger-like branches of estuaries, that is, in drowned river valleys and their drowned tributaries. In most of them there is a tidal channel along the axis of the estuary. The axial channel increases downstream in depth from 3 to 30 m. or more and in width from 370 to 750 m. or more. Tidal scour excavates the axial channel and keeps it deep and free of oysters and sediments. At its sides the axial channel is flanked by fringe reefs. Their surface slopes gently

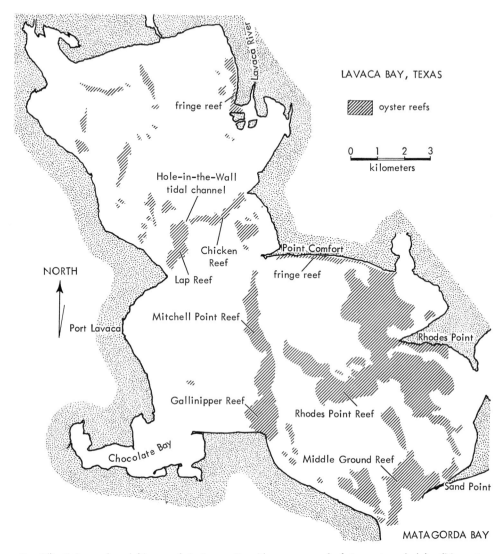

LAVACA BAY, TEXAS

oyster reefs

0   1   2   3
kilometers

Lavaca River

fringe reef

Hole-in-the-Wall
tidal channel

Chicken
Reef

NORTH

Lap Reef

Point Comfort

fringe reef

Port Lavaca

Mitchell Point Reef

Rhodes Point

Gallinipper Reef

Rhodes Point Reef

Chocolate Bay

Middle Ground Reef

Sand Point

MATAGORDA BAY

Fig. J57. String reefs and fringe reefs in Lavaca Bay, Texas, composed of *Crassostrea virginica* (Moore & Danglade, 1915).

from the shore toward the shoulder alongside the channel. The densest oyster populations are on the shoulders close (275-650 m.) to the edge of the channel (Fig. J55).

The fastest tidal currents are in the channels. As the water spreads during a rising tide from the channel over the adjacent shoulder it must slow down very much. Where it slows down most of the planktonic oyster larvae drop down and settle out, giving rise to the fringe reefs.

STRING REEFS

String reefs have fairly narrow crests, which may become exposed for a width of 10 m. when the tide is low. Crests are straight or curved as garlands. Many are arranged in an en echelon series maintaining nevertheless a straight alignment. Others are more loosely arranged and form various odd-shaped islands at low tide.

A straight series of crests is Panther Reef in San Antonio Bay, Texas, which is 10 km.

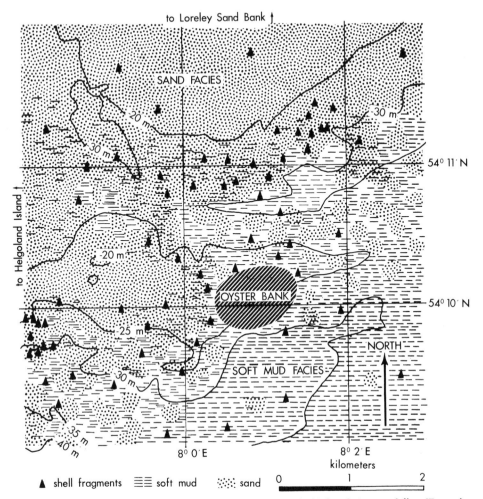

Fɪɢ. J58. Patch reef of Helgoland in the North Sea, composed chiefly of *Ostrea edulis*. (From data given by Caspers, 1950.)

long and originates at Panther Point on the bayward side of Matagorda Island. Tidal channels separate several looped crests called Long Reef, Halfmoon Reef, Grass Island Reef, and Pier Reef. They form a string reef 12 km. long across Aransas Bay from St. Joseph's Island to Lamar Peninsula on the Texas mainland. The longest string reef known is nearly 42 km. long in a straight line and forms a discontinuous barrier separating Atchafalaya and East Côte Blanche Bays on the Louisiana shore from the open Gulf of Mexico. All these were built by *Crassostrea virginica* (Fig. J56).

No true string reef is built by euhaline oysters. String reefs are found only in brackish lagoons or as barriers between sea and brackish bays.

Most string reefs are at right angles to the nearest shore, as first noted by Gʀᴀᴠᴇ (1901). Actually, the string reefs are normal to the direction of tidal currents. These in turn are guided by the shape of the lagoon and the placement of the passes to the open sea. Lagoons that are long and straight and have straight shores on both flanks tend to establish regular tidal currents parallel to the long flanks. In such situations string reefs develop best and become arranged at right angles to the nearest shore. Many of them become partial barriers across the lagoon. Good examples are string reefs in the lagoons along the Texas coast (Fig. J57).

PATCH REEFS

Patch reefs grow far from shore and have irregular but fairly compact outlines. Their sizes and locations depend mainly on availability of an appropriate substratum.

A good example is Helgoland Oyster Bank in the German Bight of the North Sea about 15 km. east of Helgoland Island and 50 km. from the nearest mainland shore. It covers an area of 800,000 sq. m.; its larger east-west axis is 1,150 m. and its north-south axis 750 m. long (CASPERS, 1950). The reef is 23-28 m. below sea level and is built almost exclusively by *Ostrea edulis* LINNÉ. The reef is now much depleted by over-fishing but nevertheless has about 1.5 million full-grown and half-grown oysters (Fig. J58). ZENKEVITCH (1963, p. 449) reports 14 million oysters on the Gudaut bank of the east coast of the Black Sea.

The distribution of sedimentary facies and biocoenoses around Helgoland Oyster Bank is characteristic for oyster banks of nearshore oysters, and generalizations of the local situation are in order. At the north is Loreley sand bank, less than 10 m. below sea level, composed of coarse sand and gravel. The sand bank is all that is left of a former island.

Sea bottom slopes gently from the sand bank southward and the bottom sediments become successively finer grained. Thus the bottom sediments are arranged in successive facies belts surrounding the sand bank. The oyster bank is at the deeper, down-slope end of the sand facies. Farther down the slope the bottom sediments consist of soft muds.

These muds are too soft for oyster larvae to find a suitable substratum for attachment. In contrast, the oyster bank sits on firm sand mixed with soft mud, mollusk shells, and their fragments.

On the muds south of the oyster bank the oysters are replaced by the mussel *Mytilus edulis* LINNÉ, because the mytilid mussels have a competitive advantage over oysters on a soft mud bottom, because they can attach themselves by many separate byssus threads anchored all around on seaweeds and other objects. Vice versa, oysters have the advantage wherever the substratum is firm and has shells or shell debris for the oyster larvae to settle on (ZENKEVITCH, 1963, p. 442).

Up the slope the oysters are replaced by other, unattached bivalves most of which are forms that plow through or dig in the sand. Many of them are venerids. They have a competitive advantage over oysters because they can adjust to shifting sands, whereas oysters must have a stable and firm substratum. Thus oysters are confined to a specified facies belt, where waves and currents are too weak to induce shifting of the sands and too strong to let muds accumulate. This belt must be a fair distance away from shore and down the sea-bottom slope.

Salinities there stay below 34‰ because of the slightly brackish western European Continental Coastal Water Current. The current is high in nutrients and turbidity, follows the coast northward and northeastward, and originates from the influx of many large rivers in Belgium, The Netherlands, and West Germany.

PHYLOGENY

CONTENTS

BEGINNINGS

ALLEGED OYSTERS FROM PALEOZOIC

No Paleozoic fossils are known today that are definitely identified as remains of oysters or that can be shown to be their direct ancestors. In the past, however, several authors have described remains claimed by them to be oysters or ancestors of oysters. The following list, possibly not complete, enumerates them (Fig. J59).

Ostrea costata STEININGER, 1831, from Eifel Mts., W. Europe. Description inadequate, no illustration (see STEININGER, 1834, p. 366).

Ostrea matercula DE VERNEUIL, 1845, from Magnesian Limestone (Perm.) near Itchalki on the banks of Pyana River, Gorkiy (formerly Gouvernement Nishniy Novgorod), Russian Soviet Federated Socialist Republic (VERNEUIL, 1845, p. 330-331, pl. 21, fig. 13a-c; NECHAEV, 1894, p. 188, pl. 7, fig. 1-2). This enigmatic form is now placed questionably in *Annuliconcha* NEWELL, 1937, family Aviculopectinidae, Pectinacea.

Ostrea nobilissima DE KONINCK, 1851, from Visé Limestone (Visean), at Visé, Belgium. First, DE KONINCK (1851, p. 680, pl. 57, fig. 10) described it as *Ostrea*, but later (DE KONINCK, 1885, p. 201-202, pl. 40, fig. 1-5) he made it the type species of *Pachypteria* DE KONINCK, 1885, family Aviculidae.

Pachypteria has also been found in Carboniferous limestones of Derbyshire and Yorkshire, Eng-

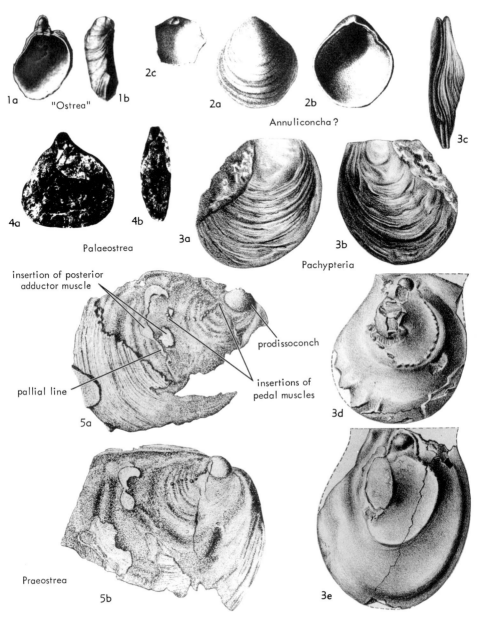

FIG. J59. Alleged oysters from the Paleozoic.

1. *"Ostrea" patercula* WINCHELL, 1865, from sand-stone at base of Burlington Ls., Miss., at Burlington, Iowa, USA; *1a,b,* inside and ?posterior or ?anterior views. (Original of figure by courtesy of L. B. KELLUM, Museum of Paleontology, Univ. Michigan.)

2. *Annuliconcha? matercula* (DE VERNEUIL, 1845) from Magnesian Ls., Perm., of P'yana River, Gorkiy, USSR; *2a-c,* lower valve, outside and inside views; valve of young specimen (de Verneuil, 1845).

3. *Pachypteria nobilissima* (DE KONINCK, 1851) from Visé Ls., L.Carb.(Visean), at Visé, Belg. (*3a-e*) (de Koninck, 1885).

4. *Palaeostrea sinica* GRABAU, 1936, from Maping Ls., Perm., of Kwangsi, China (*4a,b*) (Grabau, 1936).

5. *Praeostrea bohemica* BARRANDE, 1881, from Kopanina F., U.Sil., of Jinonice-Butovice, Hájek, Bohemia, Czech.; *5a,b,* internal molds of RV (Kříž, 1966).

land. A second species, *P. ostreiformis* MAILLEUX, was described from the Frasnian of Belgium and a third species *P. gevini* TERMIER & TERMIER, 1949, from yellow sandy marls (late Visean) of Kerb en Neggar, 65 km. NW. of Aouinet Legra in the western Sahara. The genus was discussed by GEVIN (1947), DECHASEAUX (1948), and TERMIER & TERMIER (1949). The TERMIERS placed it in the Pseudomonotinae. The Saharan species has a small attachment area on the right umbo; the valves are mostly equivalve, but one of the left valves collected has greater inflation; the hinge is rectilinear and lacks teeth, has a depressed resilifer at the anterior end, and the single muscle insertion is subcentral; radial ribs and two auricles are present in the young. Although this animal was attached, it cannot be classed with the Ostreidae according to DECHASEAUX and hardly can be an ancestor. Such similarities between *Ostrea* and *Pachypteria* as are visible must be convergence features caused by attachment. According to NEWELL (personal communication, 1963), *Pachypteria* is a junior synonym of *Pseudomonotis* BEYRICH, 1862, family Aviculopectinidae, Pectinacea (see p. N342). It is attached by its right valve and has an aragonitic shell; it is unlikely that it is an ancestor of oysters.

Ostrea patercula WINCHELL, 1865, from sandstone at base of Burlington Limestone (Miss.) at Burlington, southeastern Iowa (Fig. J59,1). Described by WINCHELL (1865, p. 124) and figured by WHITE (1884, p. 288, pl. 34, fig. 1-2). Possibilities of contamination were good at Burlington, which was at the time a major Mississippi River port and railroad terminal (letter from B. F. GLENISTER, February, 1962). Figure and description are insufficient to recognize this form with certainty as a mollusk, let alone to place it in the Ostreidae. The type was deposited at the University of Michigan, but cannot be found today (letter from L. W. KELLUM, February, 1963).

Ostrea prisca HOENINGHAUS, 1829, from Visé, Belgium. This is a *nomen nudum* in both publications (1829, p. 14; 1830, p. 237).

Palaeostrea sinica GRABAU, 1936, from Maping Limestone (Perm.) of Kwangsi, China. GRABAU (1936, p. 284-286, pl. 28, fig. 1) claimed that this fossil shell cannot be distinguished from a Mesozoic oyster and made the separation purely on a stratigraphic basis. No information on the muscle imprint was given and the material at hand consisted of a single valve resting on rock matrix and a fragment of the umbonal part of another fossil (Fig. J59,4). It appears to be wholly unidentifiable. This species is the type species of *Palaeostrea* GRABAU, 1936, by original designation.

Praeostrea bohemica BARRANDE, 1881, and *P. bohemica* var. *simplex* BARRANDE, 1881, from Late Silurian (Kopanina Formation) of Karlstejn (formerly Karlstein), southwest of Praha, and Lochkov and Dlauba Hora, Bohemia, Czechoslovakia. BARRANDE (1881a, p. 147, pl. 3, fig. 1-2,

and pl. 3, fig. 3-4, respectively; 1881b, p. 233-234) (Fig. J59,5). The type species is *P. bohemica* by monotypy, because *simplex* was regarded merely as a variety by BARRANDE. JIŘI KŘÍŽ of the Charles University, Praha, recently monographed the genus (KŘÍŽ, 1966). According to him, the genus is the only one classed in the Praeostreidae KŘÍŽ (1966) and assignable to the superfamily Mytilacea RAFINESQUE (1815). This Silurian-Lower Devonian form cannot be considered as an ancestor of the oysters.

EARLIEST KNOWN FOSSIL OYSTERS

The most ancient fossils known today that are undoubtedly oysters are Carnian (Late Triassic).

Before 1880, several authors had described bivalves from earlier (Middle Triassic) beds which they, following general practice, claimed were oysters and described as *Ostrea* or *Ostracites*. For example, GOLDFUSS (pt. 2, p. 1833-44) figured 10 species from the Muschelkalk (M.Trias.) of Germany as *Ostrea* and ROEMER (1851, p. 312, pl. 36, fig. 19) described *Ostrea willebadessensis* as a new species from an oolitic limestone layer of the upper Muschelkalk (Ladin.) quarried on the road from Cloister Willebadessen to Altenheerse, about 3 km. south of the town of Driburg on the southeastern foothills of the Teutoburger Wald, State of Nordrhein-Westfalen, West Germany.

Later studies of various supposed oysters from the Muschelkalk of Germany, however, led to the realization that these supposed oysters were attached by their right valves, rather than by left. The first one to emphasize this observation and to conclude that these bivalves could not be true oysters was NOETLING (1880, p. 321-322). His stand has found approval by many authors since 1880 (PHILIPPI, 1898, p. 617; WAAGEN, 1907, p. 172-175; Cox, 1924, p. 65-66) and is generally accepted today.

These observations led BITTNER (1901, p. 72) to propose the genus *Enantiostreon,* and many Mid-Triassic bivalves which had been called oysters before were transferred to it. SCHMIDT (1928) reviewed the entire fauna of the Muschelkalk of Germany and regarded *Ostrea willebadessensis* ROEMER (1851) as one of the many variants of *E. difforme* (VON SCHLOTHEIM, 1820). He

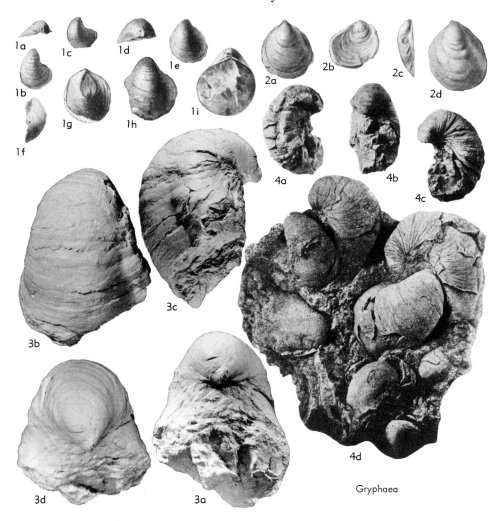

Fig. J60. Earliest fossil oysters known; *Gryphaea* from Triassic sea of Arctic.

1. G. keilhaui Böhm, 1904, *Myophoria* Sandstone, Carn., Bjørnøya Island; *1a-i,* ×1 (Böhm, 1904).
2. G. skuld Böhm, 1904, *Myophoria* Sandstone, Carn., Bjørnøya Island; *2a-d,* ×1 (Böhm, 1904).
3. G. chakii McLearn, 1937, Pardonet Formation, Carn.-Nor., Sikanni Creek, B.C., Canada; *3a-d,*

×1 (Stenzel, n. Specimen by courtesy of Shell Oil Company of Canada, Ltd.).
4. G. arcuataeformis Kiparisova, 1936, Carn., Korkodon River, far eastern Siberia; *4a-d,* ×1 (Kiparisova, 1938).

placed the 10 supposed oysters described by Goldfuss from the Muschelkalk in the following genera: *Enantiostreon* Bittner, 1901, family Terquemiidae; *Placunopsis* Morris & Lycett, 1853, ?Terquemiidae; and *Pseudomonotis* von Beyrich, 1862 [=*Prospondylus* Zimmerman, 1886], Pseudomonotidae. To these one might add *Atreta* Étallon, 1862 [=*Dimyopsis* Bittner, 1895], Plicatulidae.

Plicatulidae and Spondylidae living today are attached by their right sides, but have strong interlocking hinge teeth. On the other hand *Enantiostreon, Placunopsis, Pseudomonotis,* and *Atreta* are devoid of interlocking teeth although they are attached by their right valves as determined by the position of the adductor muscle insertions.

The removal of the various Mid-Triassic

bivalve species from true oysters to other families has left only a few species that, for lack of incisive investigation, are still carried as *"Ostrea"* by some authors. In short, no bivalves are known from Mid-Triassic or older beds that can be demonstrated to have been attached by their left valves and can be assigned to the oysters with complete certainty.

However, NAKAZAWA & NEWELL (1868) have described two species of bivalves from the Permian of Japan which they place confidently in the family Ostreidae and questionably in the genus *Lopha sensu latissimo*. Unfortunately, the specimens available are too poorly preserved to allow one to distinguish the right from the left valves with complete confidence. Thus their generic and even their familial taxonomic positions are uncertain.

Only three oyster genera are known from the Upper Triassic. However, not all of them are closely related to each other and thus they prove that even before Late Triassic time oysters had attained considerable evolutionary divergence and that we have not yet discovered any fossil remains representing the very earliest oysters and the missing ancestors that were the links connecting the Late Triassic genera. Or, the Late Triassic oysters are really not so closely related, because they are polyphyletic, and the differences between the Late Triassic genera are rather the consequences of their polyphyletic origins rather than the results of evolutionary divergence from a common ancestor.

GRYPHAEA LAMARCK, 1801

The various species from Triassic deposits described as *Gryphaea* and examined independently and repeatedly by several authors have diagnostic generic features of *Gryphaea* and are quite correctly placed in this genus (Fig. J60). One of them, namely *G. arcutaeformis* KIPARISOVA (1936), is quite similar to *G. arcuata* LAMARCK (1801), type species of the genus, as was pointed out by KIPARISOVA (1936, p. 100-102, 123-125, pl. 4, fig. 1-2, 4, 6-10; 1938, p. 4, 33-34, 38, 46, pl. 7, fig. 17-21, pl. 8, fig. 1-2, 11) and affirmed by VYALOV (1946).

Triassic species of *Gryphaea* have been found at the following places, arranged from east to west. ——1) Kolyma River drainage basin in Magadan-skaya Oblast [Province], far eastern Siberia, USSR. Described by KIPARISOVA (1936, 1938) and VYALOV (1946).——2) Bjørnøya [Bear Island], south of Svalbard [Spitsbergen], Norway. Here Triassic gryphaeas were discovered for the first time (BÖHM, 1904).——3) Ellesmere Island, Arctic Canada (KITTL, 1907; TROELSEN, 1950; TOZER, 1961; TOZER & THORSTEINSSON, 1964).——4) Borden and Prince Patrick Islands, Queen Elizabeth Islands, Arctic Canada (TOZER, 1961; TOZER & THORSTEINSSON, 1964).——5) Rocky Mountain foothills along the Peace and Pine Rivers, east-central British Columbia (McLEARN, 1937; WESTERMANN, 1962) and the region of the Sikanni Chief River, B.C., Canada (McLEARN, 1946; 1947). ——6) Cedar Mts. in east corner of Mineral County, west-central Nevada, USA. Here S. W. MULLER discovered *Gryphaea* in the Luning Formation (Carn.).——7) Gammaniura in the mountain group of the Monte Judica, about 40 km. west of Catania, eastern Sicily. Several species of *Gryphaea* have been described by SCALIA (1912) from an oolitic limestone exposed there. Although the species as described by SCALIA are mostly ill-founded, there is little room for doubt that the genus is correct. No one seems to have investigated this locality since 1912. The rock may be part of the autochthonous Mufara Formation (U.Trias.) or part of the Lavanche Olistostrome, a chaotic allochthonous sheet of blocks which are Cretaceous to early Miocene in age (letter from Paolo Schmidt de Friedberg of Novara, March 17, 1964). The supposed Triassic age of the gryphaeas may be erroneous.

Most of the Triassic gryphaeas have a circumpolar distribution and are from the Triassic sea that occupied the Arctic region. Localities in British Columbia and Nevada are from deposits in oceanic passages connecting at the north with the Triassic Arctic realm. The Sicilian locality must be set aside as dubious until it has been reinvestigated.

It is probably a safe conclusion that the Triassic home and place of origin of *Gryphaea* was the sea that during Late Triassic time occupied the Arctic region and the seaways that opened into it. The genus first showed up during the Carnian.

LIOSTREA DOUVILLÉ, 1904

Small, smooth oysters that probably are representatives of *Liostrea* have been found in many places in Rhaetian deposits (Zone of *Rhaetavicula contorta* (PORTLOCK, 1843)) of Europe. Such an oyster has been reported from the upper part of the Sully Beds (L.Rhaet.) at Cadoxton, Glamorganshire, southern Wales. It has been inadequately described and only a single view of one broken valve was figured as *"Ostrea*

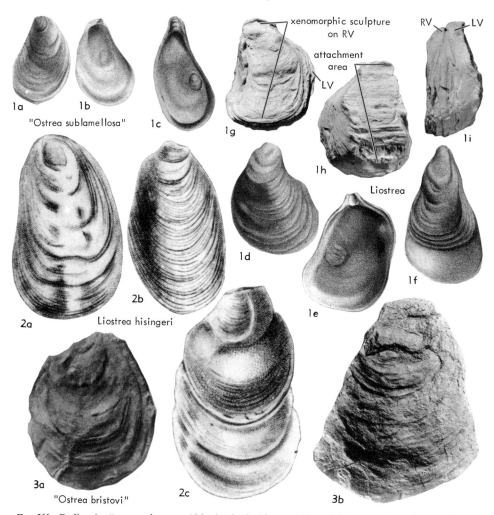

Fig. J61. Earliest fossil oysters known: Triassic-Liassic *Liostrea hisingeri* (NILSSON, 1832) from northwest Europe.

1. *"Ostrea sublamellosa"* DUNKER, 1846, Lias., Germany (West).—*1a-f.* Specimens from vicinity of Halberstadt, ×1 (Dunker, 1846).—*1g-i.* Specimens from vicinity of Hildesheim, Hettang., Zone of *Scamnoceras angulatum,* both valves seen from right, left, and front, ×1 (Stenzel, n. Specimen by courtesy of F. TRUSHEIM, Hannover).

2. *L. hisingeri* (NILSSON, 1832), L. Högenäs Series, Hettang., Skåne, Sweden; *2a-c,* ×2 (Lundgren, 1878).

3. *"Ostrea bristovi"* RICHARDSON, 1905, Sully Reds, low. Rhaet.; Glamorganshire, Wales; *3a,b,* ext., ×0.8, ×1 (Richardson, 1905; Arkell, 1933).

Bristovi ETHERIDGE ms." by RICHARDSON (1905, p. 422, pl. 33, fig. 4). The species remains nondescript, notwithstanding its listing by ARKELL (1933, p. 97). Seemingly, these oyster remains are stratigraphically the earliest in England and Wales. One or more species have been described from the Kössener Schichten (Rhaet.) of the Alps and are widely distributed in homotaxial formations of the Rhaetian in the Alps and Carpathians (DIENER, 1923; KUTASSY, 1931) (Fig. J61).

Triassic liostreas have been found (KIPARISOVA, 1938, p. 33, pl. 7, fig. 16a-b; VYALOV, 1946, p. 27-28, pl. 1, fig. 1a,b) near the Arctic Circle in far eastern Siberia

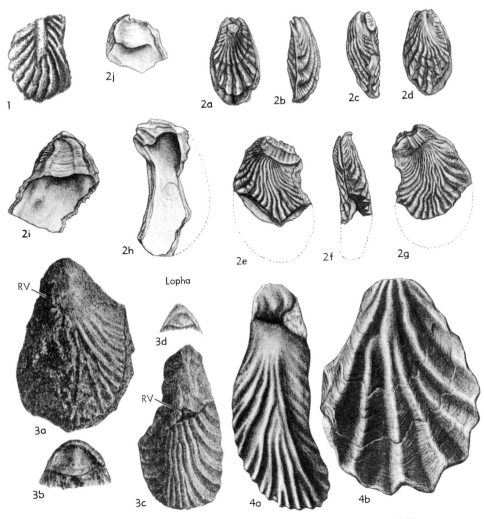

FIG. J62. Earliest fossil oysters known; Triassic *Lopha* from Mesogena seas of Alps.

1. *L. medicostata* (WÖHRMANN, 1889), *Cardita* Oolite, Carn., ×2 (Wöhrmann, 1889).
2. *L. montiscaprilis* (KLIPSTEIN, 1843), St. Cassian Formation, Carn.; *2a-j,* ×0.9 (Bittner, 1912).
3. *L. calceoformis* (BROILI, 1904), *Pachycardia* Tuff, Carn.; *3a,c,* ×2; *3b,d,* enl. (Waagen, 1907).

4. *L. haidingeriana* (EMMRICH, 1853), Kössen Formation and its homotaxial equivalents, Rhaet.; *4a,b,* ×1 (Martin, 1860).
[Assignments of stratigraphic stages according to DIENER and KUTASSY.]

at the Andesite Springs of the headwaters of the Agidzha River, a tributary to the Zyryanka River, Yakutsk Autonomous Soviet Socialist Republic. The liostreas are accompanied there by *Entomonotis ochotica* (KEYSERLING in MIDDENDORF, 1848) in interbedded gray limestones and calcareous shales (Norian).

It is probable that the stratigraphically earliest liostreas were Norian and were restricted to the Triassic sea that occupied the Arctic realm. The earliest *Liostrea* in Europe seems to be Rhaetian. Beginning with the Rhaetian *Liostrea* was widespread in Europe and the Mesogean territories.

What species names are to be applied to the Rhaetian liostreas in Europe is debatable (see p. N1103 under *Liostrea*).

LOPHA RÖDING, 1798 (SENSU LATISSIMO)

Recognition of *Lopha* found in Triassic deposits is difficult. Externally the lophas are quite similar to some species of *Enantiostreon, Atreta, Placunopsis,* and *Pseudomonotis.* The former are attached by their left and the latter by their right valves. The distinction cannot be made unless location of the adductor muscle insertion is clearly visible. However, most specimens found in Triassic deposits have tightly closed valves (Fig. J62).

The following lophas are known from the Triassic.——1) *L. calceoformis* (BROILI) from the Pachycardien tuff (Carn.) of the Seiser Alp (BROILI, 1904, p. 195, pl. 23, fig. 10-11). Originally described from only two considerably corroded left valves it was redescribed from another alpine locality by WAAGEN (1907, p. 116, pl. 34, fig. 37-38) and critically reinvestigated (Fig. J62,3). All valves described are unsatisfactory as to preservation.——2) *L. haidingeriana* (EMMRICH) from the Kössener Schichten (Rhaet.) from several localities in the northern Alps, including the type locality of the formation, 3 km. from the Bavarian border in northeastern Tirol, Austria (EMMRICH, 1853, p. 377). The same species was described as *"Ostrea" marcignyana* MARTIN (1860, p. 90, pl. 6, fig. 24-25) from Rhaetian arkoses [Grès et Schistes à *Avicula contorta*] at Marcigny-sous-Thil, on the left bank of the Armançon River, and at Montigny-sur-Armançon, Départment Côte d'Or, eastern France (Fig. J62,4). MARTIN (1865, p. 248) later conceded that his species name was a junior subjective synonym of *L. haidingeriana* (EMMRICH, 1853), this correction has been accepted by later authors (DIENER, 1923; KUTASSY, 1931). The species has been found in southern Bavaria and adjoining Austria, western Switzerland, the Carpathian Mountains of Poland, the Bihar Mountains of northwestern Rumania. Very similar remains have been found in Burma (HEALEY, 1908, p. 37, pl. 5, fig. 17-19).——3) *L. mediocostata* WÖHRMANN, 1889, p. 201, pl. 6, fig. 5) from the "Cardita Oolith" (Carn.) from the Salzberg near Hall in Tirol, Austria (Fig. J62,1).——4) *L. montiscaprilis* (KLIPSTEIN, 1843, p. 247, pl. 16, fig. 5) from the Kassianer Schichten (Carn.) of the Monte Caprile [or Zissenberg] in the Lombardy Alps, Italy. It was redescribed by WÖHRMANN (1889, p. 200, pl. 6, fig. 1-3) and BITTNER (1912, p. 70, 74-75, pl. 6, fig. 14-18) (Fig. J62,2). The latter pointed out that WÖHRMANN had confused the right with the left valve. The species has been reported from Tunisia, Tripolitania, Sicily, Bavarian Alps, Slovenia, Hungary, the Jordan River valley, Singapore Island, and the Luning Formation of Nevada, USA.——5) *L. parasitica* (KRUMBECK, 1913, p. 47-48, pl. 3, fig. 4-7) from the Fogi Beds

(U.Trias., probably Nor.) of western Buru Island, Molucca Archipelago, Indonesia.——6) *?L. blanfordi* (LEES, 1928) was described as *"Ostrea"* *(?Exogyra)* from the Elphinstone Formation (Nor.) in the Elphinstone Inlet in Muscat and Oman, at the north tip of the Arabian Peninsula. This is a U-shaped plicate form known only from left valves.——7) *?L. tinierei* (RENEVIER, 1864, p. 80-81) was described as *"Ostrea"* from the right bank of the Tinière River, opposite the hamlet Placundray near Les Chainées above Villeneuve at the east end of Lac Léman, Canton Vaud, Switzerland. It is a U-shaped form recalling *?L. blanfordi* (LEES) found in Rhaetian beds.

Triassic lophas are restricted to the Mesogean and Pacific realms. They appeared for the first time during the Carnian. They are commonly associated with crinoids, echinoids, brachiopods, sponges and corals, even compound corals. They lived in warm euhaline waters.

Although much of the Triassic and earlier deposits are yet to be searched for oyster remains, data at hand indicate that the earliest, Carnian oyster genera were *Gryphaea* (home and probable place of origin: the Triassic Arctic realm) and *Lopha (sensu latissimo)* (home and place of origin: the Triassic Mesogean and Pacific realms). *Liostrea* appeared later, in the Norian, in the Triassic Arctic realm. Indications that *Liostrea* might have evolved from *Gryphaea* are certain anatomical similarities and its later appearance and early restriction to the Arctic realm, the home of *Gryphaea.*

Gryphaea did not spread from its home area until the early Liassic. *Gryphaea* and *Lopha* at first did not live side by side, except in the oceanic connecting passage (Luning Formation of the Cedar Mts. in Nev., USA), where Arctic and Mesogean faunas intermingled.

The great geographic distance between the places of origin of *Gryphaea* and of *Lopha* are believed to be indicative of diphyletic origins of these two oyster genera.

DIPHYLETIC ORIGINS

Phylogenetic chains documented by innumerable species described from countless stratigraphic levels are fairly common in the oysters so that one is tempted to build up long phylogenies from these chains. One must be cautious in selecting the links between such chains, however.

In going from one link to the next, one

must select the one species that is both in correct stratigraphic sequence and among all available species morphologically the most similar to its predecessor species. Internal anatomical and morphological features are the more important ones in this procedure (STENZEL, 1959), and one must avoid a purely provincial outlook when searching for a missing link, because oysters are capable of spreading from one depositional basin to another.

In this fashion the Gryphaeidae can be traced from the Triassic species of *Gryphaea* to their descendants living today: *Neopycnodonte cochlear* (POLI) [*=Ostrea cochlear* POLI, 1795, v. 2, p. 179, 255, 261], living circumglobally in cool, deeper, euhaline waters, and *Hyotissa hyotis* (LINNÉ) [*=Mytilus hyotis* LINNÉ, 1758, p. 704, no. 207] and three more, congeneric species, all living geographically separated from each other in shallow, warm tropical, euhaline waters. The Gryphaeidae, always euhaline, never were rich in genera at any one time, and only two genera survive today.

Similarly it is possible to trace the Lophinae from the Triassic species of *Lopha (sensu latissimo)* to their descendants living today: *L. folium* ecomorph *cristagalli* (LINNÉ) [*=Mytilus crista galli* LINNÉ, 1758, p. 704, no. 206] a euhaline, warmwater species, living in the western Indo-Pacific Ocean, which is the type species of the genus, and several other species. Thus, two separate phylogenetic stems of oysters are traceable from their very first appearance in the Late Triassic to today. All known oyster species, fossil and living, are either offshoots or parts of these two separate phylogenetic stems. As concerns tangible evidence documented by fossil remains, the oysters are diphyletic.

The problem reduces itself to the question whether the very first oysters, that is, pre-Late Triassic ones, were truly monophyletic or were diphyletic. Because remains of oysters that ancient never have yet been discovered, the problem must be attacked by other means than fossil remains and becomes subject to speculation. Certain anatomical features of the Gryphaeidae are significant in the question and are discussed below. Other differences between the Gryphaeidae and the Ostreidae are given in the systematic portion (see p. N1096).

PELSENEER (1896, pl. 3, fig. 4-5, pl. 4, fig. 7-8; 1911, p. 94, pl. 9, fig. 8-9) was the first to discover that in two living oyster species now placed in the Gryphaeidae by me the intestine passes through the pericardium and through the ventricle of the heart itself. This arrangement has been confirmed by HIRASE (1930, p. 38), RANSON (1948b, p. 5), THOMSON (1954, p. 161), and HAROLD W. HARRY (by anatomical dissection, 1966) so that it is reliably documented in both genera. In contrast, all representatives of the Ostreidae that have been dissected have the intestine bypassing the pericardium on the dorsal side of the latter. Peculiar as it may seem, the arrangement found in the Gryphaeidae is not unusual for the Bivalvia. The great majority of the Bivalvia have the same arrangement (PELSENEER, 1906, p. 223).

This being so, one is forced to conclude that it is the original arrangement in the Bivalvia and in the ancestors of the oysters as well. Thus it is the primitive arrangement in the oysters, as PELSENEER (1911, p. 93-94) pointed out, and the arrangement of the intestine bypassing the pericardium is an evolutionary achievement of later date or an "advanced" feature.

Also, the living Gryphaeidae are probably nonincubatory, because they all have a large promyal passage. No direct observations on incubation or on ejection of eggs are really available for this family, but the promyal passage has been demonstrated in all Gryphaeidae which have been dissected and its presence has been confirmed by HARRY (dissections made in 1966). The tie-in of promyal passage with nonincubatory propagation in the oysters appears to be reliable, although reasons for it are obscure.

In the Ostreidae, some genera, notably *Ostrea,* are incubatory. Again, the vast majority of Bivalvia are nonincubatory and only a scattering of genera have progressed to incubation. Nonincubatory propagation must be the primitive situation, so general considerations demand, and the Gryphaeidae are primitive in this respect. In contrast, the Ostreidae are in part incubatory and must be regarded, in part at least, as progressive in regard to propagation.

By and large, the living representatives of the Gryphaeidae have retained more of the primitive features than the rest of the oysters. This being so, one is perhaps justified in expecting that most, and perhaps all, features by which the Gryphaeidae differ from the Ostreidae are primitive and were present in the very oldest, Triassic, representatives of *Gryphaea,* even if those primitive features did not leave any recognizable traces on the fossil shells. These primitive features are: 1) nonincubatory mode of reproduction, 2) presence of a promyal passage, 3) penetration by the intestine of the pericardium and ventricle of the heart, 4) orbicular outline of the insertion of the adductor muscle, and 5) position of this insertion comparatively closer to the hinge than in the Ostreidae. The last two are visible on fossil specimens and for that reason assume extraordinary importance in any phylogenetic inquiry.

The adductor muscle has many functions indispensable to the survival of the individual oyster (see p. *N*999). For that reason it is so large and powerful and must be capable of reacting instantaneously. In order to accomplish all these functions it must be copiously and unceasingly supplied with fresh oxygenated blood. However, heart and circulatory system in the oysters are rather poorly organized, and delivery of fresh oxygenated blood is hardly efficient. This inadequacy has been improved upon in the oysters through the anatomical position of the heart: it is placed as close to the muscle as possible and the artery supplying fresh oxygenated blood to the muscle (posterior aorta of AWATI & RAI, 1931, p. 59, fig. 31, and YONGE, 1960, p. 48) is as short as possible.

In other words, heart and adductor muscle are functionally tied together in the oysters. If for some reasons of adaptation the adductor muscle has to shift position within the shell cavity, heart and pericardium must move with it irrespective of the location of other organs in the shell cavity. For instance, if there is need for the adductor muscle to shift toward the venter, that is, away from the hinge, the heart and pericardium would shift in exactly the same way so as always to stay adjacent to the muscle, whereas the intestine would remain unaffected and be left behind. If this shift were to take place, heart and pericardium would become disengaged

from the intestine. They would follow the muscle and leave the intestine behind so that the intestine would come to be on the dorsal side of the pericardium. For some reason or other this is what happened early in the phylogenetic history of the Ostreidae but not in that of the Gryphaeidae, which retained the original, primitive, arrangement of the oyster ancestors. The reason for this shift of the muscle is unknown. One might speculate that leverage of the muscle becomes better the farther away it comes to lie from the fulcrum, the hinge axis. The shift must have enhanced leverage of the muscle, and functions of the muscle must have been made easier and more efficient.

Whatever may have been the reason behind the shift of the muscle, it is obvious that the anatomical position of the muscle is clearly marked on fossil specimens by the position of the muscle imprints. Or, fossil oysters that have muscle imprints comparatively close to the hinge (see STENZEL, 1959, p. 28, fig. 19, for exact measurements) must have had the intestine passing through the pericardium and ventricle of the heart while they were alive. All Gryphaeidae, including Triassic species, have such a position of the muscle imprints. This must be expected of them, if the observations on the anatomy of their descendants, the living species, are correct, and if our conclusion that the arrangement found in the living species is a conservative trait is correct.

On the other hand, the Triassic species of *Lopha,* that is, the oldest known representatives of the Ostreidae, have muscle imprints that have crescentic to reniform outlines and are placed closer to the ventral shell margin. It is to be expected that these oldest known representatives of the Ostreidae had the intestine passing by the dorsal side of the pericardium. In other words, this evolutionary achievement was already an established fact in the Ostreidae during Late Triassic time and must have happened before that time.

During Carnian time, when first the oysters showed up as recognizable fossils, already two separate stems were discernible: Ostreidae represented by various species of *Lopha* and Gryphaeidae by various species of *Gryphaea.* The two stems were already far apart in anatomy and other important features. So far no earlier intermediate forms are known to us which one might interpret as ancestors of both stems. Thus further investigation of the phylogeny has to rest on speculation.

There are two possibilities to consider: 1) derivation was monophyletic, that is, both stems descended from a single common ancestor which would have to be classified as an oyster; 2) derivation was diphyletic, that is, each stem descended from a different genus, but the two ancestral genera were not oysters. The two ancestral genera were closely related to each other and are to be found in the same extinct family. At present, neither possibility can be claimed proved or even provable. However, I believe the second possibility is more likely.

NEWELL (1960, p. 81) has indicated that some of the late Paleozoic and Early or Middle Triassic Pseudomonotinae resemble true oysters so closely as to suggest that the former may indeed be ancestral to the latter. Accordingly, the oysters may have arisen either from a genus of the Pseudomonotinae (monophyletic hypothesis) or from at least two genera of the same subfamily (diphyletic hypothesis). If the monophyletic hypothesis is correct, there would have to have been a long chain of transitional species starting with an ancestral pseudomonotine genus, evolving into an ancestral oyster genus, progressing by dichotomy and divergence in the two stems discussed above. This phylogeny must have been long drawn out, because in the Carnian the divergence between the two stems, at the end of the phylogenetic chains, was quite large. Because of its length it is not likely that the chain of transitional oyster species can have entirely escaped detection by searching paleontologists.

On the other hand, if the diphyletic hypothesis is correct, much of the divergence between the two stems was established by evolution within the confines of the subfamily Pseudomonotinae. Thus no lengthy chains of transitional oyster species were necessary to lead from the two pseudomonotine ancestor genera to their respective oyster descendants. Because each chain was short they could have escaped detection easily. This assumption seems to fit the facts discovered so far.

HYPOTHETICAL ANCESTORS

The roots of the phylogeny of oysters are obscure, because they are not documented by fossils or perhaps because there are difficulties in recognizing certain fossils as oyster ancestors. New discoveries of fossil remains may yet expand our knowledge. The most promising area for such discoveries is probably far eastern Siberia, where beds older than Carnian may yield recognizable remains of oyster ancestors.

Until such finds are made we must reexamine ideas concerning characteristics and appearances of these ancestors, thus turning to hypotheses and assumptions. These are presented here with full realization that ultimately they may be refuted. At the present stage of knowledge, however, new hypotheses and ideas to stimulate investigation are needed badly. This is the purpose of the following hypotheses.

We may postulate that ancestors of the Pseudomonotidae and kindred groups were sedentary animals anchored by byssus threads. Because they were not firmly cemented to their substratum but only anchored by fairly long byssus threads, they were subject to waves and currents and were tossed about by them on occasion. During such times the most stable, and therefore the safest, position for them was to lie on their sides (pleurothetic attitude). Hydrodynamics of their environment made more stable and safe bivalves possessing one convex valve and the other one flat, suited to lie on the seabottom. The adaptive advantage of such different convexities of the two valves is that an animal exposed to water currents and anchored by byssus threads but otherwise lying loose with the flat valve on bottom is more stable than would be an animal with two equally convex valves or one resting with the convex valve on bottom and the flat valve on top. This inequality of the valves induced some shifting of internal organs. Then it became heritable and genetically fixed. It so happened that the common ancestor of the superfamilies Anomiacea, Limacea, Pectinacea, Pteriacea, and Ostreacea came to lie on its right side, and the right valve became the flat one.

JACKSON (1888, p. 547-548, pl. 7, fig. 19) convincingly showed that the unequal convexities of valves of oysters, with the left valve normally more convex than the right, are not merely a mechanical result of the individual oyster's attitude during its growth, as HYATT assumed, but must be under genetic control. This important discovery proved that oysters fit quite well with other superfamilies in this respect and it is likely that the feature was inherited by the oysters and kindred superfamilies from common ancestors that were byssiferous and pleurothetic on the right side.

Although the other superfamilies mentioned are pleurothetic, with the right valve on bottom, oysters are cemented onto their substratum and pleurothetic with the left valve on bottom. This singular exception is explained below by the hypothesis that the oyster ancestors were originally pleurothetic with the left valve on top but later reversed their attitude in giving rise to the oysters.

Waves and currents exerted a considerable pull which was transmitted through the anchoring byssus threads to the byssus gland and tissues surrounding it. The pull tended to push these tissues forward, toward the anterior adductor muscle. Restrictions thus exerted on this muscle led to its gradual shift toward the hinge. In this position the muscle became no longer as effective as the other muscle and atrophied by stages. On the other side, the posterior adductor muscle remained unobstructed and was free to shift to its most advantageous position and to take over more and more of the functions of the other muscle. Ultimately the animal became monomyarian (SHARP, 1888; DOUVILLÉ, 1907, p. 97; 1913, p. 430). The presence of a byssus comprised an obstacle to free growth of the valve margins, and a byssal notch had to develop at the place where the byssus passed from the valve margins.

The hypothetical ancestor of the Ostreacea and kindred superfamilies is believed to have been byssiferous, monomyarian, pleurothetic, inequivalve (right valve flat, left convex), and to have been sedentary, anchored by byssus threads. It was nonincubatory; internally, the intestine passed through the pericardium and ventricle of the heart. It was pleurothetic on the right side, had a byssal notch at the margin of the valves, and the byssus threads were fairly long.

Cementation of the shell to its substratum probably was not so difficult an innovation, because no great changes of internal organization were required. This is attested to by the fact that several branches of these monomyarian bivalves attained cementation independently. Cementation may have been preceded by a gradual shortening of the byssus threads. Adaptational needs may have led to a shorter and shorter byssus, until ultimately the shell was so closely held that it could shift no more with reference to its substratum and the growing valve margins came into contact with the hard substratum. At this point cementation became easy.

Possibly cementation came about through adaptation of the ancestral bivalves to life on hard substrata. As long as they lived attached to sea weeds and other flexible and yielding objects, they needed many long byssus threads which could find many widespread points of anchorage. When the animals began to occupy hard, inflexible, or unyielding substrata, a few anchorage points and short byssus threads were sufficient and better.

Once cementation became a permanent way of life, byssus gland and foot became useless to the adult bivalve and were retained only in the larvae. The foot in the adult stage disappeared completely except for a pair of Quenstedt muscles (see p. *N*965). The shell itself was free to grow where the byssal notch had indented the valves, and the notch disappeared.

There remains the question of reversal of the pleurothetic attitude in the ancestor of the oysters. The monomyarian bivalves enumerated above, including the oysters, have so many anatomical features in common that it appears beyond doubt that they are of monophyletic derivation. If most of them are pleurothetic on their right sides, it would seem that this attitude is the original one among them, inherited from their common ancestor, and that the reversed attitude of the oysters is an innovation imposed upon the original attitude of the oyster ancestor. From these considerations the hypothesis is derived that oysters descended from ancestors that were byssiferous and pleurothetic on their right sides.

Perhaps reversal of the ancestral pleurothetic attitude has something to do with cementation. One innovation may have favored or entrained the other. Among bivalves that have one flat and one convex valve, a great adaptive mechanical advantage accrues to bivalves cemented with the larger, more capacious, more convex valve on bottom. The upper valve has to be lifted up by the animal every time it opens its shell.

As long as these bivalves were only loosely attached by byssus threads, and had fairly light shells, hydrodynamics of the environment precluded their lying with the convex valve on bottom. However, as soon as they became cemented, hydrodynamics of their environment lost influence and mechanical advantage could prevail. If oyster ancestors already had valves of unequal convexity before they adopted cementation as a way of life, thicker shells and cementation of the more convex valve at the bottom must have been an advantage to them.

Several different branches of this group of monomyarians came to achieve cementation independently, but only those that accomplished it with the more convex, left, valve on bottom became oysters. The others, cemented with the right valve on bottom, never became biologically as successful as the oysters.

If several different branches of this group of monomyarians were able to achieve cementation by the right valve independently, it stands to reason that it was equally well possible for more than one branch to do so with the left valve. In other words, a polyphyletic origin of the oysters is not impossible.

ATROPHY OF HINGE TEETH

Among the Bivalvia, mechanically strong shells equipped with strong hinge plates carrying large hinge teeth and corresponding sockets are developed in taxa that move about actively and plow through the sand or mud of the substratum. Such taxa have for their locomotor organ a large turgid foot activated by hydrostatic pressures of the blood in the foot produced by powerful muscles compressing the blood lacunae in foot and mantle. To accommodate a large foot when it is withdrawn into the shell, these animals must have an inflated, copious shell and must be able to open their valves wide enough to extrude the turgid foot.

When these animals plow through the sand, silt, or mud of the substratum, the valves, opened wide, easily could be wrenched out of juxtaposition so that quick shutting of the shell would be made impossible unless there are structures that effectively keep the two valves in their proper juxtaposition and that guide them when they are being shut. Such structures are of two kinds: 1) interlocking serrate valve margins, which are an auxiliary but less effective means of guidance for the valves, and 2) hinge teeth and sockets situated on a support, the hinge plate, which are a very effective means of preventing the valves from wrenching out of correct juxtaposition.

Teeth and sockets must be large and strong and their height (or rise from the hinge plate) must be sufficiently large to keep the valves in correct juxtaposition, even when the animal opens its valves to their fullest. In turn, high and strong teeth and sockets require a strong hinge plate for their support.

Animals that neither plow through the substratum nor burrow into it do not really need a large active foot. Their foot either can atrophy or must modify to take over functions other than plowing or burrowing. The oysters have followed the former of the two pathways of evolution. They have no foot when they are adult, although their larvae have a very active foot. Adult oysters need not open their valves very much, because they lack a foot and have no need for teeth, sockets, and hinge plates. Consequently, oyster larvae, which have a foot with a byssus gland, hinge plates, teeth, and sockets, lose them very soon as they undergo metamorphosis and begin to assume the adult shell form.

Reduction of teeth and sockets are fairly easily accomplished during evolution of the oysters from their immediate ancestors, because the latter themselves already had weak dentition and were largely immobilized byssiferous monomyarian bivalves.

Quite a different evolutionary path was taken by the rudists. Although they became attached like oysters, their hinge retained teeth and sockets and even enlarged their height. However, rudists descended from mobile bivalves with prominent teeth

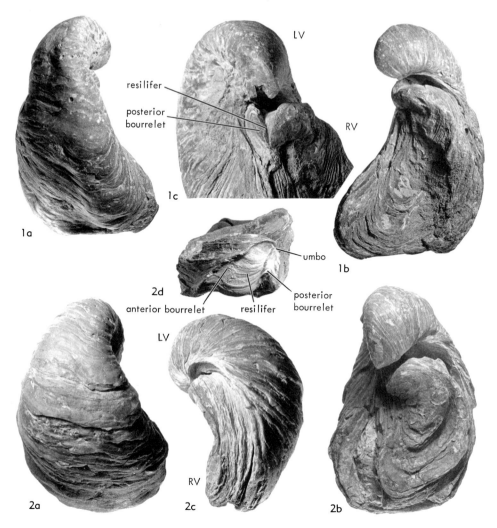

FIG. J63. The *Gryphaea* homeomorph *Aetostreon imbricatum* (KRAUSS, 1843), from Sundays River beds of Uitenhage Series, L.Cret.(Neocom.), Sondags River, Cape Province, Republic of South Africa (Stenzel, n).

1a,b. LV and RV views, ext., *ca.* ×0.6.—*1c.* Umbonal view, *ca.* ×1.
2. "*Gryphaea*" *imbricata* (KRAUSS) of SHARPE (1853, pl. 23, fig. 3) (BMNH no. LL 15898).—

2a-c. LV, RV, and edge views, ext., *ca.* ×0.6.—*2d.* RV ligamental area, *ca.* ×0.1.
[Photographs by courtesy of J. D. TAYLOR and British Museum (Natural History), London.]

and sockets. It was easier for them to retain and readapt their teeth and sockets than to suppress them and to evolve entirely new structures ensuring proper closures of the valves.

EVOLUTION OF GRYPHAEA AND HOMEOMORPHS

The phylogeny of *Gryphaea* has aroused considerable interest and there is an exten-

sive literature on this and related subjects. Good lists of references of this literature are found in SCHÄFLE (1929) and BURNABY (1965). However, much of the literature is repetitious, and several misconceptions have gained currency and even acceptance, without sufficient critical reexamination.

FEATURES IN COMMON

The following features are common to the gryph-shaped oysters, that is, *Gryphaea*, some of its descendants, and its homeo-morphs (Fig. J63, J64, J65; see J73, J74, J80, J81, J83, J84, J87).

1) Exceptionally small attachment area at the tip of the left umbo,

2) Spirally inrolled growth pattern of left valve,

3) Large and highly convex, spirogyral left valve and small, flat or concave, oper-culiform right valve lacking a conspicuous umbo,

4) Growth squamae closely appressed to contour of left valve,

5) Lack of chamber formation in um-bonal part of left valve,

6) Lack of umbonal cavity under hinge area of left valve, and

7) Presence of radial sulci dividing left valve.

Most of these seven characteristics are sur-ficial or external.

Many authors have commented on the extraordinarily small size of the attach-ment area at the tip of the left umbo in gryph-shaped oysters (JACKSON, 1890, p. 317, pl. 24, fig. 22-24). From SWINNERTON'S (1939, p. xliv and lii; 1940, p. xcviii, fig. 8; 1964, p. 419-420) data one can calculate the numerical average (2.1 mm.) of the maxi-mal diameter of the attachment area and the average size (less than 4.5 square mm.) of the area (Fig. J35). These figures per-tain to *Gryphaea arcuata* LAMARCK, 1801, from clay shales in the Zone of *Scamnoceras angulatum* (VON SCHLOTHEIM, 1820) in the Granby Limestones (latest Hettang., Lias.) south of Granby, Lincolnshire, England.

In *Gryphaea* itself no individual has been found to lack an attachment area. How-ever, some species in other homeomorphous genera are regularly free of an attachment area (e.g., *Odontogryphaea thirsae* (GABB, 1861) (Fig. J64) from the Nanafalia For-mation (Sparnacian) of the northern Gulf coastal plain (Ala. to Mexico) and possibly other congeneric species).

Evidently, larvae of nearly all gryph-shaped oyster species were like those of common oysters living today in that they could not grow into adults without first becoming attached to a firm substratum at the proper time. After growing over the substratum for a short period of time, the gryph-shaped oysters built their shells steeply upward and away from the sub-stratum. This happened early in life so

that the attachment areas remained very small in most individuals.

Perhaps, the substratum was so small a piece that the postlarval oyster soon finished spreading over all of it and had to grow freely upward. Or perhaps, the gryph-shaped oysters were genetically predisposed to growing steeply upward from their bases after they had spread over a small area on the substratum. The latter idea is borne out by certain individuals of *G. arcuata* grown one on the other, such as the one figured by SCHÄFLE (1929, pl. 2, fig. 8) and copied by many authors. In this example the younger one of the two had really much more of a substratal area available than it managed to spread over. It is also possible that the gryph-shaped oysters did not suc-ceed to spread over larger areas of their substratum, because they were in competi-tion with encroaching mud or algal mats covering the substratum around them while they grew.

Whatever the situations may have been, gryph-shaped oysters were able to grow to final size without using a large attachment area on a firm substratum. Some species (e.g., *Odontogryphaea thirsae*), had evolved even further and had attained the ability to grow from planktonic larva to final size without ever becoming attached.

The highly incurved growth pattern of the left valves has always attracted atten-tion. THOMPSON (1917, p. 534) surmised that the spiral curve of the lower valve in *G. arcuata* approaches a logarithmic or equiangular spiral. This suggestion has been followed up by TRUEMAN (1922, p. 260-261), MACLENNAN & TRUEMAN (1942), JOYSEY (1959, p. 313-314), HALLAM (1959), and BURNABY (1965). These authors have found several flaws, when they tried to fit specimens to mathematically constructed logarithmic spiral curves. It is not surpris-ing that such flaws exist, rather it is sur-prising that individuals of *Gryphaea* in growing freely came so close to fitting a logarithmic spiral.

In some of the homeomorphs the curve of the left valve is quite regular; in others, notably in *G. arcuata* the regularity is ob-scured by many narrow transverse grooves and rounded ridges caused by the expanded end of the many growth squamae as seen in thin section (see Fig. J73,1). Finally, there are those that have nearly rectilinear

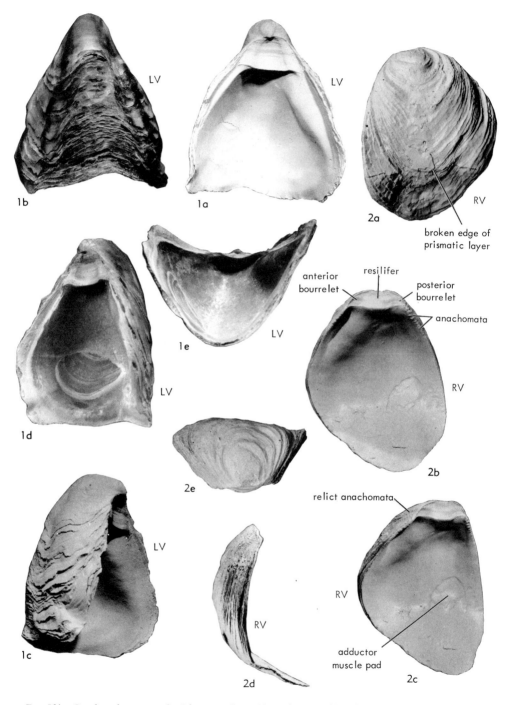

Fig. J64. *Gryphaea* homeomorph *Odontogryphaea thirsae* (Gabb, 1861) from oyster bed of Nanafalia Formation, Wilcox Group (low. Eoc.), near Shoal Creek, south of Camden, Wilcox County, Ala., USA (Stenzel, n).

stretches of the periphery and have one such stretch leading over to the next one by a bend, comp. *Texigryphaea roemeri* (MAR-COU (see Fig. J87,*1c*).

The spiral growth pattern is difficult to investigate with sufficient precision. Only in the anteroposteriorly compressed later Jurassic Gryphaeas and the early Cretaceous Texigryphaeas is the periphery convenient to define, because these have a rounded keel. The others, including *G. arcuata*, have hardly a keel so that the course of the spiral is difficult to follow. Also, the spiral is not two-dimensional, making it difficult to obtain precise results. JOYSEY and HALLAM pointed out other difficulties.

The left valves are highly convex in both directions from dorsum to venter and from anterior to posterior margin. In the latter direction they are commonly more convex than in the former. Their right valves are flat and in many species they are concave in one or another direction. The distribution of the concave areas can be highly complex in some species.

Growth squamae of the left valve generally are tightly appressed to the contour of the valve so that they are rather difficult to see and the outer surface of the valve is smooth. Even in the radially ribbed Eocene genus *Sokolowia* the surface of the left valve is smooth and free of outstanding growth squamae, because crests of the ribs and interspaces between them are smooth in radial direction and growth squamae are appressed. Smoothness and appressed growth squamae are lost near the end of growth in old individuals.

Sections through the left beak of *G. arcuata* show no voids or shell chambers there (see Fig. J73,*2a*) as in many other homeomorphs. This situation is quite different in ordinary oysters. If the latter have a prominent left beak, they commonly have many irregularly spaced shell cham-

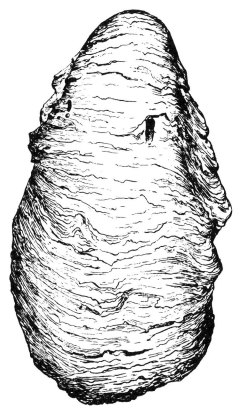

FIG. J65. *Gryphaea* homeomorph *Sokolowia buhsii* (GREWINGK, 1853) subsp. *gamma* (VYALOV), up. Eoc., Turkestanian (Trk₃) Central Asia, SSSR; LV ext., ×0.5 (Gekker, Osipova, and Belskaya, 1962).

bers filling in the space under the ligamental area of the left valve.

Oysters that have great disparity between the two valves and have a prominent left umbonal region commonly have a pronounced umbonal cavity, or recess, under the hinge plate or ligamental area of the left valve. Good examples are *Crassostrea* and *Saccostrea*. From time to time the umbonal cavity is closed off by a thin shell wall, resulting in a chamber. Continued

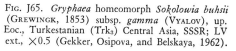

FIG. J64. *(Continued from facing page.)*

1. LV; *1a*, specimen seen from right side, whitened for photography, showing lack of attachment area at beak, ×1; *1b*, external view of same specimen, not whitened, showing traces of coloration and many growth squamae at rooflike terebratuloid fold, ×1; *1c*, oblique view of same, whitened, ×1; *1d*, oblique view of same, not whitened, showing very thin dorsal calcite cover of muscle pad partly broken out so that false convex dorsal margin of pad is suggested, ×1; *1e*, ventral view of same showing deep terebratuloid fold, ×1.
2. RV; *2a-e*, several views of specimen whitened for photography, ×1.

growth produces a series of such chambers, arranged somewhat irregularly, infilling the left umbonal region. The gryph-shaped oysters, however, have a very shallow umbonal cavity or none at all and chambers are not developed in the left umbonal region. Their left beak is solidly filled with shell material.

Radial sulci are common features in these oysters and are known from such diverse taxa as *Gryphaea (Gryphaea)*, *G. (Bilobissa)*, *Texigryphaea*, *Pycnodonte*, *Odontogryphaea*, and others. The sulci of the Gryphaeinae and Pycnodonteinae are homologous, but those of *Olontogryphaea* are only analogous to the former. These sulci are deeply incised in some species and shallow in others. Their depth is a specific to subgeneric feature. They set off flanges or separate parts of the shell which serve to concentrate and deflect the inhalant or exhalant water current and are needed as improvements to the sanitation of the animals.

PROBLEM OF FORM GENUS

Similarities between the various gryphshaped oysters appeared so persuasive to many authors that they rather uncritically classed or described diverse species under the generic name *Gryphaea* without making an effort to search for less obvious features that might set apart one group of related species from another. KITCHIN (1908, p. 79) was first to express the idea: "It is highly probable that the shells classed as *Gryphaea* do not represent a homogenetic group, but are polyphyletic in origin, including repeated offshoots from an ostrean stock."

Accordingly two avenues of procedure were open. Either accept *Gryphaea* as a form genus or investigate the various gryphshaped oysters so as to disentangle the polyphyletic groups of species in order to establish several separate monophyletic genera, each furnished with a nomenclaturally valid name. The first avenue is not recommended. Temporarily it is a convenient expedient allowing one to set aside pressing problems and a cover for the lack of incisive basic work on them. However, if the first procedure is used, repeated often, and maintained for decades, it discourages searching inquiries into phylogenetic rela-

tionships among the various genera involved in homeomorphy.

At first KITCHIN (1908, p. 77-82) followed the second avenue and proved that one of the homeomorphs, previously classed by one author as *"Gryphaea"* and by another one as *"Pycnodonta"* (DOUVILLÉ, 1904a, p. 215), is really an exogyrine related to *Aetostreon latissimum* (LAMARCK) [*=Gryphaea latissima* LAMARCK, 1801, p. 399, *=G. couloni* DEFRANCE, 1821, *=G. sinuata* JAMES SOWERBY, 1822, *=Griphea aquila* BRONGNIART, 1822, in CUVIER & BRONGNIART, p. 332]. This particular homeomorph is *Aetostreon imbricatum* (KRAUSS) (KRAUSS, 1843, p. 129; 1850, p. 460, pl. 50, fig. 2a-d) from the Sundays River Beds, Uitenhage Series (Neocom.), of the region around the Coega, Sondags, and Swartkops Rivers near Port Elizabeth, Cape Province, Republic of South Africa (Fig. J63; see Fig. J92).

The species is rather variable. Some local populations have a broader shell shape (length to height ratio fairly large) and obviously exogyrine coiling of the left beak. These have been correctly classified as Exogyras by all authors who had such specimens at hand. In common with *A. latissimum* they all have many nonappressed growth squamae on the left valve. Other local populations have shells of small length and large height looking like a tall hook-shaped *Gryphaea*. KITCHIN discovered that their attachment area was not at the tip of the left umbo as in *Gryphaea*, but was situated on the posterior flank of the beak as in the Exogyrinae. In addition, the left ligamental area has its posterior bourrelet greatly reduced in length so that it is a narrow sharp ridge rising above the general level of the ligamental area. The two features prove the species is one of the Exogyrinae.

Although in this case KITCHIN was successful in disentangling one of the homeomorphs from the form genus and in showing the true affinities of the homeomorph, he changed his mind (KITCHIN, 1912, p. 59-60) and reported:

He [Mr. PRINGLE] obtained a fine series of shells [from the Lower Liassic of Fretherne, Gloucestershire], which should prove of value in helping to elucidate the nature and origin of those ostrean forms which have hitherto been roughly

TABLE 1. *Gryph-shaped Oysters.*

Name	Subfamily & relationship to *Gryphaea*	Stratigraphic time range	Geographic distribution
Aetostreon imbricatum (KRAUSS, 1843) (Fig. J63)	Exogyrinae, homeomorph	Neocomian	East and South Africa
Gryphaea LAMARCK, 1801 (*see* Fig. J74)	Gryphaeinae, homeomorph	Late Carnian-Kimmeridgian	Nearly worldwide
Odontogryphaea VON IHERING, 1903 (Fig. J64; *see* Fig. J127)	Ostreinae, homeomorph	Late Maastrichtian-early Lutetian	Nearly worldwide
Pycnodonte FISCHER DE WALDHEIM, 1835 (*see* Fig. J80-J84)	Pycnodonteinae, descendant	Early Albian-Miocene	Nearly worldwide
Sokolowia BÖHM, 1933 (Fig. J65; *see* Fig. J121-123)	Ostreinae, homeomorph	Late Lutetian-Auversian	Transylvania and central Asia
Texigryphaea STENZEL, 1959 (*see* Fig. J87)	Pycnodonteinae, descendant	Mid-Albian-Cenomanian	Southwestern North America

classed together as *Gryphaea*. Many of these specimens show well the transitional stages between the ancestral simple oysters of flattened forms, with prolonged attached stage, and the extreme morphological type characterized by a greatly reduced attached stage and a strongly arcuate left valve. . . . He believes that the specimens comprised in this collection contribute some valuable evidence in support of this view that the features usually considered to characterize *Gryphaea* are not those of a long-lived genus but of a similar evolutionary stage which has appeared repeatedly and independently in various ostrean stocks. It is evident that in the Lower Lias alone there were two or three separate derivations for such forms. The evolution of analogous gryphaeate stages was repeated in other stocks in various Jurassic and Cretaceous horizons . . . .

Here he regarded *Gryphaea* as a stage name or form genus. His statement marks the turning point in the attitude some paleontologists took toward the problem. TRUEMAN (1922, p. 264) epitomized this attitude:

Indeed it is extremely likely that these gryphaeiform shells have been evolved repeatedly during the Jurassic and Cretaceous from species of *Ostraea* [*error pro Ostrea*] that are similar and are presumably closely related. In other words, "*Gryphaea*" is a polyphyletic group, containing species evolved along many different lines. Therefore, the name *Gryphaea* can only be applied strictly to *one* of these series, and each such series should receive a separate generic name; but until more of these characteristics are known, at least, it appears undesirable to add to the existing confusion by creating new names for each group. Indeed, as homeomorphs in some of the series are almost or quite identical, it would probably be impossible to distinguish many of the genera even though names were available.

SCHÄFLE (1929, p. 79) agreed in regarding "*Gryphaea*" as a nonhomogenetic group. ARKELL (1934, p. 58-59) quoted TRUEMAN with obvious approval.

Thus it came to pass that from 1922 onward many preferred to consider *Gryphaea* as a collection of hardly distinguishable genera, a so-called pseudogenus or form genus (MORET, 1953, p. 31, 373 footnote 2; SIMPSON, 1953, p. 183). It was rather unfortunate that from that time on no more concerted efforts were made to disentangle monophyletic groups of species from the form genus.

It is my contention that every effort must be made to disentangle the various monophyletic groups of species and to provide each group with a valid name until there are only monophyletic genera at hand. This task has largely been completed, and it is now possible, and highly advisable, to regard *Gryphaea* as a monophyletic genus tied to its well-known type species. This monophyletic genus is clearly separable from similar genera, be they its descendants or its homeomorphs, and is definable on the basis of features of anatomy, morphology, and shell structure (Table 1).

KITCHIN (1912, p. 59) maintained that among the lower Liassic gryphaeas there were two or three separate derivations from ostrean ancestors, that is, that the lower Liassic gryphaeas were polyphyletic. In more than 50 years no one has substantiated this extreme viewpoint, believed to be erroneous, and the lower Liassic gryphaeas are regarded as diversified species of a monophyletic genus.

By reason of ICZN Opinion 338, the monophyletic genus *Gryphaea* LAMARCK, 1801 (p. 398-399), has for its type species *G. arcuata* LAMARCK, 1801 (=*G. incurva* J. SOWERBY, 1815). The type species is widespread in western Europe, from southern Sweden to northern Italy and the high Tatra Mountains in southern Poland. The species ranges stratigraphically from the Zone of *Scamnoceras angulatum* (VON SCHLOTHEIM, 1820) in the Hettangian to the Zone of *Caenisites (Euasteroceras) turneri* (J. DE C. SOWERBY, 1824) in the Sinemurian. Extensive descriptions of the species were given by JONES (1865) and SCHÄFLE (1929, p. 26-37, pl. 10, fig. 7-17; pl. 11, fig. 1-4, 9).

ANCESTORS OF GRYPHAEA

DOUVILLÉ (1910, p. 118; 1911, p. 635) apparently was the first to express himself concerning the ancestry of *Gryphaea*. He suggested that the group of simple oysters of flattened shape with prolonged attachment stage found in Late Triassic and Early Jurassic beds of western Europe is the immediate ancestor of Liassic *Gryphaea*. He bestowed the name *Liostrea* to the supposedly ancestral group and used *Liogryphea* [error pro *Liogryphaea* FISCHER, 1886 (=*recte Gryphaea* LAMARCK, 1801)] for Liassic gryphaeas. This phylogenetic derivation has become a generally accepted hypothesis. For example, PHILIP (1962, p. 337) and SIMPSON (1950, p. 153) have accepted it.

Nonetheless, it never was more than an assumption without good proofs. The bases for the assumption, not explicitly stated, seem to be two: 1) *Liostrea* is found in western Europe in beds older than the *Gryphaea*-bearing Lower Liassic beds and in them it is the sole oyster genus present in this region. If one looks for an ancestor of *Gryphaea* in western Europe, *Liostrea* appears to be the only oyster genus available. 2) *Liostrea* has a simple, flattened shape which, it seems logical to assume, is more primitive than the spiral-shaped *Gryphaea*. To this day it remains to be seen whether these two ideas are sound.

PRINGLE and KITCHIN (1912, p. 59-60) claimed to have discovered good proof for this hypothetical phylogenetic derivation.

They claimed to have found in lower Liassic beds at Fretherne transitional evolutionary stages connecting the two genera. Thus they opened a lively discussion of these supposedly transitional stages of evolution. TRUEMAN (1922), in one of the first papers to use statistical data and variation diagrams in paleontology, strongly supported these claims and published a graph in their support (TRUEMAN, 1922, fig. 5). The graph shows the amount of incurving of the left umbonal region, called number of whorls by TRUEMAN, in each of five collections coming from five different stratigraphic levels. The lowest one is based on a *Liostrea,* the others on *Gryphaea arcuata* going progressively higher up in the stratigraphic section. MACLENNAN & TRUEMAN (1942) furnished additional, improved graphs based on specimens from Loch Aline, Argyllshire, Scotland. The two papers were acclaimed by many, for example, by PHILIP (1962, p. 334), and regarded as the ultimate paleontologic proof that successive evolutionary stages connected *Liostrea,* the ancestor, with *Gryphaea,* the descendant, in the stratigraphic section of the lower Liassic of the British Isles.

Objections were raised early by SCHÄFLE (1929, p. 76, and footnote 2) who explicitly took issue with TRUEMAN in stating that in southwestern Germany no transition between the two genera can be found, although one could hardly deny close relationships between them.

HALLAM reexamined statistically the supposed proof presented by TRUEMAN (1922) and MACLENNAN & TRUEMAN (1942) with the aid of specimens from various localities in the British Isles. He found no such transition connecting the two genera and concluded: "While the presumption must remain that the Jurassic *Gryphaea* evolved in some way from an *Ostrea*-like ancestor a convincing evolutionary series has yet to be demonstrated" (HALLAM, 1959a, p. 107). This contradiction added much to the interest in the question and more papers (GEORGE, 1953; HALLAM, 1959b, 1960, 1962; JOYSEY, 1959, 1960; PHILIP, 1962; SWINNERTON, 1932, 1939, 1940, 1959, 1964; TRUEMAN, 1940; WESTOLL, 1950) discussed several aspects of the problem. Particularly, the formulae and statistical methods employed so far were thoroughly criticized and improved.

It was recognized that the amount of incurving of the left umbonal region as expressed through the number of whorls was not a good measure, because it depended on the age or size of the individual oyster. The larger the individual oyster, the more of a coil or coils it has.

The progressive increase in number of coils from lower to higher beds noted by Trueman (1922) does not necessarily indicate an increase in the tightness of the coiling, it may merely indicate a progressive increase in the size or age of the individuals, whereas the tightness of coiling was the feature whose evolution supposedly had led from *Liostrea* to *Gryphaea.*

Finally, the ultimate, most refined statistical approach was used by Burnaby (1965), who was able to use Hallam's original samples. Burnaby proposed a new method of measuring tightness of coiling, thereby avoiding the influence of size on the results. His study showed that there is a slight trend from lower to higher beds toward less tightly coiled left valves in *Gryphaea arcuata.* The trend expresses itself in a reduction of a few millimeters in the length of the spiral periphery of the left valve as measured from the tip of the umbo to an arbitrarily selected point on that periphery at which the radius of the valve spiral is 2.0 cm. Thus the increase of apparent coiling noted by Trueman (1922) is best explained by the increase in size of the shells toward the upper part of the section.

Because Burnaby proved that there is no evolutionary tendency to tighter coiling within the species *Gryphaea arcuata,* the stratigraphically lowest specimens are no less tightly coiled and no nearer to *Liostrea* than specimens in higher beds. Therefore, no likelihood of a transition between *Liostrea* and *Gryphaea* is inferred. This conclusion reinforces Schäfle's observation, and the hypothesis of derivation of one genus from the other remains unproved and unlikely.

In all these discussions of the hypothetical derivation of *Gryphaea* from *Liostrea* various authors have made no use of the Triassic gryphaeas, although some have mentioned them. Schäfle (1929, p. 78, footnote

1), who is one of these, refused to regard the Triassic gryphaeas as true *Gryphaea.* The reason for neglect of the Triassic gryphaeas presumably was that no Triassic *Gryphaea* is known in western Europe and everyone has sought to find the ancestor in beds beneath the Liassic *Gryphaea*-bearing beds of this region.

Since Trueman's early work, more and better Triassic gryphaeas have been described by Kiparisova (1936; 1938), McLearn (1937), Vyalov (1946), and others. All of these fossils came from areas difficult to reach, in the present-day polar region. No wonder they were ignored by those working on the ancestry of *Gryphaea.*

Triassic gryphaeas are smaller than the large forms of *Gryphaea arcuata* from the Liassic of Europe, but otherwise they are similar and have features diagnostic of the monophyletic genus *Gryphaea.* In particular do *G. arcuataeformis* Kiparisova (1936, p. 100-102, 123-125, pl. 4, fig. 1-2, 4, 6-10; 1938, p. 4, 33-34, 38, 46, pl. 7, fig. 17-21, pl. 8, fig. 1-2, 11) from late Carnian to Norian beds on the left bank of Korkodon River, a right-hand tributary of Kolyma River in Magadanskaya Oblast [Province], far eastern Siberia, USSR, and *G. chakii* McLearn (1937, p. 96, pl. 1, fig. 8) from the Schooler Creek Group (Carn.-Nor.) in foothills along Peace River in east-central British Columbia, Canada, come close in morphology to *G. arcuata* Lamarck, 1801, from the Liassic of Europe. To disregard the evidence they present or to deny them a place in the monophyletic genus *Gryphaea* would be a serious error. They are the ancestors of the lineage of *G. arcuata,* or are very close to the real ancestors (Fig. J60).

The first appearance of *Gryphaea arcuata* in the fossil record of western Europe was investigated thoroughly by Schäfle (1929, p. 32-35, fig. 5). He was obviously puzzled by the abrupt appearance of this species, without any antecedents known to him, in the "Angulatenschichten," that is, in the Zone of *Scamnoceras angulatum* (von Schlotheim) at the top of the Hettangian Stage, early Liassic (*cf.* Arkell, 1933, p. 117). Hallam (1959a, p. 106-107) and Burnaby (1965, p. 258) recognized that wherever *G. arcuata* first appears in the

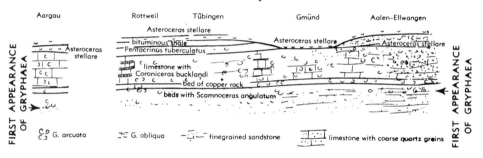

Fig. J66. First stratigraphic appearance of *Gryphaea* in western Europe (Schäfle, 1929).

fossil record of western Europe, it does so with dramatic suddenness (Fig. J66).

An abrupt appearance of a marine species within a section of continuously deposited marine beds speaks against local or provincial evolution of the species in the same depositional basin; rather, it is indicative of immigration. It is to HALLAM's credit that he was first to suggest (HALLAM, 1962, p. 574), as one of the alternatives in the problem of the origin of *Gryphaea arcuata,* that the genus might have migrated swiftly into the west European area following gradual evolution elsewhere.

The Triassic homeland or place of origin of *Gryphaea* is the Arctic region: the Kolyma River basin in far eastern Siberia, Bjørnøya Island, the Canadian Arctic Archipelago, and British Columbia, Canada. This sea basin was landlocked on all sides except in the region of British Columbia and Nevada, where it connected with the Pacific basin. In this sea basin, the first gryphaeas probably evolved from pseudomonotid or other ancestors. There *Gryphaea* made its recorded appearance in the Carnian. During the Triassic an isthmus separated the Arctic sea from the sea in western and central Europe. The isthmus was in part responsible for the widespread deposition of evaporites in western and central Europe during the Triassic. During early Liassic time the sea broke through the isthmus, establishing connection between the two seas and allowing marine faunal exchanges to occur. It was then that *G. arcuata* immigrated from the Arctic region into the western and central European sea. The spread of this species into the new area was rapid, and its first fossil record in the *Scamnoceras angulatum* Zone marks the exact time it happened. Possibly

other species of *Gryphaea* immigrated in the same way.

The environments in this sea were suitable to *Gryphaea* so that the genus became widely distributed there and produced many species and individuals. The Early Jurassic time was one of great flowering of the genus in western and central Europe.

Gryphaea is one of the two oyster genera to make an early appearance in the fossil record. There are no fossil oysters known that antedate it. To judge by its modern descendants, *Neopycnodonte* and *Hyotissa,* it must have had a "primitive" anatomy. For these reasons it is believed that it is an ancestral form of the oysters, or at least of one branch of the stock, the Gryphaeidae. Contrary to widespread opinion, the spiral *Gryphaea* shape probably is not an "advanced" feature, and *Gryphaea* possibly did not descend from a simple, flat *Ostrea*-like ancestor. It may have descended from a spirally shaped genus of the Pseudomonotinae or a related stock.

ADAPTATION AND ENVIRONMENT

SUBSTRATUM

Features that distinguish gryph-shaped from other, normal oysters must be results of adaptation to a special and rigorous environment obviously different from the usual environments in which oysters grow today. Because there are no gryph-shaped oyster species now living, it is not feasible to draw conclusions from modern species in order to explain the gryph shape. It must be concluded that either the special environment does not exist any more or that present-day oysters do not occupy it anymore, if indeed it exists. It seems unlikely that a formerly widespread sea-bottom environment that lasted for a long time has vanished utterly. Perhaps, other more efficient animals have displaced the oysters from that environment.

Gryph-shaped oyster species show less variability of shell shape than normal oysters. Elimination of the unfit as expressed by shell shape must have been rather rigorous among gryph-shaped oysters.

A feature of the environment in which gryph-shaped oysters thrived is its euhaline salinity, for they were euhaline or very nearly euhaline bivalves. *Gryphaea* itself was associated in many places with ammonites, corals, echinoids, or other animals indicative of euhaline waters. Its direct descendants, the Pycnodonteinae, were strictly euhaline and are so to this day, as can be ascertained from those now living. The genus *Sokolowia*, found in upper Eocene (=Biarritzian) deposits of Transylvania, Rumania, and central Asia, was associated in the region of Căpuș-Gilău of the Transylvanian basin with reliable indicators of euhaline waters, such as nummulites, echinoids, and brachiopods (MÉSZÁROS & NICORICI, 1962).

Taxa that are either strictly euhaline (e.g., Exogyrinae) or polyhaline to nearly euhaline (e.g., *Ostrea s.s., Flemingostrea*) have furnished offshoots that became homeomorphs of *Gryphaea*. On the other hand, taxa that are normally brackish-water oysters (e.g., *Crassostrea*) have not furnished them. In other words, oysters living in brackish-water lagoons or estuaries never evolved into *Gryphaea* homeomorphs. *Gryphaea*, its descendants, and its homeomorphs were always adapted to living in euhaline or nearly euhaline waters of epicontinental seas. There they evolved, because that is where the particular environment they became adapted to existed. The environment is not found in brackish lagoons and estuaries.

Sedimentary rocks in which fossil remains of gryph-shaped oysters are found furnish an important clue to the special environment in which they lived and to which they were well adapted. Nearly all of them lived on sediments of low energy levels, that is, on sediments deposited in fairly quiet waters. Very few have been found growing on pebbles, whereas such overgrowths are not rare among common oysters.

SCHÄFLE (1929, p. 75) pointed out that Jurassic gryphaeas are rarely found in sands, especially pure, not muddy sands. Sediments with which gryph-shaped oysters are most commonly associated are clays, marls, chalks, limestones, and glauconite marls. Originally, before their diagenesis and consolidation, these sediments were either soft, water-logged oozes or contained various amounts of small pellets composed of minette iron ore or of authigenic minerals of the glauconite group. These pellets are the modified excrements of small mud-eating animals. Most of the sediments are dark-colored, because they contain finely divided iron-sulfide minerals (FeS_2, pyrite, or marcasite). For example, *Texigryphaea roemeri* (MARCOU) is found in countless numbers in the Grayson Clay (Cenoman.), a dark gray, very sticky, carbonaceous, calcareous blocky clay shale. *Sokolowia buhsii* (GREWINGK) forms a layer (KOCH, 1896) in a thick, bluish gray, silty clay marl (late Lutet.). Abundance and great proliferation of species of *Gryphaea* during the Liassic of western and central Europe coincides with occurrence of abundant dark, carbonaceous clays in the stratigraphic section. In later Jurassic deposits such clays are much less abundant, and so are the gryphaeas.

Gryphaea, its gryph-shaped descendants, and its homeomorphs lived mostly on sea bottoms composed of soft, water-logged oozes with small fragments of shells, and their distinguishing features are adaptations to this environment. DOUVILLÉ (1911, p. 635) was the first to recognize this. TRUEMAN (1940, p. 81) mentioned it and WESTOLL (1950, p. 490) recognized that *Gryphaea* was adapted for life on unconsolidated sea floors.

ATTITUDE DURING LIFE

The broader, more bowl- or basin-like gryph-shaped oysters (with length/height ratio fairly large, near 1) always had been believed to have lived on the sea bottom with their left valves on bottom and their flat right valves nearly horizontal (ZEUNER, 1933b, p. 308). A different life attitude was postulated for the narrow canoe-shaped *Gryphaea arcuata*. Concerning this species, PFANNENSTIEL (1928, p. 390, 408) claimed that the longest diameters of the shells were kept close to horizontal and that the left valves were beneath, resting on their posterior flanges, on their posterior sulci, and on the main bodies of left valves. Their right, or opercular, valves opened out like doors, their hinge axes being "more or less vertical."

ZEUNER (1933a) conducted experiments with fossil *Gryphaea* shells lying on a very fine-grained sediment in a flume trough. He showed that currents would tend to bury the shells part way in the sediment with their left valves on bottom and their greatest diameters nearly horizontal. The shells would come to lie with their hinge axes and flat right valves nearly horizontal. Partly sunk into the soft substratum, they would lie immobilized in quite stable positions.

ZEUNER pointed out that PFANNENSTIEL probably had been thinking of a firm unyielding substratum into which the *Gryphaea arcuata* shells could not sink and on which the shells would roll around at the slightest water current. The attitude PFANNENSTIEL described was the one that might be assumed by shells free to roll around on a firm, level substratum. In that case, the shells always would be in labile positions, but would become less labile if they came to rest at a slight posterior tilt on their posterior flanges as outlined by PFANNENSTIEL. In no case could the hinge axes become vertical, however. The premise of an unyielding substratum is rather questionable, and the attitude postulated by PFANNENSTIEL is incorrect. ZEUNER's explanations are well substantiated and must be accepted with only minor modifications.

SIMPSON (1953, p. 284, fig. 35) showed another interpretation of the life attitude of *Gryphaea arcuata* in which both the longest diameter of the shell and the flat right valve are vertical but the hinge axis is horizontal. The interpretation is probably a misunderstanding of data given by WESTOLL (1950). The attitude would be extremely unstable.

It is now believed that the animals lived partly sunk into the soft ooze lying with their valve commissures nearly horizontal (Fig. J67). In this position, the animal's weight, shell and all, was about equal to the weight of the ooze it displaced so that the *Gryphaea* essentially floated in or on

the soft ooze. The animals must have done vigorous self cleansing (see p. *N*1001) to keep the surrounding ooze from encroaching. The fact that the incurved left beaks of the gryph-shaped oysters did not enclose any fluid-filled chambers and had no umbonal cavities perhaps may have been an adaptation facilitating correct distribution and balance of the load the animals exerted on the ooze so that the growing valve margins surrounding the inhalant and exhalant water currents could remain above the level of the ooze. All the gryph-shaped oysters had in principle the same attitude in life.

ORIGIN OF GRYPH SHAPE

The origin of the gryph shape among oysters was first explained by HENRI DOUVILLÉ (1911, p. 635), whose very brief exposition has been adopted, modified, and expanded by later authors, notably by PFANNENSTIEL (1928, p. 385-386), SCHÄFLE (1929, p. 74-77), and TRUEMAN (1940, p. 81-82). DOUVILLÉ's explanation is well founded, and the expanded exposition given below is based on it.

Euhaline oysters adapted to ooze-covered bottoms have much larger regions accessible to them than brackish-water oysters, for brackish-water regions form only a narrow belt along coasts of continents, whereas ooze-covered sea bottoms stretch far and wide. However, ooze-covered sea bottoms generally are devoid of firm substrata on which oysters can grow. Such firm substrata as are available are mainly small shells or their fragments. This sort of bottom contrasts with the brackish and shallow-water environments in which mangrove roots, large plant debris, or large pieces of shells are plentiful for oysters to grow on. The gryph shape is a mechanical consequence of and an adaptation to this sort of environment, that is, to life in normal-salinity marine sea waters, deep enough to be removed from strong wave action and tidal currents and far enough removed from shore to be outside the influx of fresh or brackish waters, and to life on bottoms covered with soft ooze on which small particles offer the only firm substrata for growing oysters. Such ooze-covered sea bottoms normally are at somewhat greater depth in calm waters with rather

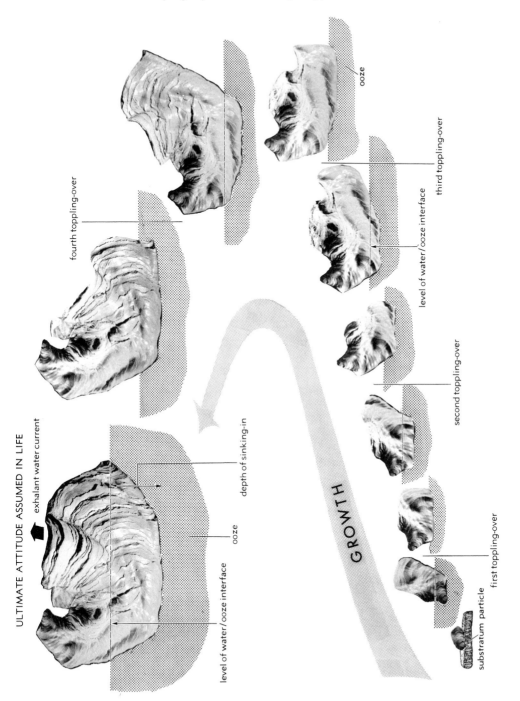

FIG. J67. Origin of the gryph shape of oysters as exemplified by an individual of *Texigryphaea roemeri* (MARCOU, 1862) from the Grayson Clay (Cenoman.), Texas, ×0.75 (Stenzel, n).

stable salinities and stable but low temperatures.

From observations made on living oyster larvae it is well known that they cannot attach themselves successfully on mud-covered surfaces, and many times it has been demonstrated that oyster larvae are selective as to the substratum to which they choose to affix themselves. The larvae of gryph-shaped oysters found extensive soft, ooze-covered bottoms, on which they were unable to settle, except wherever small particles offered suitable substrata. The larvae and postlarval oysters became adapted successfully to such substratum conditions so that it became possible for gryph-shaped oysters to occupy large areas of sea bottom that were unfavorable to colonization by other oysters. Successful they were as is attested to by the vast number of individuals of many species.

The very young postlarval oyster, later to become gryph-shaped, first grew over the small substratum and then grew upward steeply, at an angle of circa 60-70° from the horizontal (Fig. J67). By growing upward so early and so steeply the young oyster managed to bring that part of its shell margins over which the inhalant water current enters the mantle cavity into a position where the sea water was better supplied with oxygen and less polluted with mud and hydrogen sulfide. The water film at the immediate interface between water and bottom sediment was poor in oxygen, because hydrogen sulfide solutions rose from the dark-colored ooze at the bottom. Most of the ooze contained fair amounts of finely divided organic matter, probably derived from dead plankton and the feces and pseudofeces of the oysters themselves, and the water in the pore spaces of the ooze had hydrogen sulfide in solution (Fig. J67).

As the young oyster built up its shell steeply and was attached to a small particle, which in turn rested on a soft and oozy support, the young oyster shell soon reached an unstable attitude and toppled over on its ventral side. As it fell over, the left beak with the small substratum particle attached to it was lifted up off the ooze. The moment when the young oyster shell started to topple over depended on the size or rather weight of the small particle

to which the oyster had become attached, the yield point of the ooze beneath, the size and weight of the oyster, the strength of water currents, vigor of the shell movements made by the oyster, and many other factors. Some young oysters must have reacher an unstable attitude sooner than others, and specific differences in shell shapes may have had their influence.

After the young oyster had toppled over, it continued to build up its shell steeply and thereby shell growth changed direction. This process was repeated several times periodically, and in this fashion the lower valve repeatedly changed direction of growth so that it became convexly curved and its beak became incurved.

The periodicity of this process is well shown on many individuals of *Texigryphaea roemeri* (MARCOU, 1862) from the Grayson Clay (Cenoman.) of Texas (Fig. J67). In many individuals the shell is not continuously and regularly curved, but shows intermittent angular adjustments in direction of growth. Each angular adjustment followed a toppling-over. The specimen figured exhibits three major angular adjustments and some minor ones. The attachment area at the tip of the left beak is roughly triangular and 6 by 10 mm. in size. The original substratum is no longer preserved and had disappeared already before the shell was finally buried in sediment, as attested to by a small colony of Bryozoa, which grew over a small part of the attachment area with many apertures clearly visible. The sedimentary rock in which this species is found is a dark gray, very sticky clay shale containing the finely divided iron sulfide minerals (FeS_2) marcasite or pyrite. The Grayson Clay contains millions of specimens of this species, hardly ever in clusters, and very many of them show angularity of growth.

Other gryph-shaped species show a smoother, more uniformly curved growth. In these species the periodicity of toppling over followed by building up was more frequent and resulted in countless small intermittent angular adjustments of growth which produced a more uniformly curved convex lower valve. It is conceivable that in some species the two, toppling over and building up, were combined into a smooth and continuous process. The very smoothly curved left valves of many species of *Pycnodonte* may owe their origin to such a smoothly functioning process.

Many factors may exert an influence on the growth process and may tend to produce smoothness of the lower valve. For instance, vigorous repeated valve movements by the animal may keep the oyster, with its shell, from ever attaining much of an unstable attitude and may initiate adjustments of attitude quite often. Also, the ratio of weight of body and shell to that of the small substratum piece, attached at the left umbo, is important in this connection.

The very smoothly curved left valves and the growth squamae closely appressed to the contour of that valve are adaptations to life on ooze. They reduce to a minimum any hindrance to a passive adjustment or slippage of the animal's position on the ooze. Growth squamae extending out ending in frills, such as are found in common oysters, would prevent adjustments of position.

During the periodic process of toppling over and building up, the distribution of weight undergoes a gradual shift. The incurved left beak is produced gradually and comes to lie successively higher and higher above the plane of the valve commissure. In this fashion a counterweight to the growing ventral shell margin is produced. The fact that the umbonal cavity of the left beak in many gryph-shaped species is filled in continuously with solid shell-layers and that the left beak has no chamber building must be of considerable influence in making the umbonal region a more effective counterweight as pointed out by SCHÄFLE (1929, p. 76). The gradual shift in weight distribution tends to slow down the frequency of toppling over and to reduce the size of the adjustment of the direction in which the left valve is growing. These changes and the increase in size of the shell cavity result in a gradual reduction of dorsoventral curvature of the left valve. The valve grows in a spiral rather than along a circular path; at first, the radius of curvature is small, but it gradually increases in size so that the spiral opens up concomitantly. Ultimately, a stage may be reached by an old oyster when no further adjustments of the attitude of the shell are necessary. At this stage, the shell may grow quite irregularly (see SCHÄFLE, 1929, fig. 4).

Extreme coiling of the left beak has been suggested by TRUEMAN (1922, p. 265; 1940, p. 84) as the cause of extinction in the phyletic lineage culminating in *Gryphaea arcuata.* Many individuals of this species supposedly had continued to grow until they reached a stage when the coiled apical part of the left valve started to press against the opercular right valve beneath. Thus such advanced individuals encountered gradually more and more difficulty in opening their shells to feed and propagate and they became extinct. This idea of hypertely has been adopted by some later authors although no attempt to furnish proof has been made.

One should be able to find many specimens with valves tightly locked, if this manner of death had been widespread enough to cause extinction. So far only one author (HENDERSON, 1935, p. 558) has claimed to have found such specimens. These were collected at Hock Cliff, Fretherne, on the left bank of the river Severn, 9 miles south of Glouster, England, but the specimens have not been illustrated. Other paleontologists who have collected at this locality have not made such a claim, so considerable doubt remains as to the reliability of the observation. So far all shells illustrated in the literature or available for study show a gap between the coil of the left beak and the top of the right valve beneath, provided sufficient care has been taken to remove the sediment matrix completely. The smallest gap seen is 1.5 mm. but in most individuals it is 3 mm. or more.

Many gryph-shaped species, especially those of *Texigryphaea,* have the left beak coiled in a three-dimensional spiral so that the tip of the left beak comes to lie well behind the umbo of the right valve, where this valve is thickest, and allows plenty of space for the valves to open. However, in *Gryphaea arcuata* the three-dimensional arrangement is much less marked, and the spiral comes close to lying in one plane. Nevertheless, even in this species some individuals have the tip of the left beak turned aside so that the thickest part of the right valve at the hinge is not directly opposite the tip of the left beak, if the hinge axis is used as a reference axis. In addition, very old and large individuals of gryph-shaped species tend to cease growing in a strictly regular spiral and during growth their hinge axis tends to shift ventrally away from the coil of the left beak. This shift allows the left ligamental area to grow in height chiefly and provides more clearance between the two apical parts of the shell. This sort of growth pattern is different than that postulated by TRUEMAN for extreme individuals of *G. arcuata,* but has been clearly depicted by SWINNERTON (1940, p. lvi, fig. 7d), whereas TRUEMAN's postulate has not been illustrated.

It is commonly not realized to what slight extent live oysters open their valves during feeding. Data gathered from *Crassostrea virginica* by W. J. DEMORAN of the Gulf Coast Research Laboratory, Ocean Springs, Mississippi, in August, 1962, show that a living oyster, with shell 106 mm. high, opens its valves 3.5 mm. at their ventral margins, at the best. The largest size of *G. arcuata* listed by SWINNERTON (1940, p. lxxxvi) is 51.5 mm. and the largest specimen available here, graciously sent by Prof. SYLVESTER-BRADLEY, is 75 mm. high. *G. arcuata* has a much smaller shell cavity and a smaller right valve in proportion to the height of its shell than *Crassostrea.* Consequently, it would open its valves to a lesser extent than a *Crassostrea* of the same shell size. It is evident that the largest *G. arcuata,* 75 mm. high, would have to open its valves less than 2.5 mm. at the ventral margins, that is, at the side opposite the coil of the left beak. Directly under the coil the amount of clearance must have been much smaller and must have been in proportion to the distances from the hinge axis, around which the right valve pivots. Calculations show that a clearance of less than 0.6 mm. between the coil of the left valve and the top of the right valve at its thickest point would suffice to let the largest *G. arcuata* open its valves with ease and comparably to living *C. virginica.*

Another argument against TRUEMAN's idea is that the onset of the impinging of the coiled apical part of the left valve onto the top of the right valve beneath must have been rather gradual. During this protracted period of time, the incessant opening and closing of the valves, which live oysters do during feeding, must have made the two valves rub against each other. The rubbing

must have resulted in considerable wear of the areas affected and must have resulted in increasing the clearance between the two valves there. It is hardly conceivable that growth and progressive coiling of the lower valve could have been faster than the wear produced by the rubbing together of the two valves, particularly so in old and large individuals. It should be noted that no specimens of *Gryphaea arcuata* have been described showing friction scars on the apical regions of the valves. On the contrary, the individuals are neatly preserved and intact at those places.

TRUEMAN (1940, p. 84) himself discussed one objection to this idea. He pointed out that the sealing of the shell would first show up in aged individuals of the species, and would, therefore, have no harmful effect on survival of the species. Sealing of the shell could have been the direct cause of extinction only if the extreme coiling were to show up in nearly all individuals before they could participate in reproduction. However, living oysters attain sexual maturity as early as 21 days after fixation, and young oysters have been found to contain ripe eggs when they were merely 1.25 by 1.20 cm. in size (see p. *N1002*). Of course, no one has ever found much of a coil developed on such small individuals of *Gryphaea arcuata.*

In summary, it is not likely that *Gryphaea arcuata,* or any other extremely coiled gryph-shaped oyster species, suffered from hypertely and became extinct because of starvation or inability to reproduce caused by impinging valves. It is more probable that this species became extinct because it was in direct competition with at least four other species of the genus living in the same seas.

GRYPHAEA AND DESCENDANTS

After gryphaeas migrated during the early Liassic from the Arctic realm into the sea covering western and central Europe, the chief flowering period of the genus began. Numerous species evolved; their fossil remains are found in the European Jurassic wherever the sedimentary environments were favorable. Among the latest species in that area were *Gryphaea dilatata* J. SOWERBY, 1816 (see ARKELL, 1932-36, p. 160-170, text figs.) from Corallian beds of England, also recorded as far east as Lithuania (KRENKEL, 1915, p. 300-301) and the Negev Desert in Israel, and *G. lituola* LAMARCK, 1819 (see LEMOINE, 1910, fiche 201-201a) from Oxfordian and Kimmeridgian beds of France and England. The disappearance of *Gryphaea* in the highest Jurassic beds of western Europe is probably not so much caused by extinction as by lack of suitable environments in this area.

ARKELL (1934, p. 64) noted several separate branches or lineages among the Jurassic gryphaeas and named them *Bilobata, Dilatata,* and *Incurva.* However, he did it in a manner admittedly unacceptable in formal zoological nomenclature. The three names remain unavailable. Chains of successional species were recognized early (SCHÄFLE, 1929, p. 79; ZEUNER, 1933b, p. 317; DECHASEAUX, 1934). Some of the phylogenetic sequences are reliably reconstructed, because successive species are connected by forms transitional between species (CHARLES, 1949; CHARLES & MAUBEUGE, 1952a; 1953a; 1953b). Four lineages were recognized among species in the stratigraphic section from Hettangian to Bajocian of the Paris Basin, including Belgium, northern France, and Luxembourg by CHARLES (1949) and CHARLES & MAUBEUGE (1952a). However, they refrained from giving formal subgeneric names to them, though they selected a type species for each. According to CHARLES & MAUBEUGE these lineages originated during the Jurassic, but it is quite possible that some of the differentiation took place in the Late Triassic. In any case, they furnish no argument whatsoever in favor of polyphyletic origins.

Among later species of *Gryphaea* in Jurassic deposits, certain traits seemingly foreshadow transitions to *Texigryphaea* and *Pycnodonte.* These traits are a broad and deep posterior sulcus dividing the left valve, an ill-defined smooth rounded radial keel running down the main body of the left valve, a slight amount of compression of the left valve in anteroposterior direction, a three-dimensional spiral growth pattern, a more opisthogyral left beak, and sharp radial grooves or gashes on the right valve. These are traits that are considerably enhanced in *Texigryphaea* and *Pycnodonte* of the Cretaceous. Why should such traits make their foreshadowing appearance on

Late Jurassic species of *Gryphaea,* unless *Texigryphaea* and *Pycnodonte* are the direct Cretaceous descendants of *Gryphaea?*

HILL & VAUGHAN (1898), in reviewing Lower Cretaceous gryphaeas of the Texas region, described species most of which are nowadays classified as *Texigryphaea* (STENZEL, 1959). They were strongly convinced that the texigryphaeas were descended from Jurassic gryphaeas (HILL & VAUGHAN, 1898, p. 32).

In a phylogenetic diagram (HILL & VAUGHAN, 1898, p. 65, fig. 2) they showed only Cretaceous species from the Texas region. For this reason Jurassic ancestors were omitted and the oldest known species from Cretaceous rocks of the Texas region, named *"Gryphaea" wardi* HILL & VAUGHAN, 1898, was entered as the group's primogenitor.

"Gryphaea" wardi (HILL & VAUGHAN, 1898, p. 49-50, pl. 1, fig. 1-16; STANTON, 1947, p. 27-28, pl. 14, fig. 1-3, 6-11, 13) is from the Glen Rose Limestone (Alb.) of western Travis County, central Texas. Its stratigraphic level is about 350 feet (=100 m.) below the top of the formation, which is truncated by a transgressive regional disconformity at the base of the overlying Walnut Formation (mid-Alb.). The latter formation contains countless *Texigryphaea mucronata* (GABB, 1869) [called *G. marcoui* by HILL & VAUGHAN; it is not the *G. mucronata* of HILL & VAUGHN]. The 350 feet of beds between the two is barren of fossil oysters, so that HILL & VAUGHAN's statement that "the two undoubtedly grade into each other" cannot be correct.

HILL & VAUGHAN's diagram and statements have misled later authors to accept *G. wardi* as the ancestor of *Texigryphaea,* and as proof that the latter descended from a line of simpler ostrean ancestors, meaning ancestors that were more primitive than *Texigryphaea* and were *Ostrea*-like, not showing any strong homeomorphy with *Gryphaea s.s.* (KITCHIN, 1912, p. 593; ARKELL, 1934, p. 60; TRUEMAN, 1940, p. 83; GEORGE, 1962, p. 11-12). These authors considered *G. wardi* as one of the last species in a line of simpler ostrean ancestors. ARKELL identified it as a *Catinula,* a genus that had its acme in the Jurassic.

All shells of *G. wardi* are quite small, much smaller than any full-grown *Texigryphaea* from beds above the Glen Rose

Limestone. The largest at hand is 29.6 mm. long and 25.5 mm. high. They have orbicular muscle imprints, vesicular shell structure in both valves, vermicular ridgelets on the commissural shelf of the upper valves, and corresponding pits on that of the lower valves. These features securely place the species in the subfamily Pycnodonteinae and remove it from *Catinula* and from *Gryphaea*. The attachment area of the species is large; in the largest specimen at hand it is 28 by 24 mm. Because of the large attachment area on the left valve, there are equally large areas on the right valve covered with xenomorphic sculpture. The shell rises steeply from the attachment area as in many of the Gryphaeidae. There are scattered rounded radial ribs on the left valve visible where the valve rises steeply from its attachment.

"*Gryphaea*" *wardi* is a neotenous species of *Pycnodonte (Costeina)* and takes its place with the other radially costulated pycnodontes such as *P. (C.) costei* (Coquand) (1869, p. 108, pl. 26, fig. 3-5, pl. 38, fig. 13-14) from Santonian beds of Martigues, France. It is not considered to be a direct ancestor of *Texigryphaea*, but is a separate lineage of the Pycnodonteinae.

Whatever its generic assignment may be, it is not a simple or primitive ostrean, but a fairly advanced species of the Gryphaeidae. As such it differs much from *Liostrea*, the alleged ancestor of *Gryphaea*, so that the two can not be united as a persistent oyster stock of similar primitive oysters from which *Gryphaea* and its homeomorphs are supposed to have evolved by iterative evolution. Rather, *Gryphaea* itself is the ancient monophyletic root-stock of one of the two branches of oysters, namely the Gryphaeidae; several mutually unrelated genera similar to *Gryphaea* have evolved at various times either from *Gryphaea* itself (*Gryphaea* descendants) or from a few not gryphoidally incurved unrelated genera of the Ostreidae (*Gryphaea* homeomorphs).

Texigryphaea probably evolved directly from *Gryphaea*, but the series of transitional species linking the two at the turn from Jurassic to Cretaceous is largely incomplete. *Texigryphaea* first appeared in abundance as an immigrant with the invasion of the sea that deposited the mid-Albian Fredericksburg Group (including the Walnut Formation) in southwestern North America. The last of the *Texigryphaea* species is *T. roemeri* (Marcou, 1862) from the Grayson Clay (Cenoman.) of Texas and northeastern Mexico (Stenzel, 1959, p. 22, fig. 1-2, 6, 13, 17a,b). The most extreme species is *T. navia* (Hall, 1856, pl. 1, fig. 7-10; Stanton, 1947, p. 27, pl. 19, fig. 1-2), which has the most prominent keel and is more oblique and more compressed than the other species. It is from the Kiamichi Formation (mid-Alb.) and synchronous beds in southwestern North America. Geographic distribution of the genus is distinctly provincial. *Pycnodonte* shares vesicular shell structure with *Texigryphaea*, and probably both had a common ancestor which descended from *Gryphaea*. The most ancient *Pycnodonte* species known today are *P. (Costeina) wardi* (Hill & Vaughan) and *P. (Phygraea) vessiculosa* (J. Sowerby, 1822) (see Woods, 1913, p. 374-375, pl. 55, fig. 10-14, pl. 56, fig. 1), which made its earliest appearance in the Shenley Limestone (l.Alb.) of Leighton Buzzard, Bedfordshire, and in Bed 3 of the Folkestone Beds (l.Alb.) of Folkestone, Kent, England (letter from R. Casey, July, 1961). One of the latest species is *P. queteleti* (Nyst) from lower Oligocene beds of Zuid Limburg, Netherlands (Albrecht & Valk, 1943, p. 121-122, pl. 12, fig. 405-407).

Whereas *Texigryphaea* never produced more than a few provincial species, *Pycnodonte* became prolific in species and attained worldwide distribution. In the Miocene it gave rise to *Neopycnodonte*, which is represented today by a living species of circumglobal distribution. *Hyotissa*, also living today, is probably one of the descendants of *Pycnodonte*.

The descendants of *Gryphaea*, the Gryphaeidae, had their acme in the Cretaceous, and have gradually dwindled since that time. Today, they are a small group of only two genera.

SPECIATION

Speciation, or the making of a species, takes place among oysters chiefly in two ways: by successional speciation (Huxley, 1943, p. 172-173, 385-386) and by geographic separation.

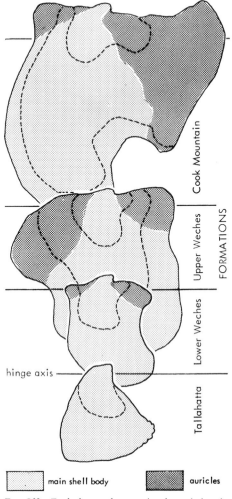

main shell body   auricles

FIG. J68. Evolution and successional speciation in the *Cubitostrea sellaeformis* (CONRAD) stock, mid-Eoc., Gulf and Atlantic Coastal Plain, N. Am., ×0.5 (Stenzel, 1949). [Outlines of LV of large specimens in each species arranged in stratigraphic sequence.]

Successional speciation is the very slow and gradual transformation of a large interbreeding population composing a species into a successor species. Time is the main factor at work. It produces a single file of successive species, and the phylogenetic diagram depicting this sort of speciation is a single unbranched line (Fig. J68).

Geographic separation divides a species of extensive geographic range by a barrier and breaks the interbreeding population of

the ancestor species into two or more separates. Then, lack of gene exchange allows the separate populations to drift apart genetically. It produces two or more descendent species from one ancestor species. The phylogenetic diagram depicting this sort of speciation has a dichotomous, trichotomous, or polychotomous arborescent configuration.

SUCCESSIONAL SPECIATION

Successional speciation among oysters was demonstrated by STENZEL (1949) on the stock of *Cubitostrea sellaeformis* (CONRAD, 1832), consisting of four widespread mid-Eocene species distributed along eastern and southern shores of North America (Fig. J68, J69). Each species is distinguished readily by its morphologic shell features. Transitions between them are not found, because the four species are sepa-

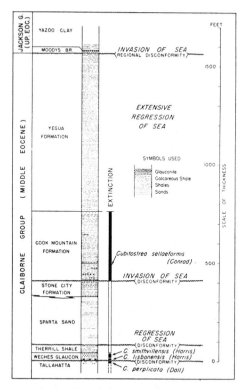

FIG. J69. Stratigraphic distribution of four successional species composing *Cubitostrea sellaeformis* stock in Gulf Coastal Plain of North America (Stenzel, 1949).

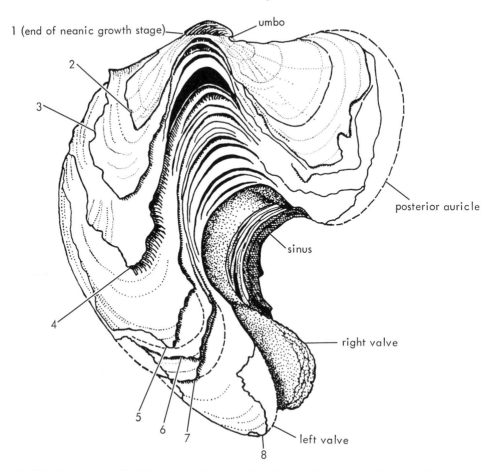

Fig. J70. Full-grown shell of *Cubitostrea sellaeformis* from Archusa Marl of Cook Mountain (Wautubbee) Formation, mid.Eoc., near Wautubbee, Clark County, Miss., USA, ×0.6 (Stenzel, 1949).

Eight major imbrications, numbered 1-8, probably indicate an age of eight years for this individual. Oblique posterior view seen at approximately 45° to hinge axis, showing very large posterior auricle and very twisted valve commissure. RV largely hidden by LV; RV stippled.

rated by nonmarine deposits containing no oysters or by recognizable disconformities, indicating absence of any deposits.

Among several distinguishing features is the size of each species. A size increase from one species to the next is demonstrable by measuring the maximal sizes (maxima of heights of the shells) obtainable in extensive collections (e.g., *Cubitostrea sellaeformis* (CONRAD, 1832), 18.2 cm. (Fig. J70); *C. smithvillensis* (HARRIS, 1919), 14.7 cm.; *C. lisbonensis* (HARRIS, 1919), 8.6 cm.; *C. perplicata* (DALL, 1898), 6.4 cm. (Fig. J43; see Fig. J116, J117)).

The lineage became extinct with *Cubito-strea sellaeformis,* which is an index fossil of the Cook Mountain Formation and its lateral contemporaneous extensions (Fig. J70).

Because successional speciation involves the entire interbreeding population composing a species, no evolutionary laggards are left and it is impossible for two such species to exist side-by-side simultaneously. If such a situation had been in existence at any time, the two species would have interbred readily and the distinctions between them would have disappeared thereby. As a consequence, successional speciation produces reliable chronologic time markers and the species can be used as index fossils in stratigraphy.

Two obvious shortcomings affect their use, however. Because transformation in so large an interbreeding population can proceed only very slowly (MAYR, 1942, p. 236), the chronologic time span of each species is fairly large. Therefore, successional oyster species have no great resolving power in stratigraphic chronology.

As illustration of this conclusion, oyster species used in Mesozoic stratigraphy and derived by successional speciation have much larger time spans than ammonites associated with them. An oyster species of this sort runs through several ammonite zones. *Gryphaea arcuata* LAMARCK, 1801, is found in at least three ammonite zones of the Liassic (CHARLES, 1949, p. 40; ARKELL, 1933, p. 120-149).

Nevertheless, STEPHENSON (1933) used species of *Exogyra s. s.* with good success as zone and index fossils to divide the Upper Cretaceous strata of the Atlantic and Gulf Coastal Plains and to trace the zone of *Exogyra (Exogyra) cancellata* STEPHENSON (1923, p. 182, pl. 50, fig. 5-6) for at least 4,000 km.

STENZEL (1949, p. 42-45) proved that all deposits containing *Cubitostrea sellaeformis* from Maryland to Mexico were part of the Cook Mountain Formation or its equivalents and belonged to the middle of the Claiborne Group. Thereby he proved that the Santee Limestone of South Carolina, which contains this species, had been misplaced in the Jackson Group.

The second shortcoming is the provincial distribution of oysters. Oyster stocks undergoing evolution as successional species must remain in a geographically circumscribed province. The *Cubitostrea sellaeformis* stock is found only from Virginia to the Gulf of Mexico, that is, along the ancient eastern and southern coasts of North America. Contemporaneous beds in the Paris or London Basins do not have it. *Exogyra cancellata* is found only along the ancient eastern and southern coasts of the Late Cretaceous North American continent, but is absent from contemporaneous deposits of the Western Interior of North America.

GEOGRAPHIC SEPARATION

This well-known cause of speciation (MAYR, 1942) affects mostly shallow-water oyster species living along shores and can become effective in several ways. Change in climate can make a stretch of the geographic range of a species uninhabitable to this oyster and divide the range into two separate parts. Or, tectonic events or falling sea level may raise a land barrier splitting the range into two.

For instance, the land bridge connecting North America with South America is a comparatively recent result of tectonic movements in Central America. Previously, the Pacific Ocean communicated with the Gulf of Mexico through a sizeable gap, and most probably a species of *Crassostrea* ranged through the gap from the western coast of North America to the Gulf of Mexico and the eastern coast of the continent. As the gap closed, the range of the species became divided, and two separate species arose from the separated populations: *C. corteziensis* (HERTLEIN, 1951) ranges from Panama to the head of the Gulf of California and *C. virginica* from Yucatan to the Gulf of St. Lawrence. The two are daughter species of a common Pacific-Atlantic ancestor species of Miocene and Pliocene age. This is a case of dichotomy by geographic separation.

An interesting case of polychotomous phylogeny is the stock of *Crassostrea gryphoides* (VON SCHLOTHEIM, 1813, p. 52), a gigantic oyster (see p. N1027) of very great west-east range during Miocene time, which has aroused interest since 1763 because of its large size (RUTSCH, 1955). This reef-forming brackish-water ancestral species has received many formal names based on local races and variants in shell shape and has been found in Miocene deposits of the Tethys Sea and its branches from Portugal, Spain, and Morocco in the west to Czechoslovakia, Austria, Cilicia in Asia Minor, and Somalia (AZZAROLI, 1958, as *Somalidacna lamellosa,* see Fig. J102) and into Japan in the east. It is evident that these occurrences from Portugal to Japan are all of the same species, because they fall into a former continuous open west-east seaway of Miocene time, the ancient Tethys Sea, along which climates were fairly uniform and warm, varying from tropical to desertic. Neither climate nor land barriers existed to prevent this species from spreading the full length of the seaway.

Toward the end of Miocene time and later, tectonic events produced several land barriers across the former seaway, breaking it up into separate sea basins, gulfs, or bays. The separate populations left behind in these broken remnants of the Tethys Sea gave rise to three, or possibly more, local species of *Crassostrea: C. angulata* LA-MARCK, 1819, on the shores of Spain, Portugal, and Morocco; *C. cattuckensis* (NEW-TON & SMITH, 1912) [=*C. madrasensis* PRESTON, 1916] on Indian Shores; and *C. gigas* (THUNBERG, 1793), China, Japan, to Sakhalin Island (see p. *N*1037-*N*1038). At least two of these three daughter species have retained the ability to grow to extraordinary size, a feature that so characterizes their ancestor *C. gryphoides.*

The earth movements that raised the land barriers and broke up the ancient Tethys Sea were synchronous. They were part of widespread tectonic disturbances that were synchronous as far as geologic evidence indicates. Therefore, the three above-mentioned species presumably arose simultaneously. This idea is contrary to the belief that one species can give rise to three descendent species only by two successive dichotomies.

Some aspects of the geologic history of the stock of *Crassostrea gryphoides* have been discussed by DOLLFUS, DOLLFUS & DAUTZENBERG, and RUTSCH (1915; 1920, p. 465-471; 1955). Many formal species names have been proposed and are nomenclaturally available for this species. Some of the names were published in the same work by VON SCHLOTHEIM (1813). The species name used here has been selected by RUTSCH, whose decision as the first reviser must be respected according to *Code* (1961, 1964) Article 24 (a)(i).

CLASSIFICATION

CONTENTS

HISTORICAL REVIEW

LINNÉ, 1758

LINNÉ (1758) described only seven living species that today are regarded as true oysters, but he described them under three generic names: *Ostrea,* with three species of true oysters; *Anomia,* with one; and *Mytilus,* with three. On the other hand his concept of the genus *Ostrea* was very broad, including 32 nominal species now distributed among four bivalve families (DODGE, 1952).

He distinguished four subdivisions in his genus *Ostrea,* of which the fourth, namely *"Rudes vulgo Ostreae dictae"* [=the Rough ones, commonly called oysters], was obviously intended as the typical core of the genus but still contained nine species. Of these nine, only three are regarded today as true oysters, one of them is unrecognizable, and the remaining five are placed in other families. One of the three true oysters is *Ostrea edulis,* which was made the type species of the genus by ICZN Opinion 94 (Oct. 8, 1926) and placed on the Official List.

Obviously, LINNÉ's concepts of his genera *Ostrea, Anomia,* and *Mytilus* had many points needing corrections. Such corrections were made by BRUGUIÈRE and LAMARCK.

BRUGUIÈRE, 1791

Early death took the author before his work could be finished. His text is incomplete, none covering the Testacea or shelled animals, but he published 189 engraved plates. On these well-arranged plates, attempt was made to group together on each plate only those species that the author regarded as closely related. For example, pl. 187 has five figures depicting oysters that have serrate valve commissures; pl. 189 contains six figures of oysters that have a shell devoid of radial ribs but have a strongly curved left beak of either exogyroidal or gryphaeoidal spiral curvature. The pl. 178-189 have as their headings "Huitre *Ostrea*" and a line at their bottom reads: "Histoire Naturelle, Vers Testacés à Coquille Bivalve irregulaire" [Natural History, Shell-bearing Worms with irregular Bivalved Shell]. These plates depict what BRUGUIÈRE considered to be the genus *Ostrea.* It obviously included various genera of true oysters and some Malleidae and Pectinidae.

Nevertheless, BRUGUIÈRE's work foreshadows LAMARCK's (1801) classification. For instance, pl. 189 foreshadows LAMARCK's *Gryphaea,* and the species illustrated on it were given species names and listed under that genus by LAMARCK. The importance of BRUGUIÈRE's work rests on his status as a precursor of LAMARCK. His classification is based on superficial, purely external morphology.

LAMARCK, 1801

As BRUGUIÈRE's successor in Paris, LAMARCK (1801) used BRUGUIÈRE's unfinished work judiciously to propose new genera. Thus he used Plate 189 of BRUGUIÈRE (1791) to distinguish the genus *Gryphaea.* By and large, he relied on a few superficial, external features to discern genera.

Nowhere is this as apparent as in his treatment of *Gryphaea* LAMARCK, 1801. From the species described or listed by him under this genus, it is obvious that the definitive feature used to distinguish this genus from other oysters was its hooked left beak. It made no difference whether the beak was spiral gryphoidally—that is, mainly orthogyrally—or spiral exogyroidally—that is, mostly opisthogyrally. Therefore, exogyrine oysters were included in his original list of species and in his concept of *Gryphaea.*

The original list of species of *Gryphaea*
LAMARCK, 1801, included also *G. angulata*
LAMARCK, 1801, a *nomen nudum* attached
to an undescribed living southern European
species. It remained a *nomen nudum* until
1819. It was based on a freakish left valve
(DELESSERT, 1841, pl. 20, fig. 3a-c) which
had grown curved so that the umbonal
region became curved over to the right side
and was vaguely similar to the incurved left
beaks of fossil Gryphaeas. This one feature
led LAMARCK to judge that the species must
be placed in the same genus with these
fossils (Fig. J71), which had been known
for many years to pre-LINNEAN authors as
gryphites (the postfix *-ites* was customarily
used for taxa exclusively fossil). Because
LAMARCK thought now to have discovered a
living species of *gryphites,* he decided to
coin a generic name without that postfix
and invented the new generic name *Gryphaea*. However, LAMARCK's *"Gryphaea"*
angulata is now *Crassostrea angulata* (LA-
MARCK).

LAMARCK (1801, p. 400) also described
the new genus *Planospirites* and its sole
species *P. ostracina,* based on a single right
valve collected by FAUJAS SAINT-FOND (1799
[1802?]) at St. Pietersberg south of Maastricht, Zuid Limburg, Netherlands, from
the "tuffeau de Maestricht," a crumbly,
porous lime grainstone of Maastrichtian
age (see Fig. J96). There, the two valves
of the species are found very rarely in
natural juxtaposition, and left valves are
much rarer than right ones. For that reason only a right valve was obtained and
described by LAMARCK, and he was so
puzzled by it that he described it as a
univalve shell, that is, a gastropod, under
"Genres incomplètement connus." The
type specimen has been figured by FAUJAS
(1802?, pl. 22, fig. 2) and by JOURDY (1924,
p. 7-8, pl. 1, fig. 1). The figure given by
FAUJAS has been called *Ostracites haliotoid-
eus* by VON SCHLOTHEIM (1820, p. 238).

Actually left valves were probably available to FAUJAS and to LAMARCK, but were
not recognized as belonging with the right
valves, because the two are so different and
occur disconnected. The left valve of an
exogyrine oyster figured by FAUJAS (1802?,
pl. 28, fig. 5) under the name *"rastellum"*
quite probably is a left valve belonging with
Planospirites ostracina (see Fig. J96,*1*).

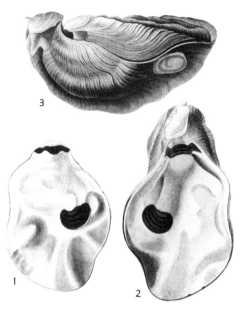

FIG. J71. *Crassostrea angulata* (LAMARCK, 1819),
species living on southwestern coasts of Europe,
?×1 (after Delessert, 1841). [Lectoholotype of
Gryphaea angulata LAMARCK, 1819.]—*1*. RV int.—
2. LV int.—*3*. Both valves viewed from posterior
side.

These uncertainties have remained unresolved, although good topotype materials
are available at Maastricht. As a consequence, authors have hesitated to use
Planospirites, although it is a good exogyrine genus.

LAMARCK, 1819

LAMARCK gave BRUGUIÈRE credit for having separated the scallops from the oysters,
both of which LINNÉ had included in one
taxon, *Ostrea* of LINNÉ, 1758, and for recognizing them as natural taxa and defining
their principal limits. He himself separated *Vulsella, Podopsis,* and *Gryphaea*
from *Ostrea,* as defined by BRUGUIÈRE, and
thereby restricted *Ostrea* more closely to its
natural limits. Restrictions and sharpened
definitions of genera involved are indeed a
major part of his accomplishments.

The oysters *(les Ostracées)* were divided
by LAMARCK into two subdivisions: 1)
ligament semi-internal, shell foliaceous, attaining often great shell-wall thickness, and
2) ligament internal, shell thin or paper-

like. Subdivision 1) contained *Gryphaea, Ostrea,* and *Vulsella;* subdivision 2) *Placuna* and *Anomia. Ostrea* itself was divided into two unnamed sections: A) Valve margins simple or undulate but not plicate and B) valve margins distinctly plicate, corresponding mainly to the Lophinae of today.

To *Gryphaea* were assigned 12 species, one living and 11 extinct; to *Ostrea* section A) 32 living and 18 extinct species; to *Ostrea* section B) 16 living and 15 extinct species. Altogether, LAMARCK described or listed 94 nominal species now known to represent oysters, but many species were overnamed. For instance, *Ostrea edulis* LINNÉ, 1758, was described under at least four species names, and *Crassostrea virginica* (GMELIN [1791]) received three names: *borealis, canadensis,* and *virginica,* all three under *Ostrea* section A). Because of the great variability of oysters, such cases of overnaming were to be expected.

More serious were errors in generic assignments. *Anomia anomialis* (LAMARCK, 1819) from the Calcaire Grossier (Lutet.) near Paris, France (CLERC & FAVRE, 1910-18, pl. 28, fig. 102-105, and pl. 29, fig. 106-107) [=*A. tenuistriata* DESHAYES, 1832, v. 1, p. 377] was originally described as *Ostrea* section A) *anomialis. Pycnodonte vesicularis* (LAMARCK, 1806) from the Chalk (Campan.) at Meudon near Paris, France, was described under two names, as *Podopsis gryphoides* LAMARCK, 1819 (CLERC & FAVRE, 1910-18, pl. 14, fig. 46-47) and as *Ostrea* section A) *vesicularis* LAMARCK, 1806 (CLERC & FAVRE, 1910-18, pl. 26, fig. 94-95). Another oyster was described as *Chama laevigata* LAMARCK, 1819 (CLERC & FAVRE, 1910-18, pl. 36, fig. 263); it is the same as *Ch. haliotoidea+Ch. conica* JAMES SOWERBY, 1813 (v. 1, p. 6, pl. 25, fig. 1-5 and p. 69, pl. 26, fig. 3) from St. Mary Donhead, Wiltshire, Eng., probably from Cenomanian beds. It is best to call it *Planospirites haliotoidea* (SOWERBY). In this case, LAMARCK probably was influenced in his generic assignment by SOWERBY's work.

RAULIN & DELBOS, 1855

The classification proposed by RAULIN & DELBOS, 1855, was restricted to oysters found in the Tertiary beds of the Aquitaine region of France. It was explicitly designed to take care of the considerable number of species found there. Accordingly, they not only recognized the three widely accepted taxa *Exogyra, Gryphaea,* and *Ostrea* as subgenera, but also introduced new minor subdivisions, which they called variously *"coupures"* or *"groupes"* or *"sections,"* wherever the number of closely related species threatened to become too large and unwieldy. In this fashion, they acknowledged one of the major functions of classification, namely, to bring order out of chaos and to provide a logical hierarchy.

The criteria used by RAULIN & DELBOS as bases for their classification were those they judged to be least variable, if one studies a variable series of shells of a given species. They were two: 1) morphology of the ligamental area *(configuration de la surface ligamentaire)* which in their definitions they called beak *(crochet)* for short, and 2) external ornamentation of the shells. They noted that the proportions of the three subdivisions of the ligamental area maintain great constancy in a series of shells of a given series, but that these proportions are much diversified among the various species. They remarked that previous authors had attached no great importance to this feature. In discussing external ornamentation, they mentioned that the two valves have either congruent or incongruent ornamentation. RAULIN & DELBOS were the first to use a structural feature that is quasi-internal (i.e., the ligament) as a basis of classification within the oyster family. Their methods were quite modern in that they took into account individual variability. These methods were the first attempt among oyster paleontologists to relinquish the typological approach to classification.

Under *Ostrea (Ostrea)* only eight sections were named. RAULIN & DELBOS resorted to names in the plural and consistently referred to them in the plural. Because of the consistent use of the plural, these names are not available for formal nomenclature *(Code* Article 8D). For each of the eight sections they designated a type species through tautonymy, and each section was defined by a brief descriptive text. Their classification of the *"famille des Os-*

tracées," consisting of the single genus *Ostrea*, is as follows:

Classification of Ostrea by Raulin & Delbos (1855)

Genus OSTREA
  Subgenus OSTREA (EXOGYRA)
    Section A: smooth species
    Section B: plicate or costate species
  Subgenus OSTREA (GRYPHAEA)
    Section A: smooth species
    Section B: plicate or costate species
  Subgenus OSTREA (OSTREA)
    Vesiculares: The two valves smooth, beak short
    Laterales: left valve smooth, concentric growth imbrications *("plis")* on the right valve
    Virginicae: left valve has slightly crinkled or plicated growth imbrications, beak much prolonged
    Edules: radiating plications on the left valve, right valve smooth, beak large and short
    Flabellulae: radiating plications on the left valve, right valve smooth, beak narrow and small
    Cornucopiae: costae on the left valve, right valve smooth
    Undatae: rounded costae on the two valves
    Carinatae: angular plications on the two valves

Judging by their type species and definitions, the eight new sections of *Ostrea (Ostrea)* may be interpreted in modern terms as follows: Vesiculares, *Pycnodonte* FISCHER DE WALDHEIM, 1835; Laterales, *Gryphaeostrea* CONRAD, 1865; Virginicae, *Crassostrea* SACCO, 1897; Edules, *Ostrea* LINNÉ, 1758, *s.s.;* Flabellulae, *Cubitostrea* SACCO, 1897; Cornucopiae, *Saccostrea* DOLLFUS & DAUTZENBERG, 1920; Undatae, dubious; Carinatae, Lophinae VYALOV, 1936. It can be seen from dates given above that RAULIN & DELBOS were largely ahead of their time and defined but did not formally name well-founded genera, which were named by others many years later.

MISCELLANEOUS AUTHORS, 1758-1886

Several generic names that are nomenclaturally acceptable and turn out to be useful were introduced in the years following LINNÉ's (1758) work. Nearly all of them, however, were incidental introductions, not based on any improved methods of description or discernment of generic features; thus they did nothing to advance methods employable in classifying oysters. The exceptions are *Exogyra* SAY, 1820, *Amphidonte* FISCHER DE WALDHEIM, 1829, and *Pycnodonte* FISCHER DE WALDHEIM, 1835.

Planospirites LAMARCK, 1801, had been the first exogyrine oyster genus to receive a name, and *Exogyra* SAY, 1820, was the second one. The former remained a dubious taxon for a long time and even was considered to be a gastropod. Also, LAMARCK failed to give an illustration of his genus. These facts militated against an understanding of the significance of this genus.

In contrast, *Exogyra* SAY, 1820, was well described and illustrated, so that it attained acceptance right away and became the leading genus of its subfamily. Additional exogyrine genera were introduced by BAYLE, 1878, but the lack of descriptive definitions of them slowed down their general acceptance by later authors.

FISCHER DE WALDHEIM was the first to note chomata on the commissural shelves of oysters. Both *Amphidonte* and *Pycnodonte* were named on the basis of chomata and their arrangement (see Fig. J80, J93). Thus he was the first to call attention to an internal shell feature important in classification. Succeeding authors paid little attention to these features until IHERING (1907) and SUTER (1917).

Gryphaeostrea CONRAD, 1865, was proposed in somewhat careless fashion. Only through the efforts of MEEK (1876, v. 9, p. 11) was CONRAD persuaded to furnish a descriptive definition. Perhaps, the situation is explainable by CONRAD's rather advanced age (62 years in 1865). At the time CONRAD seems to have been influenced by the theory of evolution and searching for genera that might be interpreted as evolutionary missing links. For these he used combination names such as *Gryphaeostrea* and *Ostreanomia*. *Gryphaeostrea* probably was so named because he believed it to be transitional between *Gryphaea* and *Ostrea*. If this was his assumption, it was erroneous. The name is rather unfortunate (see Fig. J98).

Pernostrea MUNIER-CHALMAS, 1864, was based on material misunderstood by its author, who believed it to be related to both *Perna* and *Ostrea* or at least similar to both (see p. *N*975, N1104, Fig. J78). Nevertheless, the author must be commended for trying to use the structure of the ligamental area to discern genera and their limits.

FISCHER, 1880-87

Fischer's *Manuel de Conchologie* furnishes a convenient summary of the efforts of conchologists and paleontologists since Linné (1758) and must have been an important and very useful book in its time. His descriptions of the various taxa in the Ostreidae were careful, although the classification given by him suffered by his refusal to take note of internal shell features and by his corresponding reliance on external, conchological features in nearly every case in this family. Only in the case of *Exogyra*, which he treated as a subgenus of *Ostrea s.l.*, did he use a feature that is quasi-internal. The topography of the ligamental area in *Exogyra* was very well described.

He recognized as genera: *Ostrea, Heligmus, Naiadina,* and *Pernostrea.* The following were accepted as subgenera of *Ostrea: Ostrea, Chalmasia, Alectryonia, Gryphaea,* and *Exogyra.* His classification is vitiated by the inclusion of *Heligmus, Chalmasia,* and *Naiadina,* which are now placed in other families. His scheme overemphasized the apparent (but not real) differences between *Ostrea* and *Pernostrea* and undervalued those between *Ostrea, Exogyra,* and *Gryphaea,* which he renamed *Liogryphaea.* As long as classification was based almost exclusively on external features, too few generic characteristics were taken into account to allow proper evaluation of the taxa. Such internal features as the muscle imprints were easily available on simple inspection, and so was the vesicular shell structure, but they were ignored.

Fischer started a new nomenclatural trend by proposing *Liogryphaea* as a new section of *Ostrea (Gryphaea).* It was erected for *Gryphaea arcuata* Lamarck, 1801, as type species by monotypy (see p. *N*1099). The unfortunate effect was that many oyster biologists then started using *Gryphaea* for the living genus, which correctly must be called *Crassostrea* Sacco, 1897 (Stenzel, 1947; Gunter, 1950; Gunter, 1954). The nomenclatural question

has been settled by ICZN Opinion 338 (March 17, 1955) in which *Liogryphaea* was rejected officially.

SACCO, 1897

In his great monographic treatment of the Tertiary faunas of the provinces Piemonte and Liguria in northwestern Italy, Sacco (1897a, p. 99-102, and 1897b, p. 3-30, pl. 1-9) minutely described many species and proposed four new subgenera of *Ostrea (Crassostrea, Cubitostrea, Cymbulostrea,* and *Gigantostrea),* all based on the usual external conchological features. Although Raulin & Delbos, 1855, had already separated the Edules from the Virginicae, Sacco made a clear distinction between *Ostrea s.s.* and *Crassostrea* by giving them formal subgeneric names and good descriptions (see Fig. J101, J109). His *Cubitostrea* is a well-defined monophyletic taxon, for which he gave a phylogenetic tree of the species from the mid-Eocene through Pliocene. *Cymbulostrea* does not differ from *Ostrea s.s.* sufficiently to deserve naming (see Fig. J111). *Gigantostrea* is the same as *Pycnodonte* except that it has a more irregular shell growth pattern than most well-known species of *Pycnodonte* (see Fig. J81). However, the two respective type species are quite similar in their shell growth pattern.

Crassostrea and *Ostrea s.s.* were distinguished by him on a subgeneric level only. Today we know that the biologic differences are much larger, making a generic distinction between them necessary. However, it is remarkable that a valid distinction between the two taxa can be made without the help of any neontological information, that is, purely on the basis of paleontological data.

Gryphaeostrea, which Sacco called *Exogyra (Aetostreon?),* was very well discussed by him (see Fig. J98). A phylogenetic tree was outlined for species from the Cretaceous to the Miocene. The phylogeny probably is correct. In other words, Sacco was the first frankly to outline phylogenetic trees for oyster species, and these phylogenies are a credit to his work.

DOUVILLÉ, 1886-1936

The features of oysters valuable in classification were reported in several papers by

DOUVILLÉ (1886-1936). In most of these studies oysters were treated only incidentally, but the results of his work turned out to be quite important.

In the first study (DOUVILLÉ, 1886), he examined the *Gryphaea* homeomorph *Aetostreon imbricatum* (KRAUSS, 1843) from the Neocomian of the Kingdom of Choa, now a part of Abyssinia (Fig. J63). In order to find its correct systematic position among oysters, he investigated the ligament and ligamental areas and demonstrated the structural differences in these features between *Ostrea s.l.* (=Ostreinae of today), *Gryphaea s.l.* (=Gryphaeinae), and *Exogyra s.l.* (=Exogyrinae). He traced the evolution of the exogyrine ligament and ligamental areas from the Jurassic to the Cretaceous. Thereby he laid the foundation for definition of the Exogyrinae, which has been accepted by later authors. DOUVILLÉ'S (1886) conclusions concerning *Aetostreon imbricatum* (KRAUSS, 1843) were followed by KITCHIN (1908) (see p. N1066-N1067).

In a later paper (DOUVILLÉ, 1911), he reviewed the various phyletic branches of the Exogyrinae and their stratigraphic distributions. Although this study is not strictly devoted to classification, it clearly shows his ideas of oyster taxonomy. He distinguished *Liogryphaea* FISCHER (=*Gryphaea* of today) and listed species from the Hettangian to the Oxfordian. To judge from the species listed by him, he very well recognized *Pycnodonte,* but unfortunately used the then-current, emended name *Pycnodonta.*

In one of his last publications DOUVILLÉ (1936b) turned to the study of shell structure. He was the first to recognize that vesicular shell structure was an important diagnostic feature of supraspecific importance. Thus he laid the foundation for all future work on the Pycnodonteinae (see p. N986-N987, N1105).

IHERING, 1907; SUTER, 1917

IHERING was second after FISCHER DE WALDHEIM to pay serious attention to the chomata as a means of classifying oysters. To accommodate the many extinct and living Argentinian species of *Ostrea,* which he was studying, he proposed the subgenus *Eostrea* IHERING (1907, p. 42) to include all species of *Ostrea* having chomata on the internal valve borders. Without designating a type species he included in *Ostrea (Eostrea)* the Eogene species from Patagonia, without naming any of them specifically, and the two living Argentinian species *O. puelchana* D'ORBIGNY, 1842 (*Paléontologie* v., p. 162), and *O. spreta* D'ORBIGNY, 1846 (Mollusca v., p. 672) [=*O. equestris* SAY, 1834]. COSSMANN (1916, p. 12) pointed out that all species included by IHERING in *Eostrea* were really members of *Ostrea s.s.,* presumably because *Ostrea s.s.* has the same sort of chomata. IREDALE (1939, p. 394) designated *O. puelchana* as the type species of *Eostrea* and since this species is doubtless well placed in *Ostrea s.s.,* IHERING'S *Eostrea* is a junior subjective synonym of *Ostrea* (see Fig. J113).

SUTER (1917, p. 86), who was well aware of IHERING'S publication and COSSMANN'S criticism, sought to save the basic idea underlying IHERING'S attempt at classifying *Ostrea* by proposing the subgenus *Anodontostrea* to include all species of *Ostrea* lacking chomata. Without designating a type species he included in *Ostrea (Anodontostrea)* nine species, among them "*O. angasi* [*angassi*] SOWERBY," which later was designated as the type species by FINLAY (1928, p. 264) (Fig. J36).

This oyster, *Ostrea angassi* G. B. SOWERBY (1871, pl. 13, fig. 28), lives around Tasmania and along the south coast of Australia from the Clarence River in northeastern New South Wales to the Swan River at Perth, Western Australia. However, this species, too, has chomata (THOMSON, 1954, p. 144) and falls into *Ostrea s.s.,* so that *Anodontostrea* must be a junior subjective synonym of *Ostrea s.s.* (see Fig. J113).

Just why SUTER overlooked the chomata in this case is not known. The species has only a few chomata. Older specimens possibly have obsolescent chomata, easily overlooked. Thus all these efforts come to naught, although the underlying idea that chomata are a feature important to classification is sound.

JOURDY, 1924

General JOURDY gave in this monograph (JOURDY, 1924) an excellent summary of all knowledge about exogyras published before 1924.

His investigation of the hinge and ligamental areas of exogyras revealed that there are unmistakable peculiarities differentiating them from other oyster species and from individual oysters that fortuitously have spiral opisthogyral ligamental areas. The structural differences make it quite certain that exogyras differ profoundly from other oysters and must be a phylogenetic unit. They cannot be placed under *Ostrea,* as had been done by DESHAYES, D'ORBIGNY, CoQUAND, and others, who recognized only one grand genus *Ostrea sensu latissimo.*

JOURDY discussed evolution of the hinge structure of the exogyras from the simpler beginnings in the Jurassic to its most evolved stage in the Late Cretaceous, which he called *"stade anodonte"* and *"stade monodonte,"* respectively. The distinction is based on the appearance of an auxiliary interlocking structure, consisting of a broad, shallow, pitted or striated depression on the left and a corresponding striated protuberance on the right valve, situated on the dorsal ends of the commissural shelves near the posterior margins of the curved spiral ligamental areas. This structure had been described by STEPHENSON (1914, p. 47) (see Fig. J90).

JOURDY emphasized the structural unity of all exogyras and thereby laid the foundation for regarding them as a distinctive taxon, ultimately to be recognized as a subfamily by VYALOV (1936).

In adding information on Jurassic Exogyrinae to the seven *"groupes"* which PERVINQUIÈRE (1912, p. 173-174) had distinguished among Cretaceous Exogyrinae of France and North Africa, JOURDY came to the conclusion that eight phyletic branches are discernible among the Exogyrinae. For these he did not propose any formal names, but showed where the older names *Aetostreon, Rhynchostreon,* and *Ceratostreon* would fit in, and for each phyletic branch he named several member species.

JOURDY (1924, p. 31, 96-97, pl. 4, fig. 2 and pl. 11, fig. 1, left row), without mentioning the genus *Gryphaeostrea* CONRAD, listed and discussed a few Miocene species of that genus, particularly *G. ricardi* (CossMAN & PEYROT, 1914) from early Burdi-galian beds at Saucats, Gironde, France. He called them all *Exogyra* and claimed that they showed generic features of *Exogyra s.l.* clearly enough to prove that they should be included in *Exogyra,* although their generic position has been contested.

ORTON, 1928; NELSON, 1938; GUNTER, 1950

For many years, oyster biologists were so preoccupied with their studies of local biological phenomena in living commercial oysters that they did not participate in discussions of generic classification. Rather, they remained content to classify all oyster species as *"Ostrea."* However, during that period, one after another biological difference between the common commercial oyster species was encountered. Finally, it became obvious that numerous biological differences demonstrated the presence of two distinct groups of species and that these demanded recognition in classification as separate taxa.

ORTON (1928) was first to enumerate the biological differences between the two groups. He recognized that these differences were profound enough to demand separate generic or subgeneric names. He distinguished a "Type I" taxon and stated that it has a subcircular shell outline, large eggs, larviparous propagation [he meant incubatory, instead], and monoecious sexuality, that it spawns at temperatures around 15°C., and that its several species flourish in temperate regions. He listed only three species under "Type I," one of which was *Ostrea edulis* LINNÉ. "Type II" oyster species, he noted, have a dorsoventrally elongate shell, small eggs, and nonlarviparous adults [he meant nonincubatory]; individual oysters are dioecious (hermaphroditic) and spawn around 20°C.; the various species flourish in tropical or subtropical regions. Under "Type II" he listed three species, of which one was *O. virginica* GMELIN.

By now it has been shown that both groups are dioecious and protandric alternating hermaphrodites. However, among incubatory oysters their successive sex phases overlap so much that at first it was difficult to prove that sex phases are successive and overlap very much, rather than that they are contemporaneous throughout (see p. *N961*).

NELSON (1938, p. 55) pointed out that *Crassostrea virginica* lives in the Gulf of St. Lawrence region of Canada, where individuals growing in intertidal situations freeze solid for four to six weeks in the winter. Therefore, "Type II" oysters could not all be called tropical to subtropical. Although the geographic distributions were very poorly stated by ORTON, some real, fundamental differences distinguish geographic ranges of the two groups (see p. *N1037-N1038*).

ORTON was so impressed by these differences that he proposed radical changes in names for the species: *Ostrea edulis* to *Monoeciostrea europa* and *O. virginica* to *Dioeciostrea americana*. IREDALE (1939, p. 394) designated *O. edulis* as the type species of *Monoeciostrea*, making this taxon a junior objective synonym of *Ostrea*. STENZEL (1947, p. 173) designated *O. virginica* as the type species of *Dioeciostrea*, thus fixing it as a junior objective synonym of *Crassostrea*. In this fashion ORTON's radical nomenclature was eliminated.

ORTON's proposals had been based on an erroneous philosophy of nomenclature. Generic names need not be coined in manner reflecting generic descriptions. After all, they are merely devices for information retrieval from the scientific literature. Actually the names proposed by ORTON turned out to be misleading, because it is now known that all oysters are dioecious [=hermaphroditic], as was pointed out by GUNTER (1950, p. 440).

In summary, ORTON's work was the first attempt, and a successful one, to infuse biological information into the pool of basic data on which classifications are based. As such it was very important, and incidental data and ideas of his that turned out to be unacceptable are unimportant and were corrected very soon. To the surprise of many, the conchological and paleontological differences between the two taxa noted before ORTON's work (1928) was published were found to be quite reliable guides to classification. The result of this experience was to produce considerable confidence in the reliability and validity of conchological and paleontological investigation methods. The chief result of ORTON's studies, name-

ly, that oyster species are divisible into two natural groups was fully supported by NELSON, who added to it by showing that "Type I" oysters lacked a promyal passage, whereas "Type II" oysters have such a passage, which he called promyal chamber. He believed that the differences between the two deserved recognition in taxonomy on the genus level (NELSON, 1938, p. 55) and suggested *Ostrea* for "Type I" of ORTON and *Gryphaea* for "Type II." In selecting the latter name he followed FISCHER (1880-87). At this stage, the question of proper generic names came to the fore.

GUNTER (1948, 1950) extensively discussed this question and several other problems concerning the family with STENZEL, who had just published his nomenclatural synopsis of the Ostreidae (STENZEL, 1947) and was in a position to help. GUNTER (1948) checked STENZEL's conclusions independently and published his concurrence with them. GUNTER (1950) reiterated these nomenclatural conclusions and added much information on features that separate the two groups of oyster species. In a table he contrasted the two with the aid of ten characters, also adding several more species to "Type I" and "Type II." Through STENZEL (1947) and GUNTER (1950) the correct generic name, *Crassostrea*, for *C. virginica* (GMELIN) became widely known to oyster biologists and has now been adopted by most of them.

ARKELL, 1934; ARKELL & MOY-THOMAS, 1940

In a study of oysters of the Fuller's Earth (Jur.) in western Europe, ARKELL (1934) proposed a new, experimental system of classification and nomenclature for some of the Jurassic and Cretaceous oysters. His revolutionary system was based on then-prevailing ideas of phylogeny of *Gryphaea* and its descendants. His premise was that "it has long been realised that all the Mesozoic species of oysters conveniently called *Gryphaea* do not form a monophyletic group or genus in the ordinary sense of the word" (ARKELL, 1934, p. 58). He fully approved the ideas concerning *Gryphaea* as a polyphyletic form-genus propounded by KITCHIN and TRUEMAN (compare p. *N1062-N1078*). ARKELL considered

Experimental "Rationalised" Classification of Jurassic and Cretaceous Oysters Proposed by Arkell (1934)

OLD NAMES	"RATIONALISED" NAMES
Pycnodonta marcoui (HILL & VAUGHAN)	*Ostrea (Marcoui) gryphaea*
Pycnodonta wardi (HILL & VAUGHAN)	*Ostrea (Marcoui) catinula*
Gryphaea dilatata SOWERBY	*Ostrea (Dilatata) gryphaea*
Ostrea (Catinula) alimena COSSMANN	*Ostrea (Dilatata) catinula*
Gryphaea bilobata SOWERBY	*Ostrea (Bilobata) gryphaea*
Ostrea (Catinula) matisconensis LISSAJOUS	*Ostrea (Knorrii) catinula* mut. *matisconensis*
Ostrea (Catinula) knorrii VOLTZ	*Ostrea (Knorrii) catinula*
Ostrea (Liostrea) subrugulosa MORRIS & LYCETT	*Ostrea (Acuminata) catinula*
Ostrea (Liostrea) hebridica var. *elongata* DUTERTRE	*Ostrea (Acuminata) virgula* mut. *elongata*
Ostrea (Liostrea) hebridica FORBES	*Ostrea (Acuminata) virgula* mut. *hebridica*
Ostrea (Liostrea) acuminata SOWERBY	*Ostrea (Acuminata) virgula*
Gryphaea incurva SOWERBY	*Ostrea (Incurva) gryphaea*
Ostrea (Liostrea) irregularis VON SCHLOTHEIM	*Ostrea (Incurva) catinula*
Ostrea (Liostrea) liassica STRICKLAND	*Ostrea (Incurva) virgula*
Pycnodonta corrugata (SAY)[1]	*Ostrea (Corrugata) gryphaea*[1]

[1] Added by ARKELL & MOY-THOMAS (1940, p. 404).

it needful to reconcile nomenclature of various Jurassic and Cretaceous gryphaeas with these ideas by proposing experimental nomenclature which he claimed to be "rationalised" (ARKELL, 1934, p. 64) and better than the conventional one. His scheme is outlined in the tabulation above.

This experimental nomenclature did away with the hierarchy of old names that followed the *Code* and consisted of names for genus, subgenus, and species. The "rationalised" names proposed in their stead, admittedly under disregard of the *Code,* consisted of trinomina: 1) the single genus *Ostrea,* 2) a name representing lineages, written as a subgenus of *Ostrea,* and 3) a stage-designation. Only three stage-designations were admitted: *virgula, catinula,* and *gryphaea,* based merely on external shell form. ARKELL was convinced that the phylogeny of every lineage progressed in this fashion: Shape like *virgula* (an exogyrine species) to shape resembling that of a *Catinula,* to shape similar to a *Gryphaea.* However, no real proofs of such iterative phylogenies exist. The idea was strictly an assumption.

ARKELL & MOY-THOMAS (1940) are the only authors who gave serious consideration to ARKELL's (1934) scheme, even adding one more "rationalised" name. No other authors adopted or explicitly approved the scheme. The new subgeneric names were declared unavailable by STENZEL

(1947), because they did not fulfill Article 13(a) of the *Code* in that (after 1930) the two authors failed to give characters differentiating the taxa.

The basis of the ARKELL revolutionary system of oyster classification is presumed iterative evolution in the Gryphaeidae. Such iterative evolution never was more than an assumption, however, and therefore, it furnished an insecure basis for an elaborate classificatory system. If it can be shown that the concept of iterative evolution does not apply, the whole system collapses. It is better to stay with the conventional system, because the latter is by experience very flexible and adaptable to various interpretations of phylogeny and evolution. Rather use an imperfect but adaptable conventional system than propose a scheme that at the moment seems less imperfect, but requires profound reconstruction whenever one conceives new ideas of phylogeny and evolution.

VYALOV, 1936, 1937, 1948a

VYALOV (1936, 1937) deliberately set out to resolve the major problems of oyster classification. The principal taxonomic character which he selected for use is structure of the upper valve. In addition, he planned to distinguish sculpture patterns, form of the beaks, and general shape of the shell. Accordingly, he divided oysters into four subfamilies: 1) Ostreinae—both valves convex, lower valve smooth or costulated, upper valve smooth, containing 2 genera, 14 subgenera, and 11 sections; 2) Gryphaeinae—lower valve convex and smooth or costulated, upper valve flat or concave, containing 2 genera, 5 subgenera, and 6 sections; 3) Lophinae—both valves radially sculptured, valve commissure plicate or undulate, containing 1 genus, 3 subgenera, and 7 sections; 4) Exogyrinae—umbones of both valves spirally inrolled, umbonal spire of upper valve not produced beyond valve outline and turned downward, forward, or upward, containing 3 genera, 3 subgenera, and 4 sections. All together he recognized 61 taxa above species rank and below that of subfamily. Of these, 27 taxa were given new names.

Vyalov (1936, 1937) provided each taxon with a very short descriptive definition and a type species, so that his 27 new names were available immediately. However, for two of the older taxa, *Crassostrea* and *Gryphaeostrea,* he listed wrong type species. The former was corrected by him later (Vyalov, 1948a, p. 23, footnote).

Vyalov succeeded in collecting quite a few obscure generic names that had entered the literature before 1936. However, a few such names were omitted (viz., *Alectryonella, Odontogryphaea, Planospirites, Saccostrea*). Omission of obscure names hidden somewhere in the literature cannot be criticized severely, for no matter how thorough a search may be no assurance can be given that one has · found all available names. A serious omission, however, was *Pycnodonte,* which he did not mention in 1936, because he concluded, unfortunately incorrectly, that it is a junior subjective synonym of *Gryphaea s.s.,* as explained in a later work (Vyalov, 1948a, p. 28).

Vyalov did not follow the *Code* in every case. For example, he placed *Ostreonella* Romanovskiy, 1890, as a section under the genus *Liostrea* Douvillé, 1904, and arranged *Sokolowia* Böhm, 1933, as a section under the genus *Fatina* Vyalov, 1936. In the latter case, he evidently realized the close affinities of the two respective type species, but presumably he wished to conserve his *Fatina,* instead of subordinating it under *Sokolowia,* which has clear

priority as a name. In my opinion the two taxa are synonyms not separable even on the section level.

Vyalov's major classification method was ingenious in a way, for his use of the upper valve to diagnose major divisions of oysters is eminently effective in certain cases. For example, some exogyrine forms have spiral lower valves which are confusingly similar to those of some gryphaeine oysters; among them the upper valves are the key to their distinction. However, this diagnostic feature by itself is insufficient in other cases.

Vyalov used too few diagnostic attributes, and for that reason his classification became quite uncertain on generic and lower levels. He did not make serious use of the structures of hinges and of ligamental areas. Although he discussed outlines of muscle imprints in his later work (Vyalov, 1948a, p. 9, 17-18), he never mentioned them in his earlier work and did not recognize them as a diagnostic character, nor did he use vesicular shell structure. The chomata of oysters and vermiculate wrinkles of the Pycnodonteinae, as well, were disregarded. He interpreted the latter as "not a generic, and in the majority of the cases not even a specific character" (Vyalov, 1948a, p. 27), in spite of the fact that these features had already been discussed and used successfully by others.

Small wonder that the Pycnodonteinae were not recognized by Vyalov as a taxon and that the various genera of the Pycnodonteinae were scattered about: *Gigantostrea* as a subgenus of *Ostrea; Labrostrea* as a section of *Liostrea (Liostrea); Biauris, Circogryphaea,* and *Phygraea* as sections of *Gryphaea (Gryphaea);* and *Pycnodonte* suppressed as a junior subjective synonym of *Gryphaea s.s.*

In several other cases, disruption of a well-definable genus or of a set of related genera ensued from his classification. The genus *Crassostrea* was broken into *Ostrea (Crassostrea)* and *O. (Angustostrea).* The very closely related genera *Flemingostrea* and *Odontogryphaea* were widely separated as *Ostrea (Flemingostrea)* and as the new section *Sinustrea* of *Liostrea (Liostrea).*

The Gryphaeinae, as defined by Vyalov on the basis of external valve morphology, without regard as to vesicular shell structure, vermiculate wrinkles, and outline and position of the muscle imprint, are obviously a polyphyletic collection of various gryphaeine oysters, *Gryphaea* homeomorphs, and *Gryphaea* descendants. In this subfamily he placed *Sokolowia* (=*Fatina*), which is a *Gryphaea* homeomorph and an ostreine oyster. *Sokolowia* is closely related to *Turkostrea*, which VYALOV placed correctly as a subgenus of *Ostrea*. The placements of *Sokolowia* and *Turkostrea*, far apart, are another one of many dislocations of evolutionary lineages. Although polyphyletic origins were a well-known problem already in 1936, no attempt was made to unravel any of them. Such wide separations of phylogenetically closely related lineages are the real test of validity of a classification. The more severe they are and the greater their number, the more does a classification producing them differ from a natural or phylogenetic arrangement. Separations of phylogenetically closely related taxa normally should become eliminated and their total number, both those recognized as such and those yet undiscovered, should become smaller as successive authors improve classification by using more and more criteria and by perfecting the methods they employ. It is desirable to discern more criteria of classification, but VYALOV reduced their number in comparison with preceding authors.

VYALOV actually increased the number of disruptions because 44 percent of the taxonomic names recognized by him were newly established ones and because he used an insufficient number of criteria and failed to use internal or structural criteria so that too great a weight came to rest on purely external shell features. The fact that he distinguished 79 percent more taxa than had been recognized before him and that he had to resort to sections to subdivide subgenera clearly indicate that he made too many taxonomic divisions.

Because of the great number of taxa distinguished and newly proposed by him, Vyalov's classification was received without enthusiasm (BEURLEN, 1958; RANSON, 1943b, p. 162; 1948b, p. 2-3). Indeed, several of the type species selected by VYALOV for new taxa are insufficiently known species, so that the taxa based on them are in doubt.

In defense of VYALOV's classification it must be pointed out that he (VYALOV, 1948a, p. 5) was not at all convinced that he had succeeded in taking into account all peculiarities of the oysters or that no modifications of his classification would be required in future. According to him, the ultimate, unshakable classification cannot be built by a single investigator in a few years of study.

To summarize, VYALOV deserves credit for recognizing for the first time that taxonomic differences among the oysters are profound enough to require breaking the family into subfamilies. However, the basic philosophy of his work, namely that a single feature can serve as principal criterion for subdividing oysters, cannot stand the test.

Because VYALOV's early work (1936) was so brief (only four pages), need for amplification was obvious, and was given later (VYALOV, 1948a). The two cited works are similar in all essential features. However, the later publication included two additional taxa: *Rygepha* VYALOV, 1946, a section of *Gryphaea s.s.* and *Solidostrea* VYALOV, 1948c, supposedly a subgenus of *Ostrea*. Here VYALOV (1948a, p. 33) gave a compelling reason for including *Gryphaeostrea* in the Exogyrinae. Although other authors before him had placed that genus either near *Exogyra* or even included it in the latter, *Gryphaeostrea* had remained dubious as to systematic position. VYALOV emphasized that the umbonal part of the upper valve is spiral in an exogyroidal fashion in this genus.

STENZEL, 1947

At the beginning of his studies on oysters, STENZEL set out to acquire the nomenclatural tools needed for thorough work. He endeavored to collect all names that various authors had applied to supraspecific taxa among oysters and to ascertain their type species, as well as where these had been found and described (STENZEL, 1947). Although handicapped by poor library facilities, he managed to round up 116 such names.

Many of these were found to be vitiated on purely nomenclatural grounds, because they had been proposed in a nomenclaturally illegal fashion or because they were homonyms of earlier names, unjustified emendations of earlier names, or junior objective synonyms. All together, objections of these sorts reduced the list to 80 nomenclaturally available and potentially usable names.

JAWORSKI (1951) reviewed STENZEL's synopsis favorably and corrected his own conception of *Heterostrea* JAWORSKI, 1913, pointing out that it was not applied to an oyster, instead was a subjective synonym of *Myoconcha* SOWERBY, 1824 (see Fig. J147). BEURLEN (1958) criticized both VYALOV (1936, 1948a) and STENZEL (1947) for their extreme splitting of the Ostreidae into many supraspecific taxa. Evidently he misunderstood the intent of STENZEL's synopsis. STENZEL (1947, p. 165) had made it clear that his synopsis was intended to provide no more than the nomenclatural tools and not to judge which names were justified or not from a biological point of view. In spite of BEURLEN's objection to the surfeit of available names, he added one more (*Nanogyra* BEURLEN, 1958) to the 116 names.

At that, STENZEL failed to find about eight names published before 1947 and misinterpreted the type species of *Aetostreon*. Also, by following FRANCIS HEMMING's (1944) interpretation of what constitutes definition, description, or indication of a generic name, STENZEL withheld nomenclatural priority from *Lopha* RÖDING, 1798, in favor of *Alectryonia* FISCHER DE WALDHEIM, 1807. The new *Code* does not allow this conclusion.

THOMSON, 1954

In his study of living Australian oyster species, THOMSON (1954) furnished modern descriptions and a classification founded on both purely conchological features of the hard parts of the mantle/shell and biological features of the soft parts. Internal anatomy, as far as known, was discussed, and two identification keys—one founded on soft parts and another one on shell features (THOMSON, 1954, p. 162-163)—were constructed. The interdisciplinary approach made the study modern and highly informative.

In classifying materials at hand THOM-

SON followed RANSON (1943) and distinguished only three genera: *Ostrea* LINNÉ, *Crassostrea* SACCO (for RANSON's *Gryphaea*), and *Pycnodonte* FISCHER DE WALDHEIM (for *Pycnodonta*). If one makes allowance that the classifications of RANSON and of THOMSON are dedicated to "lumping," they are quite correct and informative.

An important part of THOMSON's (1954) paper was correction of generic names proposed by IREDALE: The type species of *Saxostrea* was given as *Crassostrea commercialis* (IREDALE & ROUGHLEY, 1933), now regarded as a geographic subspecies of *Saccostrea cuccullata* (VON BORN, 1778). The type species of *Pretostrea* was given as *Ostrea folium* LINNÉ, 1758 (=*O. bresia* IREDALE, 1939), and it was shown that *O. folium* and *Mytilus crista galli* LINNÉ are one and the same polymorphic species. Thus *Pretostrea* is a junior subjective synonym of *Lopha* and *Dendostrea* (Fig. J47; see Fig. J129, J130).

METHODS AND PRINCIPLES

Many authors have expressed dismay when they noticed that they were unable to identify and classify oysters readily and correctly. Most of them had been accustomed to use only a few selected specimens per species, that is, to employ typological methods of classification, so that they failed to have at hand sufficiently large and diverse samples of species they were studying. Several such situations gave rise to embarrassing errors. A case in point is FINLAY's (1928a,b) proposal of the generic name *Notostrea* (see Fig. J144) introduced either without enough material at hand to describe both valves or perhaps not caring to describe more than one valve. At any rate he confounded one side with the other (BOREHAM, 1965). No wonder that *Notostrea*, 40 years later, remains doubtful and valueless. The first prerequisite in oyster classification is availability of ample material.

Even when specimens are numerous, problems of species identification generally remain. An example is the question posed by THOMSON (1954): Are the morphs commonly called *Lopha cristagalli* (LINNÉ, 1758), *Dendostrea folium* (LINNÉ, 1758), and *Ostrea* (*Pretostrea*) *bresia* IREDALE,

1939, really three separate species representing different genera or are they merely morphs of one and the same polytypic species, as Thomson maintained? The first two of these three are quite common, and study specimens of them abound in most museums. Thomson (1954) claimed to have seen transitional morphs linking the three, whereas other authors have either not seen any or overlooked them, in spite of abundant materials available. The question is not yet settled, although I am inclined to assume that Thomson's idea of only one polymorphic species is sound, because his careful work is based on more extensive local observations made under favorable conditions. Similar suspicions have been voiced by other authors concerning other supposedly separate species.

Questions of species identity and synonymy are only too numerous among the oysters. Some questions have major importance in generic classification. In particular, the type species must be elucidated fully first, else the genus would remain obscure. Examples of such questions are the three species discussed above (see also under *Lopha,* p. N1157) and *Striostrea procellosa* ("Valenciennes" in Lamy), the type species of *Striostrea* Vyalov, 1936. Credit for solving the latter problem must be given to Ranson (1949, 1951). The case was reinvestigated by me at Paris and Ranson's conclusion as to the identity of *Ostrea procellosa* with *O. margaritacea* Lamarck, 1819 (p. 208), was fully confirmed (see Fig. J107, J108). Only then it was possible to form an opinion as to the taxonomic status and morphologic features of *Striostrea,* which had been proposed in 1936 without sufficient knowledge of its morphology and of the synonymy of its type species. It is hoped that ultimately, through the concerted efforts of various authors, all type species of oysters will become well described so that then it will become possible to allocate firm taxonomic positions to all generic names and genera involved. However, the *Code* allows new genera to be introduced with much less effort than it takes to elucidate the taxonomic features of their type species, so that an early end to this task is not in sight.

Classification and taxonomy of oysters, on the various hierarchical levels above species and superspecies, seems extraordinarily difficult and open to divergent interpretations. For this reason, it cannot be claimed that the classification proposed here is unshakably correct. In support of it, however, is the fact that more data, particularly of neontological and phylogenetic sorts, were gathered and built into it than in any other classification.

To improve on any extant classification one must adduce more observational data, gain knowledge of more taxonomic characters, and be able to make a keener evaluation of their significance and reliability. The best way to obtain more data is to use both paleontological and neontological observations to limits of present-day knowledge.

Not all observational taxonomical characters are equal in usefulness or importance to classification. However, one cannot know in advance which characters are more important than others. Only trial and error can establish their relative importance, which really is arranged in a hierarchy, largely obscure at the beginning, as complicated as the taxonomy that is to be built from the characters. The great problem is to guess at their respective importance before one has built a system of classification on them. Certain rules are of some help in evaluating the importance of a taxonomic character. Those used by me in study of oysters are as follows.

1) A feature is judged to gain in importance, the more its changes require shifts in the position of internal organs. For example, shifts of the intestine with respect to the pericardium seem to have great significance. All oysters in which the intestinal tract runs through the pericardium and ventricle of the heart (viz., *Hyotissa, Neopycnodonte*) must differ considerably from oysters in which the intestinal tract bypasses the ventricle and pericardium at their dorsal side (viz., all other living oysters).

2) Internal features are commonly more important than external ones. As example, the shape and position of insertion of the adductor muscle are internal features and therefore probably quite important. The reason is that a given relative position of the insertion could not be accomplished originally without other internal organs having to change shapes or outlines or having to shift their positions to make room. In other

words, rule 2) is an extension or application of rule 1). In classification of oysters, the shape and position of insertion of the posterior adductor muscle is judged to have great importance.

3) Shell structure is more important than most other features. For instance, all oysters with vesicular shell walls (viz., Pycnodonteinae) must form a closely knit taxon, whether defined as a single genus or a subfamily. Shell structure outranks other taxonomic characters. For example, *Hyotissa* has a plicate valve commissure and vesicular shell-wall structure. The former character would appear to entitle it, so to say, to be placed in the Lophinae, all of which have plicate commissures. This is rejected because shell-wall structure is considered to outrank and overrule the plicate commissure. The shell wall, vesicular in *Hyotissa*, requires that this genus be placed in the Pycnodonteinae, as RANSON (1941) has demonstrated.

4) Correlated features, that is, characters that constantly or prevailingly make appearance together, are more important than noncorrelated ones. For example, the group of oysters, mostly fossil, which are now called Pycnodonteinae, have several apparently unrelated characters that always show up together. These are: vesicular shell walls, vermiculate chomata on both valves, orbicular outline of adductor muscle insertion, and a rather high position of the muscle insertion between hinge and opposite valve margin. These various characters are unrelated in the sense that apparently no reason can be cited to explain why one character should entrain others. For instance, why should vesicular shell-wall structure have any influence on shape of the insertion of the adductor muscle? Because these characters appear to be independent of each other, although making their appearance together, they are regarded as good indicators of monophyly.

OUTLINE OF CLASSIFICATION

The following outline summarizes taxonomic relationships, geologic occurrence, and numbers of recognized genera and subgenera in each suprageneric group of oysters from suborder to subfamily. A single number refers to genera; where two numbers are given, the second indicates subgenera additional to nominotypical ones.

Main Divisions of Ostreina

Ostreina *(suborder)* (52;13). *U.Trias.-Rec.*
  Ostreacea (superfamily) (52;13). *U.Trias.-Rec.*
    Gryphaeidae (22;6). *U.Trias.-Rec.*
      Gryphaeinae (6;2). *U.Trias.(Carn.)-U.Jur.* *(Portland.).*
      Pycnodonteinae (4;3). *L.Cret.-Rec.*
      Doubtful genus (1). *Paleog.*
      Exogyrinae (11;1). *M.Jur.(Bajoc.)-Mio.*
    Ostreidae (27;4). *U.Trias.-Rec.*
      Ostreinae (14;1). *L.Cret.-Rec.*
      Lophinae (6;3). *U.Trias.-Rec.*
      Doubtful genera (7). *U.Cret., low.Oligo.-Eoc.*
    Chondrodontidae[1] (1;3). *L.Cret.(Alb.)-U.Cret.* *(Campan.).*
    ?Lithiotidae[1] (2). *L.Jur.(L.Lias.).*

---

[1] Included in Ostreacea for convenience.

SYSTEMATIC DESCRIPTIONS

INTRODUCTION

For the first time, it is now recognized that oysters, as commonly understood (Ostreidae AUCTORUM), are not a monophyletic family, but are very probably diphyletic. Accordingly, two families are distinguished in the *Treatise:* 1) Ostreidae *(sensu stricto),* emended here to exclude the subfamilies Gryphaeinae VYALOV, 1936, Pycnodonteinae STENZEL, 1959, and Exogyrinae VYALOV, 1936, and 2) Gryphaeidae VYALOV, 1936, which is composed of the three above-mentioned subfamilies.

The Gryphaeinae, furthermore, are revised as shown by STENZEL (1959). For the second family listed above the designation Gryphaeidae VYALOV, 1936, must be used according to the *Code,* although VYALOV (1936) merely proposed Gryphaeinae.

The distinction between the two families is based on 1) hinge structure of the prodissoconch, 2) shape and position of the imprint of the posterior adductor muscle, and 3) the course of the intestine with reference to the heart. Characters 2) and 3) are linked, as explained above (p. *N1095*).

Suborder OSTREINA Férussac, 1822

[*nom. correct.* NEWELL, 1965 (*ex* order Ostracés FÉRUSSAC, 1822)] [Diagnosis by N. D. NEWELL]

Monomyarian; foot and byssus lacking in adults; generally cemented by LV; shell chiefly, but not exclusively, calcitic, foliaceous; gills eulamellibranch; pallial line entire. *U.Trias-Rec.*

Superfamily OSTREACEA Rafinesque, 1815

[*nom. transl.* SCHWEIGGER, 1820, p. 712 (*ex* family Ostreacia RAFINESQUE, 1815)] [Diagnosis and other materials for this superfamily prepared by H. B. STENZEL (except Chondrodontidae, furnished by †L. R. Cox and H. B. STENZEL, and Lithiotidae, which was prepared by †L. R. Cox)]

Mantle open all along margins except at point of palliobranchial fusion, devoid of a pallial line except one genus (*Saccostrea,* see p. *N*1134), which has a disjunct line of pallial muscle insertions. Outer edge of gills fused to mantle; shell cemented to a firm substratum by left valve except in the case of a few species which never become attached to a firm substratum (see p. *N*995-*N*996); postlarval shell foliaceous, inequivalve ranging from extremely inequivalve in some genera to slightly so in others, edentulous, ligamental area divided into three parts, resilifer in middle. *U.Trias.-Rec.*

Family GRYPHAEIDAE Vyalov, 1936

[*nom. transl.* STENZEL, herein (*ex* Gryphaeinae VYALOV, 1936, *emend.* STENZEL, 1959)]

Nonincubatory. Prodissoconch hinge on planktonic larvae carries uninterrupted alternating series of equal tooth precursors and corresponding sockets (Fig. J72). Promyal passage extensive, reaching close to pallial isthmus; intestine passing through pericardium and ventricle of heart. Posterior adductor muscle orbicular in cross section placed closer to hinge than to opposite valve margin, ventral border of its insertion on LV elevated above general surface of valve. Valves highly unequal in most genera to subequal in others. Radial posterior groove on LV absent, more or less deep, or reduced to a mere flexure of growth lines, or obscured in later forms. Attachment area of LV tiny to small in most genera, average to large in some. Umbonal cavity of LV very shallow except in Exogyrinae; LV beak of highly inequi-

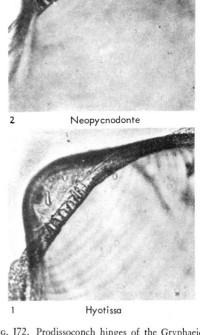

2   Neopycnodonte

1   Hyotissa

FIG. J72. Prodissoconch hinges of the Gryphaeidae, ×240 (Ranson, 1967).—*1. Hyotissa hyotis* (LINNÉ, 1758), living, IndoPacific. — *2. Neopycnodonte cochlear* (POLI, 1795), living, Atlantic, Mediterranean, and IndoPacific.

valve forms filled with solid shell matter, no chambers except in Exogyrinae. Prismatic shell layer thin in most genera, absent in a few. Ligamental area of lower valve never very high except in Exogyrinae. [Strictly euhaline and stenohaline; most genera never form true oyster reefs in which conspecific individuals grow mainly on one another.] *U. Trias.-Rec.*

Subfamily GRYPHAEINAE Vyalov, 1936

[Gryphaeinae VYALOV, 1936, p. 19, *emend.* STENZEL, 1959, p. 16; Gryphaeinae VYALOV, 1937 was placed on Official List by ICZN, Opin. 356; evidently Gryphaeinae VYALOV, 1936, had been overlooked]

Commissural shelf less well defined than in Pycnodonteinae, lacking chomata; no vesicular shell structure. *U.Trias.(Carn.)-U.Jur.(Portland.).*

Gryphaea LAMARCK, 1801, p. 398 [Official List, ICZN Opin. 338] [*G. arcuata;* SD ICZN, Opin.

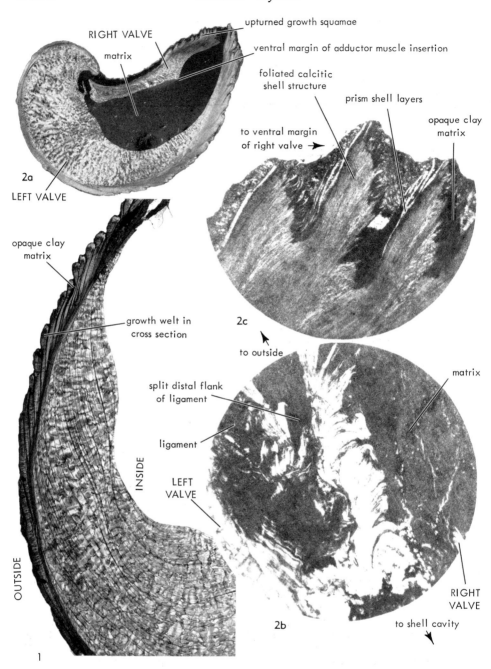

Fig. J73. Shell structures in *Gryphaea (Gryphaea) arcuata* Lamarck (1801), from Lower Jurassic (Lias.) of England (Stenzel, n).

1. Appressed growth squamae ending in terminal expansions that produce concentric growth welts on LV; thin section at right angles to hinge axis and commissural plane in ordinary light, show-ing irregular foliated shell structure on inner part and calcitic crossed-lamellar structure on outer part of valve, ×2.

2a. Thin section of shell cut at right angles to hinge

338] [=*Gryphoea* Bosc, 1802, p. 307 (*nom. van.*); *Gryphaeigenus* RENIER, 1807, p. 7 (rejected ICZN, Opin. 427); *Gryphites* VON SCHLOTHEIM, 1813, p. 52 (*nom. van.*); *Griphaea, Griphoea* DE BLAINVILLE, 1821, p. 533 (*nom. null.*); *Griphea* BRONGNIART, 1822, in CUVIER & BRONGNIART, (*nom. null.*); *Gryphea* BRONGNIART, 1823, p. 10 (*nom. null.*); *Liogryphaea* FISCHER, 1886, p. 927 (rejected ICZN, Opin. 338) (obj.); *Gryphaae* WHITE, 1887, p. 20 (*nom. null.*); *Liogryphea* DOUVILLÉ, 1904a, p. 273 (*nom. null.*) (rejected ICZN, Opin. 338); *Ghyphaea* SCALIA, 1912, p. 21 (*nom. null.*); *Lyogryphaea* COUFFON, 1918, p. 116 (*nom. null.*); *Jurogryphaea, Incurva* ARKELL, 1934, p. 62 (*nom. neg.*); *Gryphaca* JAWORSKI, 1935, p. 885 (*nom. null.*); *Rygepha* VYALOV, 1946, p. 34 (type, *Gryphaea skuld* BÖHM, 1904, p. 17; OD); *Liogriphaea* VYALOV, 1948a, p. 36 (*nom. null.*); *Liographaea* CHARLES, 1949, p. 35 (*nom. null.*); *Liogryphanaea* KRACH, 1951, p. 355 (*nom. null.*); *Gruphea, Griphea* ĆIRIĆ, 1951, p. 65 (*nom. null.*); *Cryphaea* ZAPRUDSKAYA, 1953, p. 23 (*nom. null.*)] [*non Gryphaea* FISCHER, 1886, p. 927 (=*Crassostrea* SACCO, 1897)]. Shell medium-sized to large (up to 16 cm. long and 14 cm. high), LV ranging from high and narrow (*H* [height] about 178 percent of *L* [length]) in some species to orbicular to horizontal-oval (*H* about 70 percent of *L*) to bilobate-oval (*H* about 80 percent of *L*) in others, orthocline to very slightly opisthocline, convex and capacious, ranging from highly convex to broadly convex in horizontal cross section; not compressed and devoid of median radial keel. LV with radial posterior sulcus ranging from evanescent to deeply sunken, posterior flange poorly or well set off from main body of valve (bilobate forms), and beak small, pointed, opisthogyral to nearly orthogyral, highly inrolled with tiny attachment scar in some species to less inrolled and with large attachment scar in others; LV smooth or with low smooth irregular concentric growth welts or with low smooth regularly spaced concentric undulations, mostly with appressed or rarely nonappressed growth squamae, either devoid of radial ribs or furnished with them (Fig. J74). RV concave, vertical-oval to spatulate, orbicular, horizontal-elliptic, or bilobate, truncated by hinge, without ribs or having few

narrow radial gashes or threads, and having appressed or nonappressed growth squamae. Resilifer ranging in size from as long as to 5 times length of each bourrelet of ligament, excavate in LV and flat to very slightly excavate in RV. Umbonal cavity beneath hinge plate on inside of LV largely filled in by thickened shell (Fig. J73, 2a); commissural shelf well developed but without chomata. Adductor muscle imprint orbicular to oval, with dorsal margin clearly convex and ventral edge projecting obliquely upward into shell cavity. *U.Trias.*, Boreal Prov.(B.C.-Can.Arctic-Bear I.-Far NE.Sib.)-USA(Nev.); *L.Jur.(Hettang.)-U.Jur.(Kimmeridg.)*, worldwide.

G. (Gryphaea). Small to large. LV lacking radial ribs, costellae, or threads, with evanescent to shallow radial posterior sulcus and posterior flange not detached. *U.Trias.*, Boreal Prov., Nev.; *L. Jur.(Hettang.)-U. Jur.(Oxford.)*, worldwide.——FIG. J74,3. *G. (G.) arcuata* LAMARCK, L. Jur.(Lias.), Eng.; *3a-d*, RV, various views, aragonite of adductor muscle pad leached, ×1; *3e-g*, entire shell, various views, ×1 (Stenzel, n). [Specimens by courtesy of P. C. Sylvester Bradley.] (*See also* Fig. J60.)

G. (Africogryphaea) FRENEIX, 1965, p. 32 [*Liogryphaea costellata* DOUVILLÉ, 1916, p. 58; OD] [=*Africogryphaea* FRENEIX & BUSSON, 1963, p. 1632 (*nom. nud.*)]. Small to medium-sized, outline tending to higher than long and truncate at umbo because of large attachment area and earlike extension of valves at posterior flank of hinge line. Radial posterior sulcus broad and deep; posterior flange well detached, ending in small projecting divergent lobe. LV with somewhat irregular, strong and large radial ribs. *M. Jur.(Bathon.)-U. Jur.(Callov.)*, N. Afr.(Alg.-Tunisia)-Sinai Penin.-Ethiopia-Arabia.——FIG. J74,1. *G. (A.) costellata* (DOUVILLÉ), M.Jur.(Bathon.), Massif of Moghara, Sinai Peninsula; *1a,b,e*, LV, both sides and ext.; *1c,d*, RV int. and ext. views, all ×0.7 (Douvillé, 1916).

G. (Bilobissa) STENZEL, n. subgenus, herein [*G. bilobata* J. DE C. SOWERBY, 1840, p. 4; OD] [=*Bilobata* ARKELL, 1934, p. 64 (*nom. nud.*)]. Small to medium-sized. Radial posterior sulcus deep; posterior flange well detached. No radial ribs, costellae, or threads. [*Bilobissa* is possibly ancestral to *Texigryphaea*.] *M.Jur.-U.Jur.*,Eu.——

FIG. J73. (*Continued from facing page.*)

axis and commissural plane viewed under crossed nicols, showing umbonal region solidly filled, ×1.

2*b*. Lithified ligament from same thin section (*2a*) under ordinary light, ×20. [The ligament remains attached to LV but has pulled away from RV, allowing sediment to ooze between ligament and RV.]

2*c*. Upturned growth squamae near ventral margin of RV in same thin section (*2a*) in ordinary

light, showing foliated calcitic shell structure ending in thin strips of prismatic shell layers (light-colored), ×20. [Black areas are opaque mineral aggregates, possibly pyrite or black clay.]

[Specimens courtesy of P. C. SYLVESTER-BRADLEY, Univ. Leicester, England. Thin sections made under supervision of OTTO MAJEWSKE, Shell Development Co., Houston, Texas.]

FIG. J74,2. **G. (B.) bilobata* J. DE C. SOWERBY, M.Jur.(Bajoc.), Inferior Oolite, Eng.; *2a-d,* several views of holotype, ×0.7 [photographs by courtesy of British Museum (Natural History)].

Catinula ROLLIER, 1911, p. 272 [**Ostrea knorri* VOLTZ, 1828 (=*O. knorrii* VOLTZ, 1828, p. 60) (*non O. knorri* DEFRANCE, 1821, p. 27, suppressed ICZN, Opin. 360, =*O. knorrii* VOLTZ, ICZN Official List, Opin. 360); SD ARKELL, 1932, p. 149, 180] [=*Catinulus* LISSAJOU, 1923, p. 142 *(nom. van.), O. (Knorrii)* ARKELL, 1934, p. 64 *(nom. van.)*]. Very small (up to 35 mm. high). LV deep, capacious, scoop-shaped, orthocline to opisthocline at 70 degrees to hinge axis, commonly

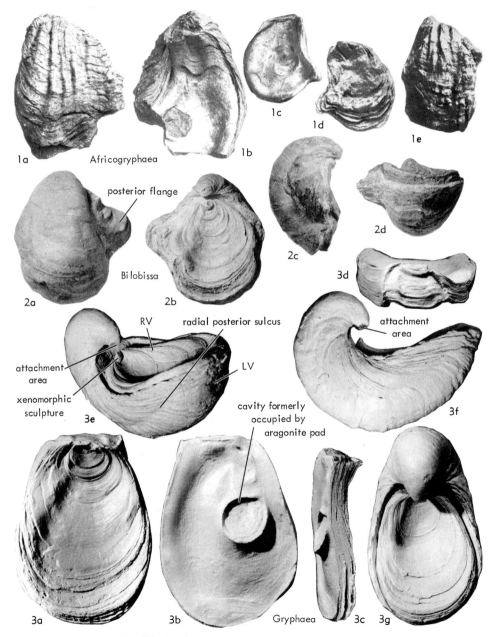

FIG. J74. Gryphaeidae (Gryphaeinae) (p. *N*1097, *N*1099).

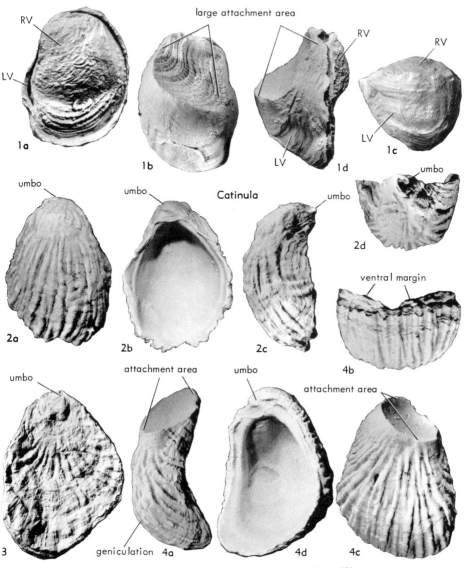

FIG. J75. Gryphaeidae (Gryphaeinae) (p. *N1100-N1102*).

higher than long (*H* 95 to 170 percent of *L*, averaging 125 percent); commonly less wide than high (*W* [width] 20 to 75 percent of *H*, averaging 45 percent); attachment area tiny to very large (10 by 11 mm., even up to 17 mm.); LVs with large attachment area growing steeply upward from substratum to form straight or convex profile without geniculation, others with tiny attachment area higher, less deep, obliquely spatulate, and less scoop-shaped, growing out at first fairly flat, but later bending abruptly and convexly to attain convex, geniculate profile; LV with many narrow, crowded, rounded, divaricating, irregular to fairly regular, discontinuous to subcontinuous, radial costae and some nonappressed growth squamae. RV flat, operculiform, fitting into LV, elliptic in those with large attachment area, grading to oblique-spatulate, pointed at umbo, smooth to ribbed, though less prominently than LV. Resilifer about twice length of each bourrelet; ligamental area mostly less high than long (2-3 mm.). Adductor muscle imprint orbicular to only faintly convex at its dorsal margin. Commissural shelf well developed, extending all around valves in some individuals, lacking chomata. *M. Jur.*, Eu.-N.Am.(La. subsurface).——Fig. J75,*1-4*. **C. knorri* (Voltz), Bathon., Schönmatt near Basel, Switz.; *1a-d,* specimen with both valves, various views; *2a-d,* LV, various views; *3,* RV, ext.; *4a-d,* LV, various views, all ×3 (specimens courtesy Naturhistorisches Museum, Basel. Stenzel, n).

[Nomenclature has been discussed by STENZEL (1947, p. 171) and SYLVESTER-BRADLEY (1952). Species lacking ribs have been included in *Catinula* by some authors, but presumably they belong to a subgenus of *Catinula* as yet unnamed. *Catinula* probably evolved through neoteny from a costellate *Gryphaea*. SYLVESTER-BRADLEY (1958) gave a statistical description of the type species.]

Deltoideum ROLLIER, 1917, p. 566 [**Ostrea sowerbyana* BRONN, 1836, p. 316 (*nom. subst. pro O.*

deltoidea J. SOWERBY, 1816, p. 111, in SOWERBY & SOWERBY, *non* LAMARCK, 1806, p. 160, =*O. delta* SMITH, 1817, p. 18); SD ARKELL, 1932, p. 149 (footnote)]. Medium-sized to large (to 17 cm. long and 21 cm. high), shell very flat, bilaterally much compressed (to 5.5 cm. wide), outline spatulate to triangular, pyriform, or crescentic; many individuals with narrow, prominent branchitellum pointed toward rear in less extreme spatulate forms, but turning up toward dorsum in extreme more crescentic forms. LV as flat as RV, with low broad smooth irregular concentric undulations and few poorly appressed growth squamae; attachment area small to large, covering much of valve. RV very similar to LV, but in some spe-

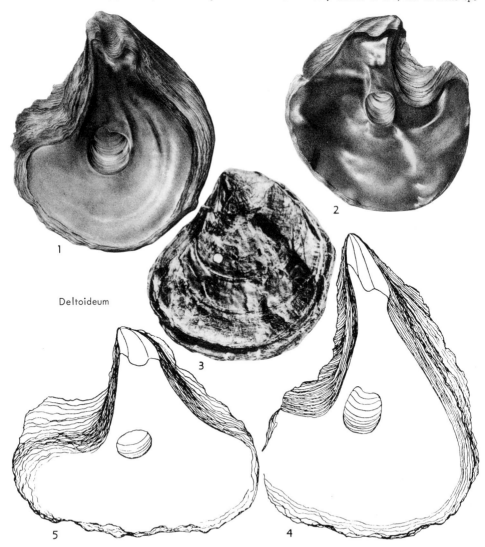

Deltoideum

Fig. J76. Gryphaeidae (Gryphaeinae) (p. N1102-N1103).

cies having short (7 mm.), narrow (0.5 mm.), rounded radial threads separated by flat wide (1.6 mm. or more) interspaces. Ligamental area flat; LV and RV resilifers very shallowly excavate; anterior bourrelet generally flatter and longer (by as much as 50 percent) than posterior one. Adductor-muscle imprint in some individuals with very slightly concave dorsal margin flanked by well-rounded convexities; position of imprint close to posterior valve margin and closer to hinge than to ventral valve margin. No chomata. Ligamental area of older individuals rather high, with subparallel anterior and posterior margins, but in many shrinking in length as age advances, so that area constricts as it grows. Some old individuals no longer grew as to size of shell cavity but merely shifted it in ventral direction leaving behind many growth foliations which flank anterodorsal and posterodorsal valve margins. [ROLLIER's claim that *Deltoideum* was attached by its RV is in error as ARKELL (1935, p. xvi) pointed out. Specimen figured by BAYLE (1878, pl. 132, fig. 1, as *Ostreum subdeltoideum* PELLAT), examined by me at the École des Mines, Paris, has large attachment area on LV and corresponding xenomorphic features on RV. ORIA (1933, p. 71-72) believed that *Deltoideum delta* and *Gryphaea dilatata* J. SOWERBY, 1816, occurred together and had to be regarded as variants of one species. Thus *Deltoideum*, if accepted as a genus, presumably would be a descendant of *Gryphaea*. ARKELL (1932-36, p. xvi-xvii) did not agree with these conclusions, but regarded *Deltoideum* as a synonym of *Liostrea*. *Liostrea*, *Pernostrea*, and *Deltoideum* are probably close relatives.] *M.Jur.(Bathon.)-U.Jur. (Portland.)*, Eu. (Eng.-France-Ger.-Pol.-USSR)-Saudi Arabia.——FIG. J76,1-5. **D. delta* (SMITH) (=*Ostreum subdeltoideum* PELLAT in BAYLE, 1878), U.Jur.(low.Kimmeridg.), LeHavre, France; *1,2,* LV int., RV int., ×0.5 (Bayle, 1878); *3,* RV with LV, ext. (holotype of *Ostrea laeviuscula* J. DE C. SOWERBY, 1825), ×0.5; *4,5,* LV ints., ×0.5 (*3-5,* Arkell, 1932-36).

Liostrea DOUVILLÉ, 1904b, p. 546 [**Ostrea sublamellosa* DUNKER, 1846; OD] [=*Liostrea* DOUVILLÉ, 1904a, v. 3, pt. 4, p. 273 *(nom. imperf.)* (type, *O. lamellosa* DUNKER, *nom. nud.,* err. pro *sublamellosa;* OD)] [=*Lioster* VOLKOVA, 1955, p. 146 *(nom. null.)*]. Small (up to 3 cm. long and 5 cm. high), subequivalve but preponderantly inequivalve; outline irregular but tending to spatulate, margins tending to converge toward umbo; some individuals slightly crescentic. LV variable, mostly capacious and highly convex in anteroposterior direction, much less so dorsoventrally; attachment area fairly large for so small a shell, correspondingly large bulging xenomorphic area on RV umbo. RV variable, gently convex to flat or slightly concave beyond xenomorphic area. Both valves carry low concentric swellings and simple irregularly spaced growth squamae, very

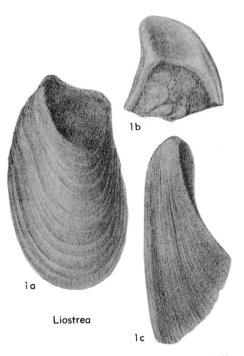

1b

1a

Liostrea

1c

FIG. J77. Gryphaeidae (Gryphaeinae) (p. *N1103*).

poorly or not at all appressed. Radial posterior sulcus of LV absent in some but present in many individuals, well developed, deep, and broad; posterior flange not detached. Sulcus appearing rather early in some individuals, that is, near attachment area. Resilifer about twice as long as each bourrelet. [Type species is highly variable and long-ranging stratigraphically (Rhaet.-Sinemur.). The following names are believed to be synonyms (SCHÄFLE, 1929, p. 16-20): *Ostrea hisingeri* NILSSON, 1832 (p. 354, pl. 4, fig. 2,3); *O. irregularis* MUNSTER in GOLDFUSS, 1833 (v. 2, pt. 4, p. 20, pl. 59, fig. 5); *O. sublamellosa* DUNKER, 1846 (p. 41, pl. 6, fig. 27-30); *O. anomala* TERQUEM, 1855 (p. 329, pl. 25, fig. 4-4b, not 5-5c); *O. liassica* STRICKLAND in TATE & BLAKE, 1876; and possibly *O. bristovi* RICHARDSON (*ex* ETHERIDGE, MS.), 1905 (p. 422, pl. 33, fig. 4). The first name has clear priority. The species has thin shell walls in northern parts of its geographic range and thick ones in the southern parts (France). Orbicular muscle adductor imprint, radial posterior sulcus on LV, lack of chomata, and absence of umbonal cavity place the genus firmly in the Gryphaeinae. It is perhaps a descendant of *Gryphaea* rather than its ancestor, as commonly assumed.] *U.Trias.(Nor.)*, Far E.Sib.; *U.Trias.(Rhaet.)-Jur.*, Eu.(Mesogaean region). ——FIG. J77,1. *L. hisingeri* (NILSSON, 1832) (=*Ostrea anomala* TERQUEM, 1855), Lias., Luxembourg; *1a-c,* ×1.4 (Terquem, 1855). [*See* also Fig. J61.]

Pernostrea MUNIER-CHALMAS, 1864, p. 71 [**Perna bachelieri* D'ORBIGNY, 1850, p. 341 (=**Ostrea luciensis* D'ORBIGNY, 1850, p. 315, no. 341; valid name determined by first reviser, STENZEL, 1947, p. 180, Code Art. 24a); SD FISCHER, 1864, p. 364]. Size medium to large. Outline oval to orbicular to rectangular. Similar to *Liostrea* but larger and has longer cardinal area. *M.Jur.,* W. Eu.(Alps-Pol.-Rumania)-Greenl. —— FIG. J78,*1-3.* **P. luciensis* (D'ORBIGNY); *1,* Bathon., Luc, France; type, ×0.3 (Boule, 1913); *2a,b,* Forest Marble of Pound Pill, Eng.; types of *P. wiltonensis* (LYCETT, 1863), ×0.3 (Lycett, 1863); *3a-d,* Callov., St. Scolasse-sur-Sarthe, France; types of *Perna bachelieri* (D'ORBIGNY, 1850), ×0.3 (Munier-Chalmas, 1864). [*See* also Fig. J15.]

[Because of page precedence *P. luciensis* is the preferred name of the type species. The fact that the generic name is constructed on the basis of a misconception should not mitigate against its use, see Article 18 (a) of Code.

Compare discussion under Multiple Resilifers, p. *N974.* If *Liostrea* and *Pernostrea* were synonymous as claimed by ARKELL (1932-36), *Liostrea* would have to fall as a junior synonym.]

Praeexogyra CHARLES & MAUBEUGE, 1952b, p. 118 [**Ostrea acuminata* J. SOWERBY, 1816, v. 2, p. 82; OD]. Shell tiny to small (largest diameter to about 7 cm.), outline variable, but commonly much higher than long (common proportion 2.5:1 or 3:1), crescentic or ovate to strap-shaped with zigzag edge along their heights in some. LV slightly convex, RV flat to slightly concave; both covered on outside by many straight very fine radial threads, which may be absent, however, and by fairly even-spaced prominent concentric growth squamae separated by wide smoother interspaces. Attachment area at umbonal tip of LV rather small. [ARKELL (1934) described the type species and corrected its involved synonymy. Because of its numerous affinities to *Catinula* CHARLES & MAUBEUGE regarded *Praeexogyra* as a subgenus of *Catinula.* They assumed that *Prae-*

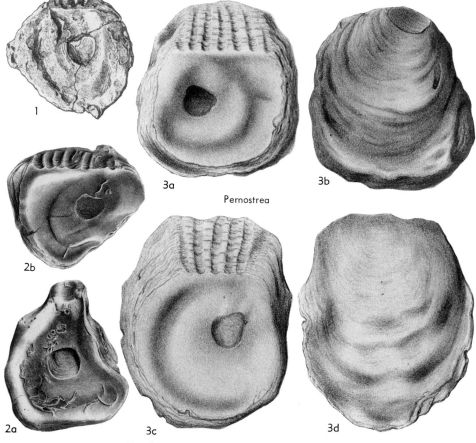

1

3a   3b

Pernostrea

2b

2a   3c   3d

FIG. J78. Gryphaeidae (Gryphacinae) (p. *N1104*).

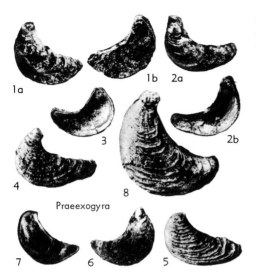

1a   1b   2a

3   2b

4   8

Praeexogyra

7   6   5

FIG. J79. Gryphaeidae (Gryphaeinae)
(p. N1104-N1105).

exogyra formed the transition between *Catinula* and true *Exogyras* of the Upper Jurassic. They pointed out that *Praeexogyra* has a resilifer that is generally arched opisthogyrally, but never twisted.] M. Jur. (Bajoc.-Bathon.), Eu. (Eng.-France-Ger.-Switz.).——FIG. J79,1-8. *P. acuminata* (J. SOWERBY), Bathon.(Lower Fullers Earth), Eng.; *1-5,8*, LV views, *6,7*, RV views (Arkell, 1934). [*1* and *2* are syntypes of J. SOWERBY Collection, British Museum (Nat. History)].

Subfamily PYCNODONTEINAE Stenzel, 1959

[Pycnodonteinae[1] STENZEL, 1959, p. 16]

Commissural shelf well defined, delimited proximally by circumferential curb; chomata short to long, branching and vermiculate; vesicular shell structure present. No prismatic shell layer except in *Neopycnodonte. L.Cret.-Rec.*

Pycnodonte FISCHER DE WALDHEIM, 1835, p. 118-119 [*P. radiata;* OD] [=*Pycnodonta* G. B. SOWERBY, JR., 1842, p. 35 (*nom. van.*); *Pycnodontes* HERRMANNSEN, 1849, p. 373 (*nom. null.*); *Pycnodunta* G. B. SOWERBY, JR., 1852, p. 259 (*nom. null.*); *Pycnondota* COSSMANN & PISSARRO, 1906, pl. 45 (*nom. null.*); *Cretagryphaea* ARKELL, 1934, p. 62 (*nom. nud.,* not available according to *Code* Art. 11d, *nom. subst. pro Pycnodonta* FISCHER); *Pychnodonta* ROMAN, 1940, p. 355 (*nom.*

[1] Pycnodonteinae STENZEL, 1959, is derived from *Pycnodonte.* The form enjoined by the zoological *Code* (Art. 29), Pycnodontinae, was avoided because the names Pycnodontidae and Pycnodontinae already were used in the Pisces. To avoid confusion and homonymy STENZEL (1959) adopted the pattern given as example accompanying *Code* Art. 55a, in this way deriving the name Pycnodonteinae.

null.)]. Small to large (up to 16 cm. long). LV mostly highly convex; attachment area small to quite large; LV umbo incurved, either rises barely above hinge line or well above it, species with low umbo subcircular or semicircular in outline, some have long straight dorsal margins continuing in auricles (as in type species), species with prominent raised umbo vertical-oval or prosocline oblique-oval to horizontal-oval in outline and lack auricles, species with either of 2 last-mentioned outlines have concave geniculate posterodorsal margins. Commissural shelf prominent; chomata either long and straight (as in *Crenostrea*) or short to long (up to 13 mm.) and arborescent and vermiculate. Radial posterior sulcus ranges from absent to broad and shallow or broad and deep. Growth squamae nonappressed to very closely appressed. Radial ribs on LV' may be absent, gently undulatory and short in some species, or small irregular unequal round-topped and well defined (as in *Costeina*). Concentric

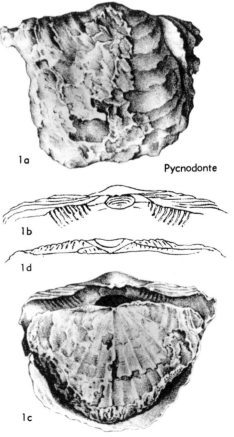

1a   Pycnodonte

1b

1d

1c

FIG. J80. Gryphaeidae (Pycnodonteinae)
(p. N1105, N1107).

1b

1a

Pycnodonte

1c

2b

2a

Pycnodonte

Fig. J81. Gryphaeidae (Pycnodonteinae) (p. N1105, N1107).

puckers or welts parallel to growth lines present or absent on LV. A few Paleogene species have 1 to 3 sharply raised hyote spines on otherwise smooth LV. RV flat to concave, commonly carries sharp radial gashes. [Many authors have assumed that *P. (Pycnodonte) radiata* (U.Cret., Crimea) is the same species as *P. (Phygraea) vesicularis* (LA-MARCK) (1806a, p. 160) from the Chalk (Campan.) at Meudon, near Paris, France. In outline, shape, and sculpture they are so dissimilar that this conclusion seems to be erroneous.] *Cret.-Mio.*, cosmop.

P. (Pycnodonte) [=*Ostrea (Gigantostrea)* SACCO, 1897, p. 14 (type, *G. gigantica* (SOLANDER) in BRANDER, 1766, p. 36; OD); *O. (Gigantostraea)* SACCO, 1897, p. 15 *(nom. null.); O. (Biauris)* COSSMANN, 1922, p. 211 (type, *O. (B.) subhippopodium* (ARCHIAC), 1850, p. 439; OD); *Gryphaea (Gryphaea) sec. Circogryphaea* VYALOV, 1936, p. 19 (type, *G. sinzowi* NECHAEV, 1897, p. 53; OD); *Biaurus, Circographaea* HAAS, 1938, p. 294 *(nom. null.); Circogryhaea* VYALOV, 1948a, p. 36 *(nom. null.)*]. LV umbo rising barely above long straight dorsal margin; auricles present; outline subcircular to semicircular; chomata arborescent and vermiculate; radial ribs absent to low, short, and gently undulatory; concentric puckers and welts absent or present; no well-defined radial riblets. *Cret.-Mio.*, worldwide.——FIG. J80,1. **P. (P.) radiata* FISCHER DE WALDHEIM, 1835, U.Cret., Crimea; *1a*, LV ext.; *1b*, LV umbo and hinge; *1c*, both valves seen from right side; *1d*, RV umbo and hinge; all ×0.5 (Fischer de Waldheim, 1835).——FIG. J81,1,2. *P. (P.) gigantica* (SOLANDER in BRANDER, 1766) (=*"Ostrea (Gigantostrea)"* SACCO, 1897), Eoc.(Barton.); Barton Cliff, Eng.; *1a,b*, LV ext. and int. views; *1c*, RV int.; *2a,b*, both valves, ext.; all ×0.5 [photographs courtesy of †L. R. Cox, British Museum (Nat. History)].——FIG. J82,1,2. *P. (P.) subhippodium* (D'ARCHIAC, 1850) (=*"Ostrea(Biauris)"* COSSMANN, 1922), Eoc.(Barton.), Trabay, France *(1a,b)* and Biarritz and vicinity, SW.France *(2a-h); 1a,b*, RV int., ext., ×1 (d'Archiac, 1850); *2a,b,e*, RV int., ext., edge view; *2c,d*, RV int., ext.; *2f,g*, LV ext., int.; *2h*, RV int.; all ×1 (Cossmann, 1922).——FIG. J82,3. *P. (P.) sinzowi* (NECHAEV, 1897) (=*"Gryphaea (Gryphaea)* section *Circogryphaea"* VYALOV, 1936), Paleoc. below Kamyshin on Volga R., USSR; *3a-c*, LV ext., int., section along mid-axis, ×1 (Nechaev, 1897).

P. (Costeina) VYALOV, 1965, p. 5 [**P. (C.) costei* COQUAND, 1869, p. 108; OD] [*nom. subst.* pro *Avia* VYALOV, 1936, p. 19, *non* NAVAS, 1912]. As *P. (Pycnodonte)* but has many long, narrow, discontinuous, dichotomous, rounded radial riblets on LV only. *Cret.*, USA(Texas)-Eu. (France)-N.Afr.——FIG. J83,2. **P. (C.) costei* (COQUAND), U.Cret.(Campan.), Maadid, Alg.; LV ext., ×0.9 (specimen collected and donated

by C. W. DROOGER, Rijks-Universiteit te Utrecht, Neth. Stenzel, n).

P. (Crenostrea) MARWICK, 1931 [**Ostrea (Crenostrea) wuellerstorfi* ZITTEL, 1864, p. 54; OD]. LV umbo prominent, rising well above hinge line; no auricles; LV outline vertical-oval; posterodorsal valve margin not geniculate; chomata strong, straight, most not branching; radial ribs short, gently undulatory; many variable prominent concentric puckers and welts on LV. [When MARWICK established *Ostrea (Crenostrea)* he gave *O. wuellerstorfi* as type species. The type specimen of that species has arborescent and vermiculate chomata so that it must fall into *Pycnodonte.*] *Oligo.*, N.Z.——FIG. J84,1. **P. (C.) wuellerstorfi* (MARWICK), up.Oligo.(Duntroon.), North Island *(1a)*, Oligo.(Forest Hill Ls.), Southland (Snowdrift quarry) *(1b-d); 1a*, holotype, LV int., ×0.5 (cast of holotype furnished by courtesy of O. PAGET, Naturhistorisches Mus., Wien, Austria); *1b-d*, LV ext., int., post., ×0.6 (specimen by courtesy of C. A. FLEMING and Mrs. A. U. E. SCOTT, N.Z. Geol. Survey) (Stenzel, n).

P. (Phygraea) VYALOV, 1936, p. 19 [**Gryphaea (Gryphaea) frauscheri* VYALOV, 1936 (=*Ostrea (G.) escheri* FRAUSCHER, 1886, p. 53, *non* MAYER-EYMAR, 1876, p. 29, =*G. pseudovesicularis* GÜMBEL, 1861, p. 659); OD] [=*Phrygaea* STENZEL, 1947, p. 180 *(nom. null.)*]. LV umbo prominent, rising well above hinge line, no auricles, LV outline vertical-oval or procline oblique-oval to horizontal-oval, 2 last-mentioned outlines inequilateral, with posterodorsal margin concave and geniculate. Chomata shorter and less elaborate than in *P. (Pycnodonte).* Concentric puckers and welts feeble; radial ribs few or absent. Growth squamae mostly very closely appressed and surface of LV rather smooth. *Cret.-Mio.*, worldwide.——FIG. J83,1. **P. (P.) pseudovesicularis* (GÜMBEL), up.Paleoc., Haunsberg north of Salzburg, Aus.; *1a-g*, LV ext., both valves left side, right side, both valves left side, right side, RV ext., RV int., ×0.9 (Stenzel, n). [All specimens by courtesy of FRANZ TRAUB, München, Ger.; *1b-c* are plaster cast of specimen of pl. 2, fig. 1a,b of TRAUB, 1938.] [=*Gryphaella* CHELTSOVA, 1969, p. 62 (type, *Gryphaea similis* PUSCH, 1837, v. 1, p. 34; OD).]

Hyotissa STENZEL, n. genus, herein [**Mytilus hyotis* LINNÉ, 1758, p. 704, no. 207; OD]. Medium-sized to large (up to 28 cm. long and equally high), valves tending to be subequal and similarly sculptured, LV slightly more convex and capacious than RV, outline variable, mostly suborbicular to vertical oval, more rarely subspatulate and somewhat falcate; attachment area large to very large. Shell commissure plicate in free-growing individuals, commissural plications originating from crude irregularly dichotomous strong radial plicate ribs, tops of which are mostly well rounded and crossed by prominent nonappressed growth

FIG. J82. Gryphaeidae (Pycnodonteinae) (p. *N*1105, *N*1107).

squamae rising here and there into prominent hyote spines. Chomata long, vermiculate, and arborescent, in many places breaking up into tubercles. Distortions of shell shape and sculpture caused by large attachment area common and extensive. [The oldest known species is *H. semiplana* (J. DE SOWERBY, 1825) (v. 5, p. 144, pl. 489, fig. 3), U.Cret., W.Eu., common at Ciply, Belg. The genus is strictly euhaline, stenohaline,

and stenothermal, a member of the compound-coral biocoenosis. Only about 4 geographically separated, closely related species live today. The type species has been placed erroneously in *Lopha* by some authors; full credit for recognizing its true affinities must go to RANSON (1939-41, 1941).] *U.Cret.(Turon.)-Rec.,* worldwide.—— FIG. J85,*1.* **H. hyotis* (LINNÉ), living, Nosi Bé, Madag.; *1a-c,* ext., int., and edge views of young individual with many tubular hyote spines, ×0.5 (Stenzel, n).——Fig. J85,2. *H. hyotis forma sinensis* (GMELIN, 1791), living, Nosi Bé, Madag.; *2a-d,* views of older specimen without tubular spines, with insertion of adductor muscle outlined by pencil in *2c,* ×0.5 (Stenzel, n). [All specimens obtained by courtesy of R. TUCKER ABBOTT,

Acad. Nat. Sci. Philadelphia.] [*See* also Fig. J27.] **Neopycnodonte** STENZEL, **n. genus,** herein [*Ostrea cochlear POLI, 1795, v. 2, p. 179 (=*Peloris gracilis+Peloriderma cochlear* POLI, 1795, v. 2, p. 255); OD]. Medium-sized (to 9 cm. high), shell walls fragile, very thin, partly translucent, outline variable, many auriculate. LV deep, capacious;

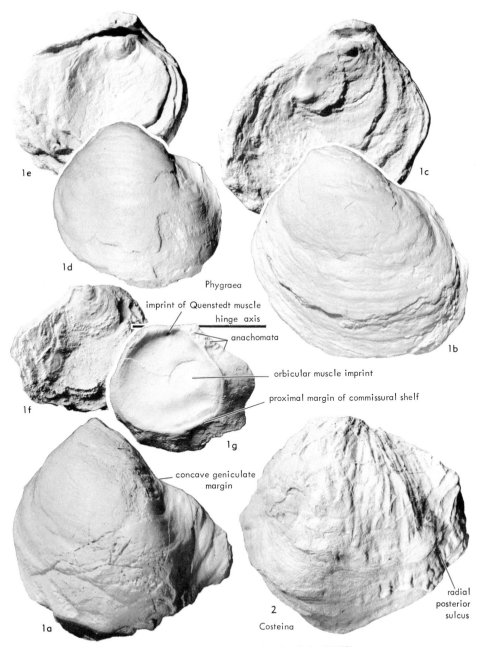

FIG. J83. Gryphaeidae (Pycnodonteinae) (p. *N1107*).

attachment area small to fairly large, situated commonly on region behind umbo; posterior half of LV rising vertically from substratum attachment so that hinge axis is at 45° to level of attachment; LV mostly smooth, devoid of imbrications, but older individuals have paper-thin foliaceous imbrications near valve margin, and these diverge at 30° to 45° from valve contour, imbrications of some individuals drawn out into long scalelike hyote and spoon-shaped extensions. Auricles on either side of hinge common, foliaceous, imbricate, extensive (to 2 cm.), irregular in outline. LV with 7 to 10 gentle, rounded, irregular radial plications of unequal length and cross section; fairly deep, well-rounded radial posterior groove sets off small posterior flange that

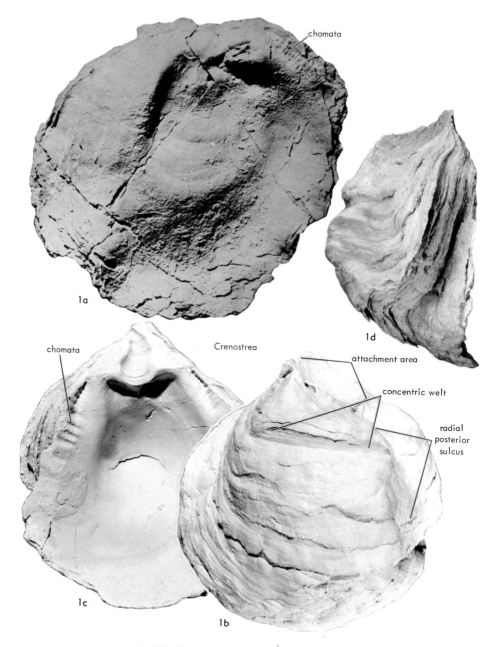

Fig. J84. Gryphaeidae (Pycnodonteinae) (p. *N*1107).

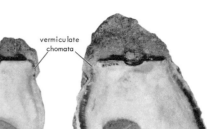

FIG. J85. Gryphaeidae (Pycnodonteinae) (p. *N1107-N1109*).

carries 2 plications, but is absent in some individuals and commonly rather difficult to discern. Commissural shelf of LV only in dorsal half of valve, disappearing completely on ventral half. Adductor muscle imprint large, its diameter about 0.25 of valve height. Chomata on LV up to 2 mm. long, smooth, very subdued, hardly detectable in some individuals, restricted to within 1 cm. distance from hinge. RV flat to concave, with scaly imbrications along margins parallel with contour of valve, showing fibrous prismatic structure where broken. Few sharp radial gashes as in *Pycnodonte*. Commissural shelf of RV nearly complete around periphery. [The type species, which is the only living species, has circumglobal distribution. It is strictly stenohaline and euhaline

in oceanic waters at 12° to 14°C. and depths of 27 to 1,500 m. The genus descended from *Pycnodonte* and Miocene species are transitional between the two genera. Anatomy of the soft parts has been described by PELSENEER (1896, 1911), and HIRASE (1930, p. 37-41).] *Mio.-Rec.*, worldwide.——FIG. J86,*1,2*. **N. cochlear* (POLI), (=*"Ostrea hiranoi* SPICER & BAKER, 1930"), dredged from 50 fathoms (91.4 m.) in Kagoshima Bay, Japan, whitened for photography except *1a; 1a-c,* LV int., int., ext.; *1d,e,* same specimen RV ext., int.; *2a-c,* another specimen, LV int., ext., both valves, all ×0.75 (Stenzel, n). [Specimens from Bernice P. Bishop Museum, Hawaii, USA.]——FIG. J86,*3*. **N. cochlear* (POLI) (=*"Pycnodonta floribunda"* MONTEROSATO,

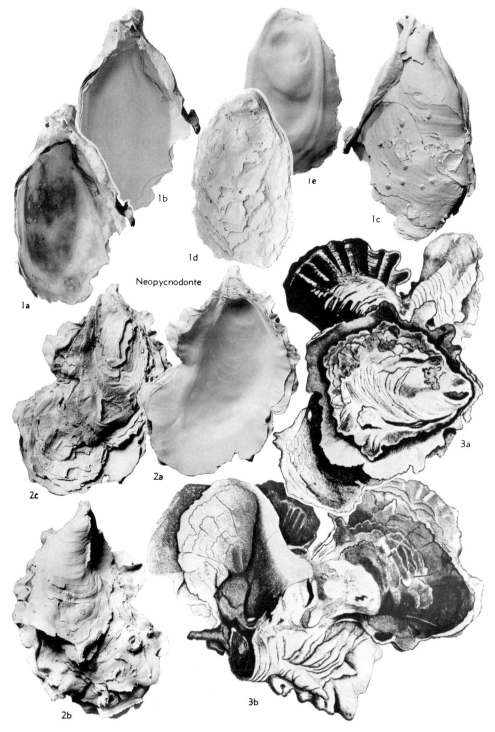

Neopycnodonte

FIG. J86. Gryphaeidae (Pycnodonteinae) (p. *N*1109-*N*1111, *N*1113).

1916, holotype) from 50-60 m. depth near Palermo, Sicily; *3a,b,* cluster, both sides, ×0.75 (Monterosato, 1916).

Texigryphaea STENZEL, 1959, p. 22 [*Gryphaea roemeri* MARCOU, 1862, p. 95 footnote (=*G. mucronata* HILL & VAUGHAN, 1898, p. 63, *non* GABB, 1869, p. 274; =*G. graysonana* STANTON, 1947, p. 28); OD] [=*Ostrea (Marcoui)* AR-

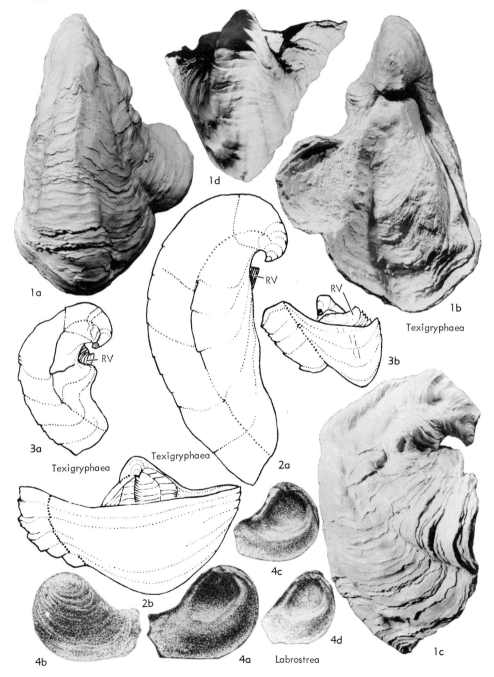

FIG. J87. Gryphaeidae (Pycnodonteinae) (p. *N1113-N1114*).

KELL, 1934, p. 64, *nom. nud.; O. (Corrugata)* ARKELL & MOY-THOMAS, 1940, p. 404, *nom. nud.*]. Small to medium-sized (to 11 cm. high), LV ranging from high and narrow or vertical-oval, bilobate-oval, or bilobate-triangular to auriculate, orthocline to considerably opisthocline, highly to broadly convex in horizontal cross section, not compressed to much compressed in obliquely anteroposterior direction, mostly provided with median radial keel; radial posterior sulcus deeply sunken and posterior flange well set off from main body of valve; beak small, pointed, mostly opisthogyral, highly inrolled and with tiny attachment area. LV generally smooth, but with some poorly appressed growth squamae and few short rounded radial ribs near keel in some species. Umbonal cavity of LV filled with solid shell deposits; no chambers. RV concave, truncated at hinge, without ribs but having some narrow radial gashes; valve margins reflexed so that it comes to lie countersunk in LV when shell is closed. Chomata straight, rarely branching, rarely vermiculate. Differs from *Gryphaea* in its vesicular shell structure, well-defined commissural shelf, chomata, deep radial posterior sulcus, and well-detached posterior flange. [When STENZEL (1959) proposed *Texigryphaea* as a new subgenus of *Gryphaea*, he failed to notice the vesicular shell structure, because the vesicles are filled in with secondary calcite. RANSON (1941, p. 64) had already recognized vesicular shell structure in one species of this genus, which he called "*Pycnodonta Tucumcarii* MARCOU." Because of this structure the taxon cannot be placed in the Gryphaeinae. Late Jurassic species of *G. (Bilobissa)* are transitional between *Gryphaea* and *Texigryphaea* so that it seems likely that the latter is a descendant of *G. (Bilobissa)*. *Texigryphaea* was stenohaline and strictly euhaline and lived on bottoms composed of clay or lime mud. It is a provincial genus restricted to southwestern North America from Kansas southward. It formed extensive shell banks composed of countless individuals.] *L.Cret.(Alb.)-U.Cret.(Cenoman.),* Mexico-SW.USA.——FIG. J87,1. **T. roemeri* (MARCOU), Cenoman.(Grayson Marl), Texas (near Spring Valley, McLennan Co.); *1a-d,* L side, R side, post., umbonal views, ×1 (Stenzel, n).——FIG. J87,2. *T. tucumcarii* (MARCOU, 1851), M.Alb.(Kiamichi F.), Texas (near Tahoka, Lynn Co.); *2a,b,* LV post. and ventral views, ×0.7 (Stenzel, 1959). ——FIG. J87,3. *T. mucronata* (GABB, 1869), mid. Alb. (Walnut F.), Texas (Tennessee Valley, Bell Co.); *3a,b,* post. and ventral views, ×0.7 (Stenzel, 1959). [*See also* Fig. J9.]

Doubtful Genus: Labrostrea VYALOV, 1945, p. 200 [**L. labrum;* OD]. *Labrostrea* was introduced by VYALOV in 1936 and 1937 as a new section of *Liostrea s.s.* with a brief definition, but the type species had not yet been described in 1936. It was

FIG. J88. *Exogyra (Exogyra) costata* SAY (1820), LV from Coon Creek Tongue of Ripley Formation (Maastricht.) of Coon Creek, McNairy County, Tenn., USA, showing perfectly preserved orbicular adductor muscle pad composed of aragonite, ×0.9 (Stenzel, n). [Pad is outlined with pencil. Specimen courtesy of Geology Dept., Louisiana State Univ.]

reintroduced by VYALOV (1945, p. 200-201), at which time the type species was described for the first time. Thus one might argue that *Labrostrea* remained a *nom. nud.* until 1945. STENZEL (1947, p. 176) argued that the definition of *Labrostrea* should be regarded as sufficient to cover its sole species and should be accepted as the description of the type species too. VYALOV (1948a, p. 26, 35) emphasized the sharply defined circumferential commissural shelf and the well set-off deep central shell cavity in the LV of this section and based his definition of it on these features. Such features are characteristic of the Pycnodonteinae and it is probable that *Labrostrea* will ultimately find its place among them. However, it is not known whether the type species has vesicular shell structure; therefore *Labrostrea* must remain of dubious taxonomic position. *Paleog., USSR.*——FIG. J87,4. **L. labrum,* Tuar-Kyr, Transcaspian Reg.; *4a-d,* LV int., LV ext., LV int., LV int., ×1 (Vyalov, 1945).

Subfamily EXOGYRINAE Vyalov, 1936

[Exogyrinae VYALOV, 1936, p. 20]

Larva attached to substratum by its left anteroventral valve margin; postlarval growth either regularly spiral throughout life or regularly spiral only for early part and then straight, as in *Ostrea*. Attachment area medium-sized to large, except in few species of *Exogyra s.s.*, some species of *Rhynchostreon,* and in *Ilymatogyra*, all of which lack attachment areas. Ligamental area very short but long in anatomical height; regularly spiralled either throughout life or for early part at least; posterior bourrelet reduced in length so that it forms narrow ridge, in many shells quite sharp, along side of resilifer groove; anterior boundary of resilifer indistinct, because sloping anterior bourrelet grades imperceptibly into resilifer. Adductor muscle imprint orbicular, or nearly so (Fig. J88). *M. Jur.(Bajoc.)-Mio.*

The Exogyrinae are classed with the Gryphaeidae, because of the outline of the adductor muscle imprint and its position close to the hinge. The first to give a good description of the peculiar features of the ligamental area in the oysters now called Exogyrinae was GOLDFUSS and first to trace its evolution was DOUVILLÉ (1886, p. 230-232).

Tribe EXOGYRINI Vyalov, 1936

[*nom. transl.* STENZEL, herein (*ex* Exogyrinae VYALOV, 1936) ="Gruppa I." MIRKAMALOV, 1963, p. 152]

Postlarval growth and ligamental area regularly spiral throughout. *M.Jur.(Bajoc.)- U.Cret.(Maastricht.).*

The Tribe Exogyrini became extinct with the end of the U. Cret. (Maastricht.). However, an Eocene species was described from Fergana, Central Asia, USSR, namely *Exogyra ferganensis* ROMANOVSKIY, 1879. This species has recently been restudied and assigned to the exogyrine genus *Amphidonta* [*recte Amphidonte*] by GEKKER, OSIPOVA, & BELSKAYA (1962, v. 2, p. 139). Nevertheless, this Central Asian species does not have a posterior bourrelet that is reduced in length to form a narrow ridge as in the true Exogyrini, and its adductor muscle insertion is clearly reniform as in the Ostreidae so that there is no doubt that it is an ostreine and not an exogyrine oyster. It is an *Exogyra*

homeomorph and is believed to be a descendant or member species of the ostreine genus *Ferganea*. It remains an incontroverted observation that all the tribe Exogyrini became extinct with the end of the U. Cretaceous (Maastricht.).

Exogyra SAY, 1820, p. 43 [**E. costata;* M] [=*Exegyra* BENETT, 1831a, p. 122 (*nom. null.*); *Exogira* MATHERON, 1843a, p. 262 (*nom. van.*); *Exagyra* FISCHER DE WALDHEIM, 1848, p. 464 (*nom. null.*); *Exogera* DESMAREST in CHENU, 1859, v. 4, p. 33 (*nom. null.*); *Exoyra* SEGUENZA, 1882, p. 180 (*nom. null.*) (obj.); *Costagyra* VYALOV, 1936, p. 20 (type, *E. olisiponensis* SHARPE, 1850, p. 185; OD); *Fluctogyra* VYALOV, 1936, p. 20 (type, *Ostrea trigeri* COQUAND, 1869, p. 119; OD); *Nutogyra* VYALOV, 1936, p. 20 (type, *O. fourneti* COQUAND, 1862, p. 229; OD)]. Small to large, mostly medium-sized (largest diameter of largest known specimen 21 cm.). Shell very inequivalve, RV flat to concave, LV tumid, convex, much larger than RV. Attachment area variable, large in most species, very small in few; spirally curved umbonal half of LV convex and rather tumid, except in species with small attachment area, spiral keel obtuse and rounded, never prominent, evanescent in old age, hardly noticeable in most individuals. LV outline orbicular to oval, margins in most species convex throughout except for dorsalmost part of posterior margin which may be rectilinear or concave; very few species somewhat falcate in outline, with much of posterior part of the LV margin gently concave (see *Fluctogyra*). Anterior part of RV exterior with many crowded concentric upturned growth squamae parallel to anterior valve margin; posterior part of RV with flat-lying foliaceous growth squamae. LV with wide commissural shelf delimited by rounded curb in species that lack chomata or by narrower rounded curb and adjoining shallow gutter in species that have chomata. Anterior set of LV chomata consisting of 3 mm. long straight parallel radial ridgelets separated by grooves, all closely spaced; posterior set consisting of 7 mm. long, transverse, vermiculate ridgelets covering entire width of commissural shelf near hinge, this part of shelf sigmoidally curved so that it forms inward and upward projecting platform covered with vermiculate chomata in some specimens (Fig. J89, J90). RV margin reflexed to form commissural shelf which allows valve to become countersunk into LV, anterior margin reflexed at about 70°. Quenstedt muscle imprint, visible only on RV, narrow, elongate, and about 1 mm. removed from posterior end of resilifer. Sculpture of LV, and to lesser extent of RV, consisting of foliaceous growth squamae alone or in combination with various patterns of radial costae which include hyote spines or rows of successive transverse bulges on crests of radial costae. *Cret.,* N.Am.(Gulf Mexico area-Atl. Coastal Plain-Utah)-S. Eu.-N. Afr.-Angola-Nigeria-W. Asia-India.

[*Fluctogyra* VYALOV, 1936, proposed as a section of *Exogyra*, supposedly is characterized by vague radial undulations. Such costules are present also on *E. (E.) erraticostata* STEPHENSON (1914, p. 49, pl. 15, fig. 4, pl. 16, fig. 1-2), which cannot be separated from *Exogyra s.s.* However, the type species of *Fluctogyra* differs from *Exogyra s.s.* in somewhat elongate falcate outline of LV margin, which seems an unimportant distinction.——*Costagyra* VYALOV, 1936, also proposed as a section of *Exogyra*, carries chomata and has a few projecting radial ribs on RV (REESIDE, 1929) (Fig. J91). Although it is very close to *Exogyra s.s.*, it may be a valid minor subdivision.——Some apparently isolated species of *Exogyra* have LV with a very small attachment area, a not so tumid, rather slender umbonal tip end, and a greater number of spiral volutions. It is puzzling that some individuals have no chomata, whereas others have well-developed ones.]

E. (Exogyra). Sculpture of LV and to a lesser extent that of RV consisting of 1) unfrilled rough foliaceous growth squamae, 2) weak discontinuous irregular radial ribs that are rounded and separated by interspaces as wide as or wider than these ribs, 3) strong continuous dichotomous radial ribs that are round-topped and separated by interspaces as wide as or much narrower than the ribs, 4) radial series of transverse rounded bumps, or 5) hyote spines developed from crests of·radial ribs. *Cret.,* N.Am.-S.Eu.-N.Afr.-Angola-Nigeria-W.Asia-India.——FIG. J89,*1.* **E. (E.) costata* (SAY), U.Cret.(Maastricht., Corsicana Marl, Navarro Gr.), Texas (San Geronimo Creek, Medina Co.); *1a,b,* shell ext. from LV and

1a   1b

Exogyra

2a

2b

FIG. J89. Gryphaeidae (Exogyrinae) (p. *N1115-N1117*).

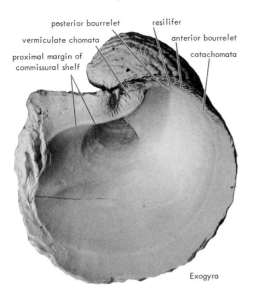

posterior bourrelet   resilifer

vermiculate chomata   anterior bourrelet

proximal margin of   catachomata
commissural shelf

Exogyra

FIG. J90. Vermiculate chomata on projecting platform at posterodorsal terminus of commissural shelf in *Exogyra (Exogyra)* *costata* SAY, 1820, from Navesink Marl, U.Cret.(Maastricht.), New Egypt, N.J., USA; ×0.8 (Stenzel, n).

RV sides, ×0.7 (Stephenson, 1941).——FIG. J89,2. *E. (E.) trigeri* (COQUAND, 1869) [="*Exogyra* section *Fluctogyra*" VYALOV, 1936], U.Cret. (Cenoman.), France (LeMans); *2a,b,* LV ext., LV int., ×0.7 (Bayle, 1878). [*See* also Fig. J24-J26, J88, J90.]

    E. (Costagyra) VYALOV, 1936, p. 20 [**E. olisiponensis* SHARPE, 1850, p. 185; OD]. Distinguished from *Exogyra s.s.* by 7 to 12 flat-topped RV radial ribs that project beyond periphery of valve; radial ribs on LV narrow, separated by wide, concave interspaces. Rib pattern highly variable. According to REESIDE (1929) some variants of type species lack radial ribs. Chomata present. *U.Cret.(Cenoman.-Turon.),* S.Am. (Colom.)-Mexico-USA(Utah)-N. Afr.-Angola-S. Eu.——FIG. J91,1. * *E. (C.) olisiponensis* (SHARPE), U.Cret.; *1a,b,* up. Turon., Alcantara near Lisbon, Port., figured types, ×0.4 (Sharpe, 1850); *1c,* Cenoman., Tunisia; ×0.4 (Pervinquière, 1912).

Aetostreon BAYLE, 1878, pl. 139 explanation [**Gryphaea latissima* LAMARCK, 1819, p. 199; SD DOUVILLÉ, 1879] [*=Actostreon* EBERZIN (ed.), 1960, p. 89 (*nom. null.*).] Medium-sized to large (up to 16 cm. high and 13 cm. long), inequivalve; LV deep, RV flat or even concave. LV with spiral, well-defined, although rounded keel, commonly surmounted at intervals by elongate knobs and broad, shallow well-marked groove, which runs subparallel with keel and is about 1-2 cm. from keel on posterior side of

valve; groove separating gently convex posterior flange from main body of LV. Both valves lacking costae, but with many poorly appressed growth squamae; LV also with low concentric swellings parallel to growth lines. Spiral beak of LV not projecting much beyond outline of RV. No chomata. RV somewhat countersunk into LV, because its anterior margin is somewhat reflexed, although not as much as that of *Exogyra s.s.*; RV with a concavity extending from umbo to branchitellum. *L.Cret.(Valangin.-Alb.),* Eu.(Spain-France-Eng.-Switz.)-Afr. (E. Afr.-Madagascar)-Caucasus.——FIG. J92,1. **A. latissimum* (LAMARCK), Apt., France (Wassy, Département Haute-Marne); *1a-c,* LV ext., LV ext., shell ext. from RV side, all ×0.7 (Bayle, 1878). (See also Fig. J63).

[STENZEL's (1947, p. 168) conclusion that *Aetostreon* must date from BAYLE in DOUVILLÉ, 1879, is no longer correct

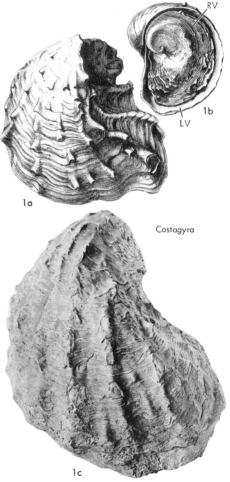

RV

1b

LV

1a

Costagyra

1c

FIG. J91. Gryphaeidae (Exogyrinae) (p. N1117).

in view of changes made in the zoological *Code* (1964). STENZEL (1947, p. 169) also erred in concluding that the correct name of the type species is *A. aquila* (BRONGNIART in CUVIER & BRONGNIART, 1822). *Gryphaea latissima*

LAMARCK (1801, p. 399) was neither figured nor described but validated by reference given to illustrations "N. Bourg. Petrif. pl. 14, no. 84, 85 Esp. foss." These illustrations in BOURGET clearly depict a broad exogyrine shell. PERVIN-

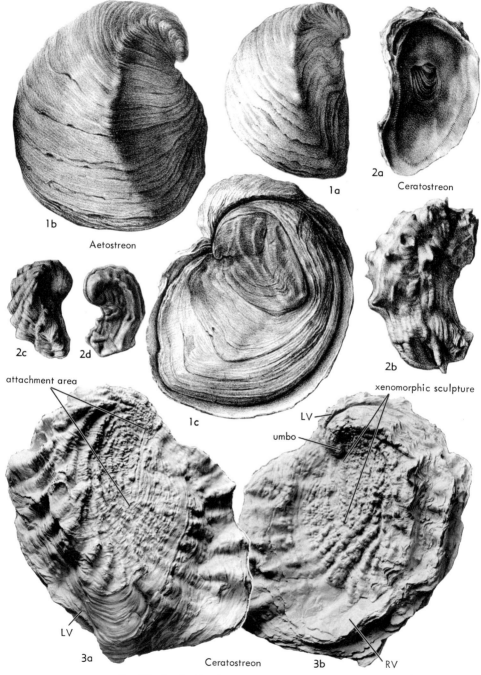

FIG. J92. Gryphaeidae (Exogyrinae) (p. *N1117-N1119*).

QUIÈRE (1910b) figured LAMARCK's two syntypes, which obviously belong to two different species, one broad and large, the other narrow and small. Only the former corresponds to the figures in BOURGET. PERVINQUIÈRE erred in stating that "the reference to the figure in Bourget applies as well to the small [syntype]." Relying on this statement STENZEL (1947, p. 169) was mistaken in concluding that the correct name of the type species is *Aetostreon aquila* (BRONGNIART). Rather, the type species is *Aetostreon latissimum* (LAMARCK, 1801) and some of its synonyms or very closely related species are *Gryphaea couloni* DEFRANCE, 1821 (=*Gryphaea sinuata* J. SOWERBY, 1822; *Griphea aquila* BRONGNIART, 1822).]

Amphidonte FISCHER DE WALDHEIM, 1829, p. 31 [**A. humboldtii* FISCHER DE WALDHEIM, 1829, pl. 1, fig. 1-4; SD FISCHER, 1886, p. 927] [=*Amphidonta* ANONYMOUS, probably BRONN, 1831, p. 335 *(nom. van.)*; *Amphiodonta* AGASSIZ, 1846, p. 18 *(nom. van.)*; *Amphitonde* SCHAUROTH, 1865, p. 166 *(nom. null.)*; *Amphydonta* VYALOV, 1948c, p. 13 *(nom. null.)*]. Similar to *Aetostreon* but with many chomata, commonly along entire periphery of both valves and spiral keel that is not prominent, well rounded, and not built up as ridge above LV surface, its growth squamae less prominent and tending to be smoother. *Cret., USSR; L.Cret.(up.Alb.),* N.Am.(Texas).——FIG. J93,1. **A. humboldtii* (FISCHER DE WALDHEIM), U.Cret.(Cenoman.),USSR; *1a-c*, LV ext., LV int., RV int., ×0.6 (Fischer de Waldheim, 1837); *1d,e*, LV ext., RV int., ×0.6 (Mirkamalov, 1964).

Ceratostreon BAYLE, 1878, pl. 133-134 explanations [*Exogira spinosa MATHERON, 1843, p. 192 (=*Ostrea matheroniana* D'ORBIGNY, 1848, p. 737; *Ceratostreon matheroni* BAYLE, 1878, pl. 134, fig. 1-2, 10-11); SD DOUVILLÉ, 1879] [=*Ceratostrea* HAAS, 1938, p. 294 *(nom. van.)*]. Medium-sized (up to 10 cm.); outline narrow, elongate and crescentically curved or comma-shaped to ovate with umbonal region forming larger end. Shell inequivalve. Attachment area generally large. Both valves keeled, but merely in sense that each has 2 different slopes which meet along spirally disposed crest that is much obscured by surface sculpture that crosses it; anterior slope narrower, descending steeply to anterior valve margin; posterior slope larger, descending more gently to posterior valve margin, flat or concave on RV, gently convex on LV. Many dichotomous, unequal, rounded costae and equally large and rounded interspaces on both valves; costae of type species short, discontinuous, and rising at their ends into prominent spines, particularly at places where costae cross keel; in other species costae are continuous and less spinous, costae in many consisting of series of contiguous transverse puckers. Chomata slender and well developed, 1-5 mm. long, situated along all periphery of the valves. *L. Cret.(Apt.-mid. Alb.),* N. Am.(Mexico-Texas-Okla.-Kans.)-S. Am.(Colom.); *Cret.(Neocom.-Senon.),* Eu.-N. Afr.——FIG. J92,1. **C. spinosum* (MATHERON), U.Cret., France (Royan, Département Charente-Inférieure); *2a,b*, LV int., ext., ×0.7; *2c,d*, shell ext. from LV and RV sides, ×0.7 (Bayle, 1878).——FIG. J92,3. *C. texanum*

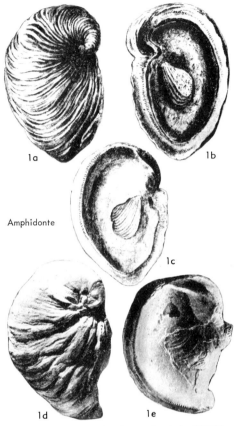

1a   *1b*

Amphidonte

1c

1d   *1e*

FIG. J93. Gryphaeidae (Exogyrinae) (p. N1119).

(ROEMER, 1852), mid.Alb.(Up. Walnut Clay), Texas (Coryell Co. near Mound); *3a,b*, shell ext. from LV and RV sides, ×1 (Stenzel, n). [*See also* Fig. J45.]

[D'ORBIGNY failed to explain why he changed the species name from *spinosa* to *matheroniana;* possibly he was aware that already an *Ostracites spinosus* VON SCHLOTHEIM, 1813, p. 73, and an *Ostrea spinosa* F. A. ROEMER, 1835, p. 58 were available. No proof is found to indicate that these homonyms are congeneric; therefore MATHERON's name is retained in preference to the others. STENZEL's (1947, p. 171) conclusion that *Ceratostreon* must date from DOUVILLÉ, 1879, is no longer correct in view of changes in the zoological *Code* (1964).]

Ilymatogyra STENZEL, **n. genus,** herein [*Exogyra arietina ROEMER, 1852, p. 68; OD]. Small (largest diameter up to 4 cm.). Shell shape resembling elevated corkscrew spiral, highly inequivalve, LV corkscrew-shaped, RV countersunk, slightly convex, operculiform. Tip of LV umbo carrying small smooth semitranslucent pointed cap set off by deep groove, cap made by postlarval calcite infilling of prodissoconch, which, because of its original aragonitic composition, has been leached and has disappeared. No attachment area present.

Umbonal part of LV devoid of ribs for about 3 mm. from tip, covered beyond by about 35 equal wrinkled dichotomous radial costules, which disappear before size of 10 mm. is reached (Fig. J32). Rounded, but well-defined keel evanesces in later growth. Many individuals with concavity on posterior slope of LV, this concavity extending in spiral to posterior valve margin where it origi-

nates from concave part of LV margins situated close to adductor muscle. LV margin of late growth stages has several tongue-shaped protrusions separated by rounded sinuses; these show best on growth lines. RV with orbicular or indented reniform outline and reflexed valve margins, because of which it is countersunk into LV opening, its spiral nucleus carrying calcitic

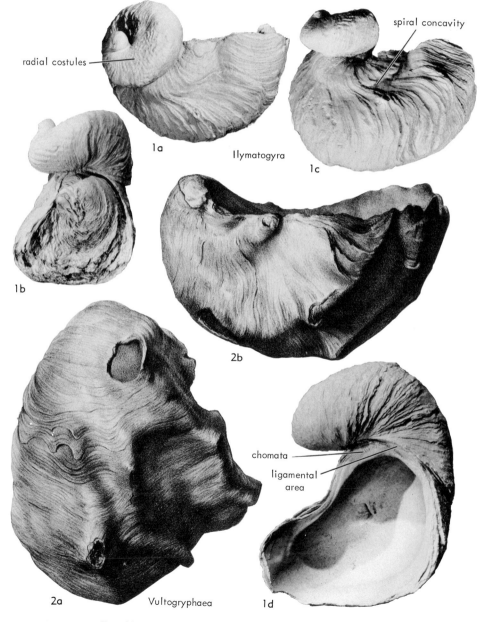

Fig. J94. Gryphaeidae (Exogyrinae) (p. *N1119-N1121, N1124*).

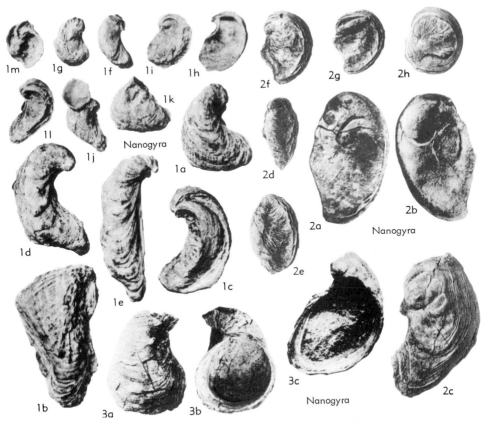

FIG. J95. Gryphaeidae (Exogyrinae) (p. *N1121-N1122*).

umbonal cap, similar to that of LV, and smooth early coil forming boss at indentation of reniform outline; remainder of RV covered by countless concentric upturned crestlike growth squamae. U. *Cret.(Cenoman.)*, NE. Mexico-USA(Texas-Okla.).——FIG. J94,1. *I. arietina* (ROEMER), Grayson Marl, C.Texas; *1a-c*, bivalved specimens, various views, ×2; *1d*, LV, ×2 (Stenzel, n). [*See* also Fig. J32.]

[The genus is monotypic. Its features indicate adaptation to an ooze bottom lacking firm fragments for attachment. The sediment in which it is found is a sticky clay now (Grayson Marl or Del Rio Clay). A detailed description of the type species was given by BÖSE (1919).]

Nanogyra BEURLEN, 1958, p. 206 [**Gryphaea nana* J. SOWERBY, 1822, v. 4, p. 114; OD] [*=Palaeogyra* MIRKAMALOV, 1963, p. 152 (type, *Exogyra virgula* GOLDFUSS, 1833, p. 33, *non Ostrea virgula* DEFRANCE, 1821, p. 26, *nom. nud.*, *=Exogyra striata* (WILLIAM SMITH, 1817) *fide* ARKELL, 1933, p. 440, footnote 9)]. Shell small (greatest diameter of large individuals 3 cm.), greatly variable in shape and outline, inequivalve; LV globular to moderately convex, its outline suborbicular or subtrigonal, elliptical, ovate, to comma-shaped, all specimens spirally twisted, but shape and degree of spirality varying considerably from one to another; LV bilobate in some individuals, because spiral groove divides it into 2 unequal lobes; spiral groove ending in sinus at valve margin located dorsal of branchitellum and producing lingulate outline that is fairly common. RV flat to partly concave or gently convex, its outline varying from suborbicular to ovate, comma-shaped, or auriform to lingulate at branchitellum. LV covered with fine radial ribs or rough concentric growth squamae that have local puckers or constrictions in some places. RV with few major growth squamae except along anterior valve margin, which has many crowded upturned growth squamae. Spiral beak of LV coils tightly over ligamental area so as to obscure it largely. In contrast to *Exogyra*, ligamental area of *Nanogyra* is much more variable as to spirality, 2 bourrelets being about equal in length, and posterior one not narrower than anterior. *Jur. (Bajoc.-Portland.)*, Eu. (Eng.-Scot.-France-Ger.-Pol.-Switz.)-India-Arabia (Yemen)-Afr. (Ethiopia-Somalil.-E.Afr.).——FIG. J95,2,3. **N. nana* (J. SOWERBY), up.Oxford.(U.Calcareous Grit), Eng.

FIG. J96. Gryphaeidae (Exogyrinae) (p. *N1122*).

(Ringstead Bay, Dorsetshire) *(2a-h, 3c),* up. Kimmeridg. [Shotover Grit Sands—Zone of *Pectinatites pectinatus* (PHILLIPS)], Shotover Hill, Oxfordshire *(3a,b); 2a-h,* RV ext., int., RV exteriors, ×1; *3a-c,* LV ext., int., int., ×1 (Arkell, 1932).——FIG. J95,*1. N. striata* (WILLIAM SMITH, 1817) (=*Exogyra virgula* GOLDFUSS, 1833, ="*Paleogyra*" MIRKAMALOV, 1963), low. Kimmeridg., France, various localities; *1a-m,* LV

ext., LV ext., shell right side, LV ext., both valves left side, left side, left side, LV int., both valves right side, LV ext., LV ext., both valves right side, LV int. on cluster, ×1 (Jourdy, 1924).

[ZIEGLER (1969) regards *N. virgula* as a descendant of *N. nana* so that it is very likely that the two are congeneric.]
[Best description of the type species and excellent synonymy were given by ARKELL (1932), p. 175-180, pl. 17, fig. 2-21; pl. 18, fig. 3-11; pl. 19, fig. 4, 4a and text fig. 48). The fact that the posterior bourrelet of the ligamental area is not reduced to a narrow strip is indicative of the primitive status of the ligament, according to BEURLEN, who regarded *Nanogyra* as the link between Jurassic *Liostrea* and Cretaceous *Exogyra*. BEURLEN pointed out (1958, p. 205) the following primitive features of *Nanogyra*: small size; under-developed spirality of the left beak; spiral of beak rarely makes one complete volution; absence of inward and upward projecting platform covered with vermiculate chomata that is an extension of the commissural shelf found in *Exogyra s.s.* It is probable that the spiral groove dividing the LV of *Nanogyra* into two unequal lobes is homologous to the spiral groove of *Gryphaea*. If correctly interpreted, this condition would speak for the derivation of the Exogyrinae from the Gryphaeinae.]

Planospirites LAMARCK, 1801, p. 400 [**Planospirites ostracina* LAMARCK, 1801 (=*Exogyra planospirites* GOLDFUSS, 1863, text v., pt. 2, p. 37, *nom. van.*); M] [=*Planospirigenus* RENIER, 1807, pl. 8, rejected ICZN, Opin. 427]. Shell medium-sized to large (greatest diameter to 15 cm.), outline oval; shell very inequivalve; LV deep and basin-shaped, RV very flat; attachment area rather large, covering entire bottom of LV which has steeply upstanding to vertical anterior wall rising from attachment area and carries few irregular wavy radical costae. RV devoid of any costae, at its anterior margin with narrow spiral band composed of thin crowded subparallel upturned successive growth squamae. LAMARCK had only separate LVs and RVs of the type species, not realizing that they belonged together. He described only the RV under the name *P. ostracina*, therefore, classing it as a univalve. *U.Cret.(Maastricht.),* W.Eu.—— FIG. J96,*1-3. *P. ostracina,* Neth.(St. Pietersberg— south of Maastricht); *1,* LV int. (type of FAUJAS-SAINT-FOND, pl. 28, fig. 5, as *Rastellum,* specimen loaned by courtesy of C. A. VAN REGTEREN ALTENA, Teyler's Museum, Haarlem, Neth., no. 5137), ×0.7; *2a,b,* RV ext., int. (topotype), ×1; *3,* RV ext. (monotype of LAMARCK, Museum Natl. d'Histoire Nat. Paris), ×0.7 *(1,2,* Stenzel, n; *3,* Jourdy, 1924). ·

Rhynchostreon BAYLE, 1878, pl. 138 explanation [**R. chaperi* BAYLE, 1878, pl. 138, fig. 1-5 (=*Gryphaea suborbiculata* LAMARCK, 1801, p. 398; *G. columba* LAMARCK, 1819, v. 6, pt. 1, p. 98; *R. suborbiculatum* (LAMARCK); SD DOUVILLÉ, 1879)] [=*Rhyncostreon* HILL & VAUGHAN, 1898, p. 25, 29, *nom. null.*]. Medium-sized (up to 13 cm. high), highly inequivalve; LV smooth and highly convex, especially so in anteroposterior cross section; LV beak narrow, so greatly elevated above hinge that valve height exceeds length (generally height is 108 percent of length, but may reach 143 percent), highly incoiled (up to 2 volutions), opisthogyral although it is more nearly orthogyral than LV beak of other genera in the Exogyrinae. Surface of LV very smooth, with faint and smooth growth wrinkles and closely and smoothly appressed growth squamae. Rounded keel present on LV only in its earliest growth stage, when about 1 cm. high; keel more noticeable in individuals having many fine radial costules, which are also restricted to earliest growth stage. Broad and shallow radial sulcus starting on LV in mature growth stage, easily overlooked. Attachment area absent or very small, at best 9 by 13 mm. in size. RV suborbicular to horizontal elliptical, longer than high, flat in dorsoventral direction, concave in anteroposterior direction and devoid of costules, with flat spiral umbo and smooth with smoothly appressed growth squamae, no reflexed growth squamae at anterior valve margin. Mature to old individuals have broad gentle terebratuloid fold at ventral margin. Adductor muscle imprint orbicular. Ligamental area very narrow. No chomata. [Although many authors use the species name *columba,* the earliest available name is *suborbiculatum.*] *U.Cret.(Cenoman.-Turon.),* N. Am.(Mexico-Texas-N.Mexico-Colo.-Minn.-Va. subsurface)-Eu. (Eng.-France-Ger.-Czech.-Pol.-Hung.-

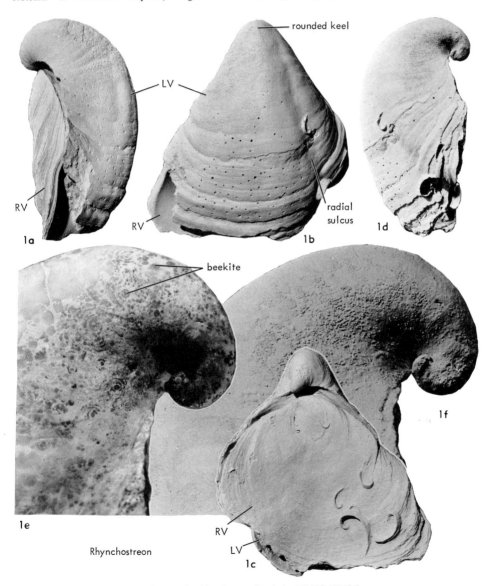

Fig. J97. Gryphaeidae (Exogyrinae) (p. *N1122-N1124*).

USSR-Spain-Port.-Sardinia-Switz.-Aus.) - Asia (India-Syria).——Fig. J97,1. *R. suborbiculatum* (Lamarck, 1801), Turon., France (Mosne, Département Indre-et-Loire); *1a-d,* whitened for photography, various views, ×0.7; *1e,* same specimen, not whitened, post. view of umbonal region showing beekite silicification centers, ×2; *1f,* same specimen, whitened, same view, showing absence of attachment area and roughness caused by local silicification, ×2 (Stenzel, n; specimen from Naturhistorisches Mus., Basel, Switz.).

Vultogryphaea Vyalov, 1936, p. 19 [**Ostrea vultur* Coquand, 1869, p. 118 (=*Rhynchostreon vultur* Bayle, 1878, pl. 141, fig. 1-2); OD]. Shell medium-sized (greatest diameter about 11 cm.), shell very inequivalve. LV beak incurved, narrow, opisthogyral, with small (4 by 7 mm.) attachment area; with rounded well-developed keel, which persists from early growth stage onto about midgrowth when it broadens and becomes one of several spine-bearing radial keels. At about midgrowth 3-6 rounded radial keels arise and become more prominent with age, bearing stout hyote spines at irregular intervals, strongest of these several keels located posteroventrally and ending in tail-like projection at branchitellum. This keel is posterior neighbor of hyote-spine-bearing keel that develops out of the early juvenile keel. RV smooth, deeply concave, carrying few widely spaced shallow radial grooves. *U.Cret.,* Eu.(France).——Fig. J94,2. **V. vultur* (Coquand), low. Chalk, France (Bonneuil-Matours,

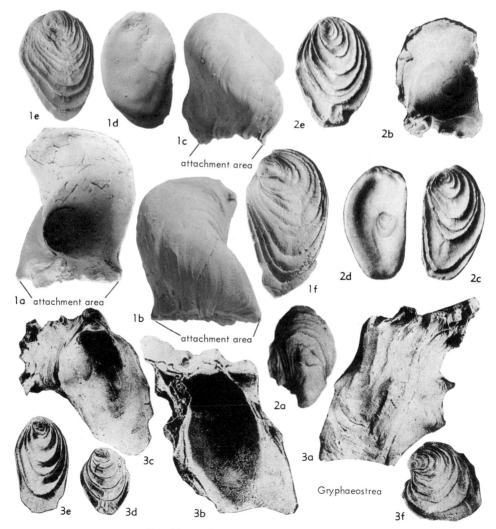

attachment area

1e 1d 1c 2e 2b

1a attachment area 1b attachment area 1f 2d 2c

2a

3c 3a

Gryphaeostrea

3e 3d 3b 3f

Fig. J98. Gryphaeidae (Exogyrinae) (p. N1125).

Département Vienne); *2a,b*, LV ext. views, ×0.7 (Bayle, 1878).

[All specimens seen were filled with hard light-colored limestone so that their interior could not be observed. The features of the ligament area of the LV were seen only obscurely in only one specimen and seemed to be as in *Exogyra*. For this reason and because of the opisthogyral turn of the beak of the LV it is believed that this is an exogyrine oyster genus. VYALOV (1936) had placed it as a new subgenus of *Fatina*, which is not closely related to it. It is a monotypic insufficiently known genus.]

Tribe GRYPHAEOSTREINI Stenzel, new tribe

[*nom. subst.* STENZEL, herein (*pro* "Gruppa II" MIRKAMALOV, 1963, p. 152)]

Postlarval shell growth and ligamental area uniformly spiral only in first part of adult life, changing by abrupt angulation to more or less rectilinear growth in later life. In later part of shell posterior bourrelet is no longer a narrow ridge but is as long and high as the anterior one. *L.Cret.* *(Apt.)-Mio.*

Gryphaeostrea CONRAD, 15 April 1865a, unnumbered errata page following p. 190, *nom. correct.* [**Gryphaea eversa* MELLEVILLE, 1843, p. 87; M] [=*Gryphoeostrea* CONRAD, 25 February 1865b (*nom. nud.*); *Gryphostrea* CONRAD, 1866, p. 3 (*nom. van.*)]. Small (less than 3.5 cm.) although some species reach 3.9 by 6.5 cm. size. LV highly convex and capacious, its commissural shelf well developed but without any chomata; deep umbonal cavity overhung by hinge plate; attachment area large in many species, small in some, restricted in position to posterior or posterodorsal flank of valve and inclined at 45° to 90° roughly to anteroposterior direction. LV beak opisthogyrally spiral; anterior wall of LV spirally curved and rising up obliquely or vertically from substratum; LV smooth, lacking radial ribs or folds except in few last surviving species; smooth growth lines give way to squamae in last stages of growth; angular to spoonshaped shelly claspers grow out periodically from growth squamae of LV to provide additional attachments. RV flat; outline oval to spatulate or triangular, devoid of claspers and chomata; its umbo flat, growth squamae simple, smooth, devoid of frills, folds, ribs, or plicae, and spaced regularly apart starting close to umbo. RV considerably smaller than LV, leaving wide (up to 0.7 cm.) margin on LV uncovered in bivalved fossil specimens. On both valves ligamental area is deep narrow spiral groove which abruptly widens and straightens out near end of individual growth. *L.Cret.* *(up.Apt.)-Mio.(Torton.),* Eu.(Eng.-France-Neth.-Belg.-Ger.-Denm.-Sweden-Switz.-Spain-Italy- Aus.-Bulg.-USSR) - Afr. (Moroc.-Alg.-Tunisia-Libya-Egypt-Congo-Somalil.-Madag.); *L.Cret.(Alb.)-Oligo.(Rupel.),* N.Am.(N.J. to Mexico).——FIG. J98,*1*. **G. eversa* (MELLEVILLE), low.Eoc.

(Thanet.), France (Paris Basin); *1a-f*, LV int., ext., another specimen LV ext., other specimens RV int., ext., RV ext., all ×2 (photographs courtesy of N. J. MORRIS, British Museum, Natural History).——FIG. J98,*2*. *G. sp.* (=*G. vomer* STEPHENSON, 1941, *non* MORTON, 1828), U.Cret. (Maastricht., Corsicana Marl), Texas(Bowie Co.); *2a-e*, LV ext., int., RV ext., int., ext., all ×1 (Stephenson, 1941).——FIG. J98,*3*. *G. plicatella* (MORTON, 1833), up.Eoc. (Jackson), Miss. (Shubuta); *3a-f*, LV ext., int., int., RV ext. (3 specimens), all ×1 (Harris, 1946).

[STENZEL (1947, p. 175) found that the type species was originally designated through monotypy. Many authors have classed the genus among the exogyras (see COSSMANN & PEYROT, 1914, p. 197; JOURDY, 1924, p. 31, 96-97, pl. 4, fig. 2; GLIBERT & VAN DE POEL, 1965, p. 56), whereas STENZEL (1959, p. 31) had argued that it must be excluded from the Exogyrinae. This opinion is no longer maintained. *Gryphaeostrea* is easily recognized as a genus, because of the configuration of the ligamental area and the regularly spaced concentric RV squamae. Only the tribe Flemingostreini has similar concentric RV squamae. *Gryphaeostrea* is the only genus of the Exogyrinae to survive the end of the Cretaceous Period. The last surviving species were *G. ricardi* (COSSMANN & PEYROT, 1914, p. 197, pl. 20, fig. 29-36) from the early Burdigal. of Saucats, Département Gironde, south of Bordeaux, France, and *G. miotaurinensis* (SACCO, 1897, p. 30, pl. 9, fig. 15-33) from the Helvetian of the Colli di Torino, northern Italy, which is reported by AZZAROLI (1958, p. 110, pl. 27, fig. 7-9) as widespread in the Miocene (Burdigal. to Torton.) of Somalia, Cirenaica, and Venetia. In North America the last species, as yet undescribed, are found in the Cooper Marl (Ludian) and Marianna limestone (Rupelian) of South Carolina and Mississippi according to F. STEARNS MACNEIL (personal communication). These last species have radial ribs on the left valve.]

Gyrostrea MIRKAMALOV, 1963, p. 152 [**Exogyra turkestanensis* BOBKOVA, 1949, p. 180; OD]. Small to medium-sized, outline quite variable, commonly oval, length to height ratio about 1:1.5, but with irregularities. LV highly convex, lacking radial keel, but surface rough with growth squamae; some species without radial ribs, few with rough rounded irregular or continuous ribs well separated from each other. RV flat, with many conspicuous growth squamae, which stand up freely on anterior half of valve. Entire shell rather rough looking. Spiral beak tending to become detached from general contour of shell by unrolling. *U.Cret.(Cenoman.-Turon.),* C.Asia (Tadzhik Basin-Gissar Mts. region-Pamir Plateau-Altai Range-Fergana-Afghanistan) - Italy - Tunisia-Palestine-Madag.-N. Am. (Mexico-Texas). [*G. cartledgei* (BÖSE, 1919) is the Mexico-Texas species.]——FIG. J99,*1*. **G. turkestanensis* (BOBKOVA), Turon., USSR(Fergana, C.Asia); *1a-c*, types of BOBKOVA, both valves, left side, right side, post. view; ×1 (Bobkova, 1961; photographs by courtesy of KH. KH. MIRKAMALOV).——FIG. J99,*2-4*. *G. akrabatensis* MIRKAMALOV, 1966, Turon., USSR(Gissar Mts., C.Asia); *2*, LV int., *3a,b*, both valves, left side, right side, *4a,b*, both valves, left side, right side, all ×1 (Stenzel, n).——FIG. J99,*5*. *G. longa* (BOBKOVA, 1961), Turon., USSR(Tadzhik Basin, C.Asia); *5a,b*, LV ext., int., ×1 (Stenzel, n). [Specimens of Fig. J99,2-5 were whitened for photography and were obtained by courtesy of KH. KH. MIRKAMALOV, Tashkent, USSR.]

[The genus was reproposed as new in MIRKAMALOV, 1966, p. 43-44. The type species was redescribed carefully by BOBKOVA, 1961, p. 114-117, pl. 25, figs. 1-4; pl. 26, figs. 1-5) who named it "*Exogyra turkestanensis* (BORNEMAN, 1935, *n. msc.*)." If the unexplained *n. msc.* stands for *nomen manuscriptum*, BORNEMAN's authorship lacks standing in nomenclature.]

SUPPOSED GRYPHAEIDAE HERE REJECTED FROM FAMILY

Acutostrea VYALOV, 1936, p. 18, was proposed as a new section of *Liostrea (Liostrea)*. It is here regarded as a taxon of the Ostreidae-Ostreinae and must, therefore, become independent of

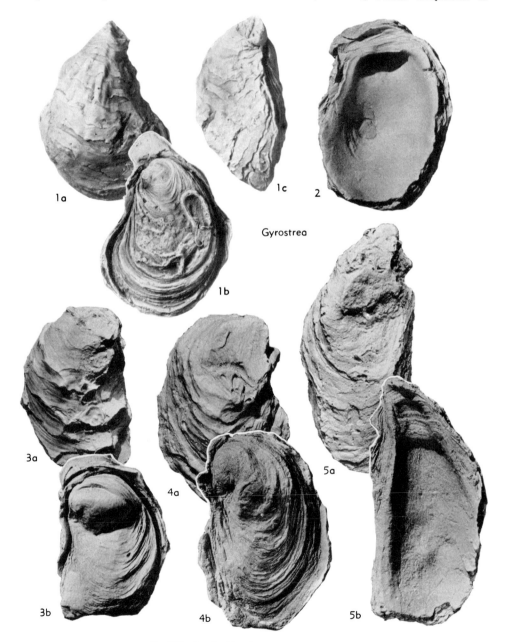

Gyrostrea

FIG. J99. Gryphaeidae (Exogyrinae) (p. *N1125*).

Liostrea, which is retained in the Gryphaeinae. See genus *Acutostrea* on p. N1128.

Anulostrea VYALOV, 1936, p. 19, was proposed as a new subgenus of *Liostrea*. It is not securely placeable in any scheme of classification, because all the known specimens of the type species have their valves closed tightly, and nothing is known of the internal features. There are indications that it may ultimately find its place among the Flemingostreini-Ostreinae-Ostreidae. See p. N1167.

Curvostrea VYALOV, 1936, p. 18, was proposed as a new section of *Liostrea (Liostrea)*. However, it is not placeable in any scheme of classification, because all the figured specimens of the type species have their valves closed, and nothing is known of the internal features. See p. N1168.

Fatina VYALOV, 1936, p. 19, was proposed as a new genus of the Gryphaeinae. However, it is a junior subjective synonym of *Sokolowia* J. BÖHM, 1933, p. 104-105, which takes precedence by reason of priority. The type species of *Fatina* was originally described as a variety of the type species of *Sokolowia* so close are their relationships. *Sokolowia* is a homeomorph of *Gryphaea*, but falls into the Ostreinae-Ostreidae without doubt, see p. N1146.

Ferganea VYALOV, 1936, p. 19, was proposed as a new genus of the Gryphaeinae. It is here regarded as a genus of the Ostreidae (Ostreinae), see p. N1143.

Kokanostrea VYALOV, 1936, p. 19, was proposed as a new subgenus of *Liostrea*. However, outline and position of the adductor muscle imprints show that the two taxa are not closely related and that *Kokanostrea* is one of the Ostreinae. It has been placed, with some hesitation, in the tribe Flemingostreini, see p. N1153.

Mimetostreon BONARELLI & NÁGERA, 1921, p. 21, was introduced as a subgenus of *Gryphaea*, but it must be placed near or in *Maccoyella* ETHERIDGE, 1892, see p. N346.

Odontogryphaea IHERING, 1903, p. 193-227, was proposed as a new subgenus of *Gryphaea*, but must be placed in the tribe Flemingostreini-Ostreinae-Ostreidae, see p. N1153. It is a homeomorph of *Gryphaea*.

Quadrostrea VYALOV, 1936, p. 18, was proposed as a new section of *Liostrea (Liostrea)*, but the holotype specimen of the type species has the two valves closed, and its interior is inaccessible. Its shape is reminiscent of *Flemingostrea*, but its systematic position remains unknown. see p, N1169.

Vultogryphaea VYALOV, 1936, p. 19, was proposed as a new subgenus of *Gryphaea*, but all the specimens seen were filled with hard limestone so that their interior could not be studied. It is here tentatively placed in the Exogyrinae, see p. N1124.

Family OSTREIDAE Rafinesque, 1815

[Official List, *nom. correct.* GRAY, 1833, p. 777 (*pro* fam. Ostreacia RAFINESQUE, 1815, p. 148; see ICZN, 1955, Opin. 356, p. 105)] [=*Les ostracées* LAMARCK, 1809, p. 317 (vernacular); Ostraceen OKEN, 1817, p. 1167 (vernacular); Ostracés CUVIER, 1817, p. 456 (vernacular); fam. Ostreacea SCHWEIGGER, 1820, p. 712; fam. Ostracea DE BLAINVILLE, 1825, p. 519 rejected ICZN Opin. 356; Ostreadea FLEMING, 1828, p. 392 rejected ICZN Opin. 356; fam. Ostraceae MENKE, 1828, p. 57; fam. Ostreae EICHWALD, 1829, p. 287; Ostracidae D'ORBIGNY, 1837, p. 100; fam. Ostreinae AGASSIZ, 1846, p. 266; fam. Ostreana BRONN, 1862, p. 474; Ostreideae EICHWALD, 1871, p. 23] [*non* suborder Ostracea DAUTZENBERG, 1900, p. 222]

Nonincubatory or incubatory. Prodissoconch hinge of planktonic larvae bearing on each valve four subequal tooth precursors and their corresponding sockets split by long smooth median gap into two equal groups. Promyal passage present in nonincubatory and absent in incubatory genera. Intestine passes by dorsum of pericardium and does not pierce heart. Posterior adductor muscle reniform or crescentic in cross section, placed nearly centrally or closer to opposite valve margin than to hinge; its insertion on lower valve not elevated. Valves subequal to highly unequal, with large to small attachment areas, no radial posterior groove. Umbonal cavity very deep to shallow to absent; chambers common. Prismatic shell layer present, may be conspicuous. [Several genera are euhaline, other brackish-water inhabitants; many form true reefs.] *U.Trias.-Rec.*

Subfamily OSTREINAE Rafinesque, 1815

[*nom. transl.* VYALOV, 1936, p. 20 (*ex* Ostreidae RAFINESQUE, 1815)]

Nonincubatory or incubatory. Promyal passage present in nonincubatory and absent in incubatory genera. Chomata present or absent; without pustules on interior surface along the valve margins, which generally are devoid of plications but rarely are irregularly plicate so that very little, if any, conformity is found among individuals. *L.Cret.-Rec.*

NONINCUBATORY GENERA

Living representatives of this informal group of oysters are nonincubatory and have a promyal passage in the exhalant chamber of the mantle cavity (Fig. J11). Their shells commonly have a more or less extensive umbonal cavity under the LV hinge plate. Extinct genera of the group are recognized

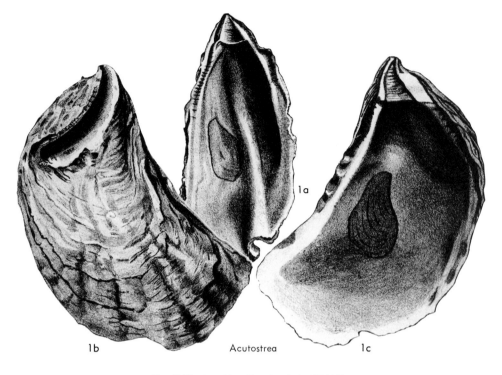

1a

1b   Acutostrea   1c

FIG. J100. Ostreidae (Ostreinae) (p. *N1128*).

by their LV umbonal cavity and similarity to living *Crassostrea*.

Acutostrea VYALOV, 1936, p. 18 [**Ostrea acutirostris* NILSSON, 1827, p. 31 (=*O. incurva* NILSSON, 1827, p. 30; *O. curvirostris* NILSSON, 1827, p. 30; *O. scaniensis* COQUAND, 1869, p. 44; *O. acutirustris* NILSSON in VYALOV, 1936, p. 18, *nom. null.*); OD]. Medium-sized (up to 9 cm. high); outline very variable, mostly elongate-spatulate, straight or more commonly falcate; subequivalve to inequivalve. LV beak commonly pointed, straight or variously curved, projecting beyond that of RV; LV ligamental area acuminate, high triangular; length of resilifer commonly twice that of adjoining bourrelet; deep umbonal cavity of LV overhung by hinge plate. LV commissural shelf with rounded gutter in which pits of chomata are located, both gutter and chomata well developed from hinge to region of adductor muscle, but fading out beyond. LV with many undulatory growth squamae and in some individuals few (about 6) wide low weak radial costae. RV lacking costae, but having concentric growth squamae. Except for chomata and smaller size resembles *Crassostrea*. *U.Cret.,* Eu.-N.Am.
——FIG. J100,*1.* **A. incurva* (NILSSON, 1827), *Actinocamax mamillatus* Zone, Sweden; *1a-c,* LV int., LV ext., LV int., ×1 (Hennig, 1897).

[HENNIG (1897, p. 11-14, pl. 1, fig. 15, 17, 21-23, 25-28) redescribed the type species, pointed out that NILSSON had described it under 3 species names, and selected *O. incurva* as the name to use. This decision of the first reviewer must be accepted. The genus is reminiscent of *Crassostrea* and is believed to be the ancestor of *Crassostrea*. VYALOV proposed *Acutostrea* as a section of *Liostrea* (*Liostrea*). However, *Acutostrea* has radial costae, as HENNIG pointed out, and an adductor muscle imprint that is clearly concave at its dorsal margin (see HENNIG, pl. 1, figs. 22, 28).]

Crassostrea SACCO, 1897, p. 15 [Official List, ICZN Opin. 338] [**Ostrea (C.) virginica* (GMELIN) [1791], p. 3336; OD] [=*Gryphaea* FISCHER, 1886, p. 927 (*non* LAMARCK, 1801) (type, *G. angulata* LAMARCK, 1819, p. 198; OD); *C. (Euostrea)* JAWORSKI, 1913, p. 192 (*obj.*); *Crasostrea* KOCH, 1929, p. 6 (*nom. null.*); *Dioeciostrea* ORTON, 1928, p. 320 (type, *D. americana, nom. subst. pro C. virginica* (GMELIN); OD); *Dioeciostraea* THIELE, 1934, p. 814 (*nom. null.*); *Angustostrea* VYALOV, 1936, p. 18 (type, *O. angusta* DESHAYES, 1824, p. 362; OD); *Grassostrea* VYALOV, 1948a, p. 23 (*nom. null.*); *Somalidacna* AZZAROLI, 1958, p. 115 (type, *S. lamellosa* =*C. gryphoides* (VON SCHLOTHEIM, 1813); M); *Crasotrea* MIYAKE & NODA, 1962, p. 599 (*nom. null.*)]. Small to very large (to 60 cm. high), outline very variable among individuals but very high, slender-spatulate forms with subparallel anterior and posterior margins seeming to preponderate. Surface rough, with many nonappressed, irregularly spaced growth squamae, simple or frilled along free ends; rounded, steep-sided radial ribs on some individuals, more common on LV than RV, such ribs tending to project beyond general outline of margins (Fig. J44); some shells with inconsistent variable radial undulating ribs ending at ventral margins with undulating valve commissure. Chambers common and LV with well-developed umbonal cavity (Fig. J13). No chomata. Adductor muscle imprint close to posterior valve margin and closer to ventral margin than to hinge; its outline with 2 fairly sharp corners, dorsal margin nearly straight. Slender-spatulate forms having LV ligamental area higher than long with subparallel anterior and posterior

boundaries, both flanked by many growth foliations (Fig. J8), such forms possessing strongly convex resilifer and convex ligamental area on RV (Fig. J14). *L.Cret.-Rec.,* worldwide.——FIG. J101,*1,* **C. virginica* (GMELIN), living, Texas coast; *1a-d,* specimens from Port Lavaca, LV ext., LV int., RV ext., RV int.; *1e-h,* specimens from Galveston Bay, LV ext., LV int., RV ext., RV int., all specimens whitened for photography, ×0.3 (Stenzel, n).——FIG. J101,*2. C. angusta* (DESHAYES, 1824) (="*Angustostrea*" VYALOV, 1936), low.Eoc. (Cuis., Sables de Cuise), France (Cuise-Lamothe); *2a,b,* LV int., RV int. (holotype of DESHAYES at École des Mines, Paris), ×0.7 (Stenzel, n).——FIG. J102,*1. C. gryphoides* (VON

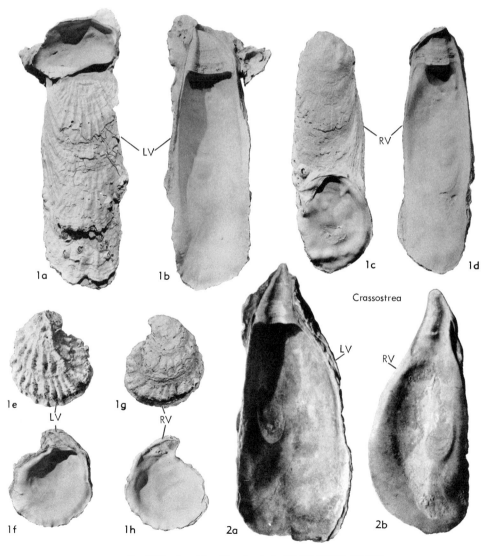

FIG. J101. Ostreidae (Ostreinae) (p. *N1128-N1129, N1131*).

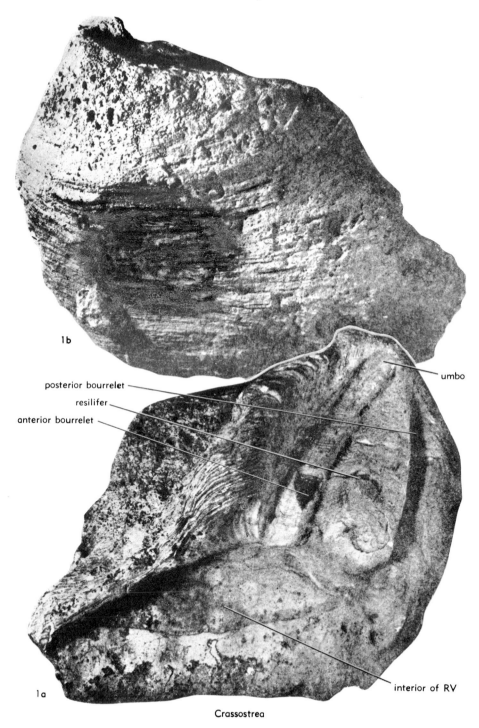

1b

posterior bourrelet

resilifer

anterior bourrelet

umbo

interior of RV

1a

Crassostrea

Fig. J102. Ostreidae (Ostreinae) (p. *N1128-N1129, N1131*).

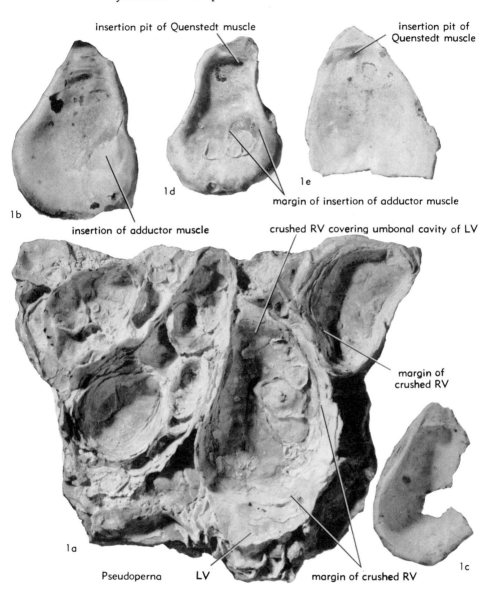

FIG. J103. Ostreidae (Ostreinae) (p. *N1131-N1134*).

SCHLOTHEIM, 1813) (=*"Somalidacna lamellosa"* AZZAROLI, 1958, holotype), Mio. (Serie del Guban), Somalia (hills on left bank of Wadi Merero); *1a,b,* RV hinge view, post. view, ×0.4 (Azzaroli, 1958). [*See* also Fig. J8, J14, J44, and J71.]

[Many authors have classed *Crassostrea* as a subgenus of *Ostrea.* However, species of these two genera will not at all interbreed in nature or in the laboratory, but within each genus the various species can be made to crossfertilize (GALTSOFF & SMITH, 1932; DAVIS, 1950). The northern species of *Crassostrea* are the only known oysters that can survive freezing solid for several weeks in winter. The genus is euryhaline and thrives in very low salinities (17 per mille) as well as in elevated ones (42 per mille) (BREUER, 1962).]

Pseudoperna LOGAN, 1899a, p. 95 (or 1899b, p. 215-216) [*nom. correct.* NEAVE, 1940, v. 3, p. 996 (*pro Pseudo-perna* LOGAN, 1899)] [**P. rugosa* LOGAN, 1899a (=*P. attenuata* + *P. orbicularis* + *P. torta* LOGAN, 1899a, + *P. wilsoni* LOGAN, 1899b=*Ostrea congesta* CONRAD in NICOLLET, 1843, p. 169); SD STENZEL, herein]. Small (to about 4 cm. high and 3 cm. long), outline highly irregular wherever neighboring individuals impinge on each other, otherwise tending to ovate and spatulate, with widest part of oval at about 0.25 to 0.3 of height above ventral margin. Attachment area very flat and large, leaving only about 1 cm. of freely grown edges on LV of larger individuals, these free edges tending to grow up vertically from substrate on anterior and posterior valve margins, but more obliquely at ventral margin of LV. Deep umbonal cavity in LV under thin hinge plate and very shallow resili-

fer groove. Free edges of LV lacking costae, erratically wavy, their growth squamae appressed. RV irregular but mostly free of costae and gently convex because of their smooth xenomorphic configuration. Chomata very small and numerous, slightly elongate at right angles to valve margin, missing in many because of abrasion. *U.Cret.,* N.Am.——Fig. J103,*1.* **P. congesta* (Conrad, 1843), Coniac.-Santon. (Smoky Hill Chalk), Kansas (Logan Co.); *1a,* fragment of flat prismatic calcite shell layer of an *Inoceramus* overgrown by individuals of *P. congesta;* large individuals are

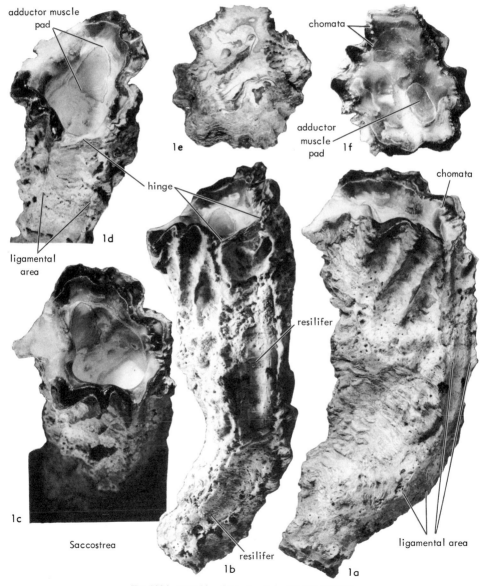

Fig. J104. Ostreidae (Ostreinae) (p. *N1134-N1135*).

bivalved, but their RVs have been crushed down into the hollow of the LV through compaction of the chalk; *1b-e,* four RV ints., ×2 (Stenzel, n; specimens by courtesy of D. F. MERRIAM, State Geol. Survey of Kansas, Lawrence, Kans.).

[It is not known which one of the two articles by LOGAN is the earlier one, both are dated June, 1899. However, all nominal species given by LOGAN under *Pseudo-perna* are junior synonyms of *Ostrea congesta* CONRAD. All specimens of the type species are firmly grown onto very flat fragments of the prismatic calcite layer of *Inoceramus* or are free specimens that have broken off from them. Most

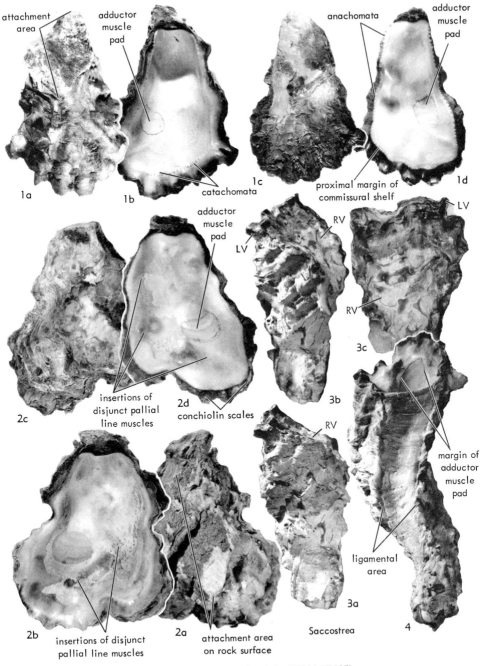

FIG. J105. Ostreidae (Ostreinae) (p. *N1134-N1135*).

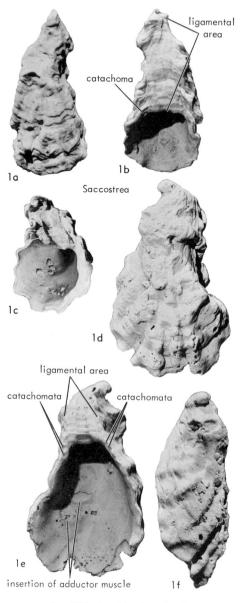

FIG. J106. Ostreidae (Ostreinae)
(p. N1134-N1135).

features of the species are believed to be caused by crowding and adaptation to growth on ooze-covered sea bottoms on which the only solid substrates available were fragments of *Inoceramus* shells. The genus is similar to *Crassostrea*, but never grows large and the shells carry many chomata. It is a monotypic genus possibly descended from *Crassostrea*. Specimens of the type species from the general type area were made available through the courtesy of Dr. D. F. MERRIAM of the State Geological Survey of Kansas.]

Saccostrea DOLLFUS & DAUTZENBERG, 1920, p. 471 [**Ostrea (Saccostrea) saccellus* DUJARDIN, 1835, p. 272 (=*O. cucullata* BORN, 1778, p. 100, =*O.*

cucullata BORN, 1780, p. 114); M] [=*Saxostrea* IREDALE, 1936, p. 269 (type, *S. commercialis* IREDALE & ROUGHLEY, 1933, p. 278; OD); *Sanostrea* MIYAKE & NODA, 1962, p. 599 *(nom. null.)*]. Small to medium-sized, outline variable, but divisible into normal ostreiform ecomorphs with spatulate or vertical-oval to irregular form and abnormal rudistiform ecomorphs with small, operculiform RV and large, slender conical to cornucopia-like LV carrying very high, slender ligamental area. LV has large attachment area, and grows preferentially on bare rock surfaces or mangrove, carries many rough nonappressed growth squamae and 10 to 30 rough irregular rounded dichotomous radial folds extending into small protruding lobes. Umbonal cavity in average very deep; LV umbonal region full of chambers and curved cross walls. RV flat, much corroded through bacterial decay, free of folds but carrying many scaly growth squamae of conchiolin. Both valves with strong chomata, which commonly encircle entire valve. Nonincubatory. A distinct pallial line of separate small muscle insertions connects Quenstedt muscle with posterior adductor muscle imprints. Differs from *Crassostrea* in its deeper umbonal cavity, strong chomata, and tendency to conical rudistiform or cornucopia-like shapes. *Mio.-Rec.,* circumglobal in tropical climates.——FIG. J104,*1;* J105,*1-4;* J106,*1.* **S. cucullata* (BORN, 1778); J104,*1,* rudistiform ecomorph living on rocks at Keppel Bay, Queensland, Australia (="*Ostrea cornucopiaeformis*" SAVILLE-KENT, 1893); *1a,b,* LV ext., two views; *1c,d,* LV, two views into umbonal cavity; *1e,* RV ext., showing corroded surface; *1f,* RV int., all ✕1 (Stenzel, n) [possible types of SAVILLE-KENT loaned by courtesy of D. F. MCMICHAEL, Australian Museum, Sydney]; J105,*1,2,* ostreiform ecomorph, living, Australia (="*Saxostrea commercialis*" IREDALE & ROUGHLEY, 1933); *1a,b,* LV ext., int.; *1c,d,* RV ext., int. (outline of adductor muscle pad and some catachomata are outlined by pencil line); *2a,b,* LV ext., int.; *2c,d,* RV ext., int., all ✕0.7 (Stenzel, n) [specimens donated by D. F. MCMICHAEL]; J105,*3,4,* rudistiform ecomorph living on rocks at Keppel Bay, Queensland, Australia (="*Ostrea cornucopiaeformis*" SAVILLE-KENT, 1893); *3a-c,* post. view, oblique post. view, right side; *4,* LV ext., showing spirally twisted ligamental area, all ✕1 (Stenzel, n) [possible types of SAVILLE-KENT loaned by courtesy of D. F. MCMICHAEL, Australian Museum, Sydney]; J106,*1,* Mio., C.France (Pontlevoy, Département Loir-et-Cher) (="*Ostrea saccellus*" DUJARDIN, 1835); *1a-f,* LV ext., int., umbonal cavity, LV ext., int., ant. views, ✕1 (Stenzel, n) [specimens donated by GEORGES LECOINTRE, La Chapelle Blanche, Département Indre-et-Loire, France; both whitened for photography]. [*See* also Fig. J7.]

[LECOINTRE (personal communication) regards *S. saccellus* (Mio., S. France) as equivalent to *S. cucullata;* not even deserving subspecific distinction. This conclusion is supported here. The spelling *cucullata* is orthographically erroneous but must be retained, because it is older by two years than the emended *cucullata*. The rudistiform ecomorphic growth pattern has evoked comment (FISCHER, 1880-87, fig. 684; KLINGHARDT, 1922, 1929) without developing explanation of its causes. The best account of living ecomorphs of the type species and their ecology has been given by MACNAE & KALK (1958). The ostreiform ecomorphs grow on sea cliffs exposed to wave action. The rudistiform ecomorphs grow in crowded situations not exposed to strong wave action. Some authors (THOMSON, 1954, p. 150) have regarded *Saccostrea* or *Saxostrea* as a junior synonym of *Crassostrea*. This is believed to be erroneous, because the two can be distinguished consistently by shell features and because eggs of *S. cucculiata* have no effect in stimulating ejaculation of sperm from ripe males of *Crassostrea virginica* and species of these two genera cannot be made to crossfertilize each other (GALTSOFF & SMITH, 1932).]

Striostrea VYALOV, 1936, p. 17 [**Ostrea (S.) procellosa,* "VALENCIENNES" in LAMY, 1929, p. 71

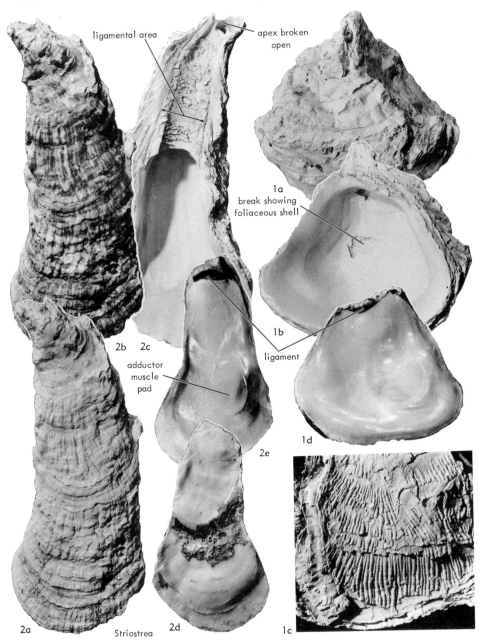

FIG. J107. Ostreidae (Ostreinae) (p. *N1135-N1137*).

(*nom. subst. pro O. multistriata* HANLEY, 1845, p. 106, *non* DESHAYES, 1830, p. 294) (=*O. margaritacea* LAMARCK, 1819, p. 208); OD]. Small to large (to 20 cm. high), with successive growth forms consisting of young ostreiform individuals represented by type specimens of *O. procellosa,* and older, larger, rudistiform individuals corresponding to lectoholotype of *O. margaritacea.* 1) *O. procellosa* form irregular, commonly flattish, generally less than 3.5 cm. wide, up to 11 cm. long and equally high, with rather variable outline approaching suborbicular and rounded-triangular. RV covered by many thin, readily dehiscent, conchiolin-rich imbricating layers that have prismatic shell structure and carry on their tops many narrow (1.3 mm. or less wide) dichotomous flat-topped radial riblets separated by narrower interspaces, riblets converging and diverging irregularly from place to place, producing shaggy appearing surface, becoming less abundant and less prominent in later growth stages. Wherever imbricating layers have peeled off, surface is smooth and carries paper-thin iridescent appressed or nonappressed growth squamae. LV irregular, with large attachment area; where grown free from substrate having crowded delicate paper-thin noniridescent nonappressed somewhat undulant growth squamae that produce very foliaceous surface. Internal face of both valves nacreous and iridescent. Adductor muscle imprint reniform, well rounded, twice as long as high, with dorsal boundary concave and fully rounded at ends; outline similar to that of *Ostrea s.s.* Position of imprint central, as in *Ostrea s.s.,* but in individuals with deep umbonal cavity tending to be more ventral. Chomata variable, absent in some, medium-sized in others, coarse and thick in many. Umbonal cavity mostly shallow, generally clearly defined. Conchiolinous fringe at margin of RV extensive (up to 2 cm. wide). 2) *O. margaritacea* growth form large (up to 20 cm. high), generally slender, rudistiform, highly inequivalve, twisted spirally, sigmoidally, or otherwise. LV tall, slender, twisted conical, with very high ligamental area; RV flattish, irregular, concave or slightly convex from dorsum to venter; its ligamental area mostly lost through corrosion. Exterior of LV, where not destroyed by erosion, crowded with delicate paper-thin noniridescent nonappressed somewhat undulating growth squamae that produce very rough foliaceous surface. Outer face of RV generally devoid of any remnants of conchiolinous prismatic shell layers, but riblets less crowded where preserved and less prominent than in *O. procellosa* form. Otherwise, RV covered with many appressed or nonappressed growth squamae without riblets. Internal faces of both valves nacreous and iridescent. Adductor muscle imprint of LV variable, from reniform and similar to that of *O. procellosa* form to nearly twice as high as long, commonly with concave dorsal margin and small well-rounded horns; its position on LV close to posterior valve

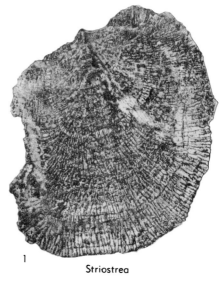

Striostrea

FIG. J108. Ostreidae (Ostreinae) (p. *N1135-N1137*).

margin; on RV farther from ventral margin than on LV. Chomata variable, absent in some individuals, as few as 3 in others, better developed on RV than LV. Umbonal cavity excellent (up to 2 cm. deep), with many paper-thin cross walls that produce chambers. Ligamental area of LV long (to 3.7 cm.) and very high (to 13 cm.), divided into much larger, sunken-in, flat-bottomed resilifer and 2 subequal flat-topped bourrelets. Well-defined groove (2 mm. wide) delimiting anterior and posterior margins of ligamental area. Nonincubatory; promyal passage extensive. Ribletbearing layers usually lost owing to their delicate nature. Differs from *Crassostrea* in its reniform adductor muscle imprints, chomata, nacreous and iridescent interior, very foliaceous shell structure, and rudistiform growth pattern. *Mid.Eoc.(Gosport Sand),* N.Am.(Ala.); *Rec.,* SE.Afr.-Madag.-C.Am. (W. coast-Baja Calif.-Panama).——FIG. J107,*1,2;* J108,*1.* **S. margaritacea* (LAMARCK, 1819), living, S.Afr.(Knysna Lagoon); *procellosa* growth form (=*Ostrea multistriata* HANLEY, 1845 (*non* DESHAYES, 1830) =*O. procellosa* "VALENCIENNES" in LAMY, 1929) (J107,*1,* and J108), rudistiform growth form (=*O. margaritacea* LAMARCK, 1819) (J107,*2*); J107,*1a,b,* LV ext., int., ×1; J107,*1c,* RV ext. showing radial riblets, ×2; J107,*1d,* RV int. showing nacreous luster, ×1; J107,*2a-e,* LV ext., LV int., RV ext., RV int. showing nacreous luster, ×0.6 (Stenzel, n); J108,*1,* very young (5 mo.) specimen showing numerous radial riblets, ×3.3 (Ranson, 1951).

[Until 1949, *O. procellosa* and *O. margaritacea* had been regarded as separate species (LAMY, 1929, p. 71,272). RANSON (1949d, p. 251) first recognized that both names are applicable to a single species which he chose to call *Gryphaea margaritacea.* Later (RANSON, 1959) he described it and figured both growth forms. My study of specimens, including the types of both, deposited in Paris collections (Mus. Natl. d'Hist. Nat.), showed that these can be arranged in a continuous growth series leading from one growth form to the other, thus proving that they belong to a single species. On describing *O. multistriata* HANLEY (1845) incidentally indicated that his material, which came from Africa, consisted of young oysters only. KOR-

RINGA (1956) described and figured both growth forms as *Crassostrea (Gryphaea) margaritacea* without mentioning the name *O. procellosa* and its identity problem. His work has yielded the best description of the biology of the type species. VYALOV (1936) cannot have been aware of the rudistiform terminal growth form when he proposed *Striostrea* as a subgenus of *Ostrea*. Because *Striostrea* is nonincubatory, has an extensive promyal passage, and has quite

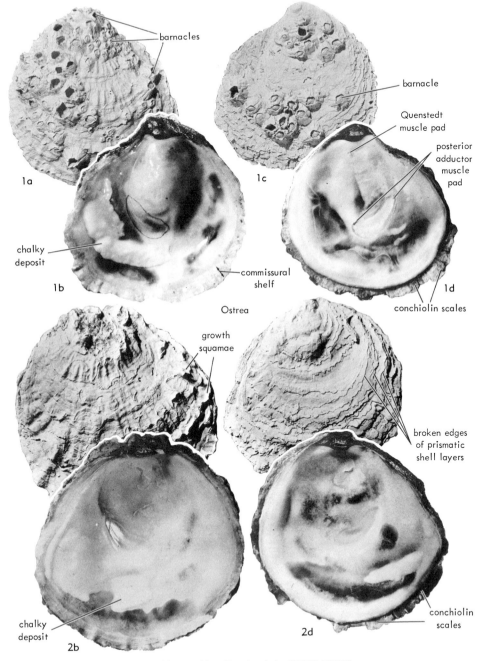

FIG. J109. Ostreidae (Ostreinae) (p. *N1138-N1139*).

different conchological features than the incubatory *Ostrea,* the two genera cannot be close relatives.]

INCUBATORY GENERA

Living representatives of this informal group of oysters are incubatory and lack a promyal passage in the exhalant chamber of their mantle cavity. Commonly, their shells have no umbonal cavity, or only a very shallow one beneath the LV hinge plate. Extinct genera of the group are recognizable by the less capacious LV and absence or shallowness of an umbonal cavity in it. The sculpture pattern is commonly similar to that of *Ostrea,* in which the LV has radial ribs and the RV lacks them.

Ostrea Linné, 1758, p. 696 [Official List, ICZN Opin. 94 and 356] [*O. edulis;* SD ICZN, Opin. 94] [=*Ostracites* Gesner, 1758, p. 39 *(nonbinom.);* *Ostreum* Da Costa, 1776, p. 249 *(nom. van.)* (obj.); *Peloris* Poli, 1791, p. 33 (type, *O. edulis* Linné; SD Gray, 1847, p. 201) (obj.); *Ostracites* Gmelin, 1793, p. 404 (rejected ICZN, Opin. 296); *Peloriderma* Poli, 1795, p. 255 *(nom. subst. pro Peloris* Poli, 1791) (obj.); *Ostracarius* Duméril, 1806, p. 168 *(nom. van.); Ostreigenus* Renier, [1807] (rejected ICZN, Opin. 427); *Ostraea* G. B. Sowerby, Jr., 1839, p. 75 *(nom. van.); Peloridoderma* Agassiz, 1846, p. 277 *(nom. subst. pro Peloriderma* Poli, 1795) (obj.); *Ostreites* Herrmannsen, 1847, p. 177 *(nom. subst. pro Ostracites* Auctt.); *Ostreola* Monterosato, 1884, p. 4 (type, *Ostrea stentina* Payraudeau, 1826 [1827], p. 81; OD); *Cymbulostrea* Sacco, 1897b, p. 12 (type, *Ostrea cymbula* Lamarck, 1806, p. 165; OD); *Ostrea (Eostrea)* Ihering, 1907, p. 42 (type, *Ostrea puelchana* d'Orbigny, 1841, p. 672; SD Iredale, 1939, p. 394); *Ostrea* (section *Anodontostrea*) Suter, 1917, p. 86 (type, *Ostrea angassi* G. B. Sowerby, Jr., 1871, v. 18, pl. 13; SD Finlay, 1928, p. 264); *Ostrea (Euostrea)* Douvillé, 1920, p. 65 *(non* Jaworski, 1913, p. 192) (type, *Ostrea edulis* Linné; OD); *Monoeciostrea* Orton, 1928, p. 320 *(nom. van.)* (obj.); *Ostroea* Tolmer, 1928, p. 91 *(nom. null.); Osrea* Tzankov, 1932, p. 78 *(nom. null.); Monoeciostraea* Thiele, 1934, p. 814 *(nom. van.); Ostrea (Ostrea)* section *Bellostrea* Vyalov, 1936, p. 17 (type, *Ostrea bellovacina* Lamarck, 1806, p. 159; OD)]. Medium-sized to large (to 18 cm. high and 20 cm. long), variable outline, but average shells tending to be roughly orbicular with hardly prominent umbones obtusely pointed and flanked by small to very large auricles or lacking them, posterior auricle, if present, much larger than anterior one. Width about 0.25 of height, resulting in rather flat shell. RV flat to gently convex, covered by many fragile flattish conchiolinous growth squamae, peripheral conchiolin fringe extensive (up to 1.5 cm. wide), so that calcareous part of

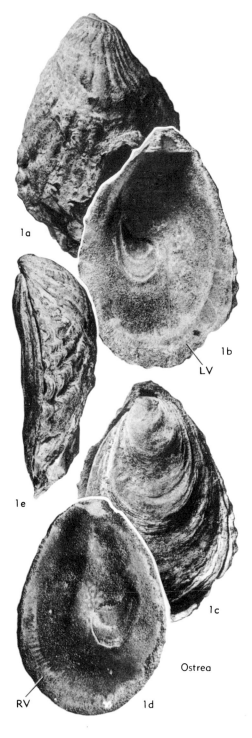

1a

1b

LV

1e

1c

Ostrea

RV   1d

Fig. J110. Ostreidae (Ostreinae)
(p. *N1138-N1139*).

RV is much smaller than that of corresponding LV, as conspicuous as in some fossil bivalved specimens. Concentric undulations absent or present, never conspicuous on RV. LV slightly convex, hardly ever deeply cupped, covered by many long unequal rounded radial ribs interrupted by freestanding frilled delicate growth squamae less abundant than those on RV; some concentric undulations present. Although radial rib patterns differ much from species to species, no hyote spines are developed on them. Ligamental areas commonly longer than high, forming triangles of long base lines. Chomata always present, but differing in prominence from species to species, few (4 or 5) in some, inconspicuous and tuberculiform (see Fig. J113), abundant and conspicuous in others, forming elongate ridgelets (see Fig. J31) arranged all around periphery. Adductor-muscle imprint reniform; both ends well rounded, with length about 4 times height; more centrally located than in other genera. LV mostly without umbonal cavity. Typical species tend to have large round flat shells with flat commissures, whereas others have small vertically elongate shells with twisted or irregularly plicate commissure. [Genus is incubatory. Diagnostic features are outline and position of adductor muscle, chomata, flattish shape of shell, absence of umbonal cavity, and different ornamentation of opposing valves.] *Cret.-Rec.,* worldwide except in polar regions.

[There is some confusion concerning the type species of *Anodontostrea* SUTER, 1917. SUTER was under the impression that the New Zealand oyster's name was *Ostrea angasi* SOWERBY (*recte angasi*), which he listed as the first species under *Anodontostrea*. It now appears (HOLLIS, 1963, p. 2-8) that name rather refers to the closely similar South Australian oyster, whereas the New Zealand oyster is to be called *O. lutraria* HUTTON, 1873.]

O. (Ostrea). No plications along valve margins; chomata few and inconspicuous. *Cret.-Rec.,* cosmop.——FIG. J109,*1,2.* * *O. (O.) edulis* LINNÉ, living, Eng.; *1a-d,* LV ext., int., RV ext., int., ×0.7; *2a-d,* LV ext., int., RV ext., int., ×0.7 (Stenzel, n) [specimens donated by E. J. DENTON, the Laboratory, Citadel Hill, Plymouth, Eng.; *1a,1c,2a,* and *2c* whitened for photography].——FIG. J110,*1. O. (O.) bellovacina* LAMARCK, 1806 (=*"Ostrea (Ostrea)* section *Bellostrea"* VYALOV, 1936), Eoc. (Thanet.; Sables de Bracheux), France (Butte de la Justice at Bracheux near Beauvais); monotype of LAMARCK at Museum de Genève, Switz.; *1a-e,* LV ext., int., RV ext., int., ant. view, ×1 (Clerc & Favre, 1910-18).——FIG. J111,*1. O. (O.) cymbula* LAMARCK, 1806 (=*"Cymbulostrea"* SACCO, 1897), Eoc. (Lutet.), France (Grignon, Département Seine-et-Oise, near Paris); *1a-c,* RV ext., LV int., LV ext., ×1 (Cossmann & Pissarro, 1904-13).——FIG. J112,*1,2. O. (O.) stentina* PAYRAUDEAU, 1826 (1827) (=*"Ostreola"* MONTEROSATO, 1884), living, Egypt (Port Said, Medit. Sea); *1a-b,* LV ext., int., *1c-e,* RV ext., ext., int., ×1; *2a,b,* LV ext., int., *2c-e,* RV ext., ext., int., ×1 (Stenzel, n) [*1a,b,d,e,* and *2a-e* whitened

1a

Ostrea

1b

1c

FIG. J111. Ostreidae (Ostreinae) (p. N1138-N1139).

for photography].——FIG. J113,*1. O. (O.) puelchana* D'ORBIGNY, 1841 (=*"Ostrea (Eostrea)"* IHERING, 1907), living, Brazil (Rio Grande do Sul); *1a-d,* LV ext., int., RV ext., int., ×0.7 (Stenzel, n).——FIG. J113,*2. O. (O.) lutraria* HUTTON, 1873 (=*Ostrea (Ostrea)* "section *Anodontostrea"* SUTER, 1917), living, N.Z.(Foveaux Strait), *2a-d,* LV ext., int., RV ext., int., ×0.7 (Stenzel, n).

O. (Turkostrea) VYALOV, 1936, p. 18 [**O. turkestanensis* ROMANOVSKY, 1878, p. 112 (=*O. strictiplicata* RAULIN & DELBOS, 1855, p. 1158); OD] [=? *Goridzella* HAAS, 1938, p. 294, *err. pro Gorizdrella* VYALOV, 1936, p. 17 (type, *Ostrea gorizdroae* VYALOV, 1937b, p. 16-18; OD); *Gorizdrella* VYALOV, 1936, p. 17 (*nom. nud.*); *Gorizdrella* VYALOV, 1948a, p. 34; *Turcostrea* VYALOV, 1948b, p. 60 (*nom. null.*)]. Differs from *O. (Ostrea)* in having strong chomata and many strong continuous, fairly narrow radial ribs on LV and tendency of ligamental area to turn in opisthogyral spiral fashion. *Low.Eoc.(Ypres.)-Mid. Eoc.(Alaisk.),* Mesogean region (N.Afr.-C.Asia).——FIG. J114,*1-4.* **O. (T.) strictiplicata* RAULIN & DELBOS, 1855 (=*O. turkestanensis* ROMANOVSKIY, 1878), mid.Eoc.(Alaisk.), C.Asia(USSR,Uzbek.); *1a-d,* both valves, left side, right side, ant. view, umbonal view, ×1; *2a,b,* both valves, left side, right side, ×1; *3,* LV ext., ×1; *4,* both valves, right side showing beekite silicification centers, ×1 (Stenzel, n) [all specimens whitened for photography and donated by D. P. NAIDIN, Moscow State Univ., USSR].——FIG. J115,*1. O. (?T.) gorizdroae* VYALOV, 1937 (="Ostrea *(Cymbulostrea)* section *Gorizdrella"* VYALOV, 1936, *nom. nud.),* Paleog., USSR (Fergana); *1a-e,* LV ext., LV ext., RV ext., LV ext., int., ×0.9 (Vyalov, 1937). [*See* also Fig. J30 and J31.]

[Valves are commonly thick-walled. This thickening is attributed to the prevailingly hot climate and calcium-rich environment. ASTRE (1922) gave an account of the variants of the type species in the western Mesogean region. YANG KIEH (1930) reviewed the evolution of *O. (Turkostrea)* into *Sokolowia.* His claim that *O. strictiplicata* is an earlier name for *O. turkestanensis* is accepted here, notwithstanding the fact that Russian authors do not (VYALOV, 1936, etc.; GEKKER, OSIPOVA, & BELSKAYA, 1962). *Gorizdrella* VYALOV, 1936, was proposed as a section of *Ostrea (Cymbulostrea)* and accompanied by a brief definition and the citation of a type species. However, at the time the type species was a *nomen nudum.* It was later described in VYALOV, 1937b. HAAS listed *Gorizdzella,* evidently an error *pro Gorizdrella;* he defined it and gave a type species indirectly by reference to VYALOV, 1936. HAAS, 1938, is thus the first nomenclaturally acceptable introduction of this taxon. To judge by the figures given in VYALOV, 1937b, *O. gorizdroae,* the type species, is based on young and small oyster specimens (Fig. J115,*1*) which cannot be classified without recourse to additional, more full-grown specimens. They might be the young of an *Ostrea (Turkostrea).*]

FIG. J112. Ostreidae (Ostreinae) (p. *N*1138-*N*1139).

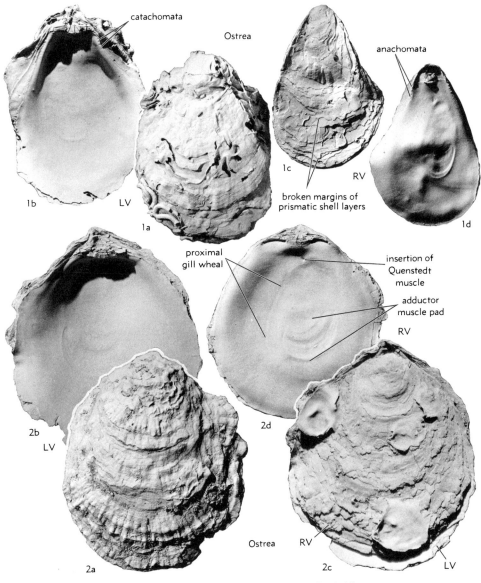

Fig. J113. Ostreidae (Ostreinae) (p. *N1138-N1139*).

Cubitostrea Sacco, 1897a, p. 99 [**Ostrea cubitus* Deshayes, 1832, p. 365; OD]. Small to medium-sized (largest dimension of LV up to 8 cm.), outline curved, crescentic to crescentic-triangular. Crescentic species generally thin-walled, with flat to slightly convex RV, those tending to triangular shapes thick-walled, with RV strongly convex on outside but almost flat on inside. LV obscurely keeled; keel crescentic, located nearer to concave posterior than to anterior valve margin, which is broadly rounded in crescentic species, angulate in triangular ones, angulation indicating position of

inhalant pseudosiphon and widest part of gills during life. Branchitellum much produced and narrowly rounded. Posterior auricle present on LV of some species. Adductor muscle imprint comma-shaped to reniform, located approximately halfway between hinge and branchitellum. LV with concentric growth squamae and high, narrowly rounded, dichotomous or intercalating, long radial ribs separated by deep, sharply rounded interspaces narrower than ribs. RV without ribs, having only appressed concentric growth squamae. Margins of LV strongly crenated by concavities beneath radial ribs, but margins of RV smooth. Short row of chomata on each side of hinge. *Eoc. (Lutet.-Auvers.),* N.Am.(Gulf Coast-Atl. Coast); *Eoc.,* Patag.; *Eoc.(Lutet.)-Oligo.,* Eu.——Fig. J116,*1.* **C. cubitus* (DESHAYES), Eoc.(Auvers.), France (Crépy-en-Valois, Département Oise); LV ext., ×1 (Stenzel, Krause, & Twining, 1957).—— Fig. J116,*2;* J117,*1.* *C. perplicata* (DALL, 1898), mid.Eoc. (U.Tallahatta F.), USA (Ala., Catons Bluff on Conecuh R., Covington Co.); J116,*2a-d,* RVs ext., ×1; J116,*2e-h,* RVs int., ×1 (all topo-

types) (Stenzel, n); J117,*1a,* both valves, right side, ×2; J117,*1c-g,* LVs ext., ×1; J117,*1b,h,i,* LVs int., ×1 (all topotypes) (Stenzel, n). [*See* also Fig. J17, J43,*1,* and J70.]

[Close relationship between this genus and *Ostrea* is proved by the rib patterns of the valves and the chomata. For this reason it is believed that the genus was incubatory. The two opposing valves differ much in size, LV extending a good distance beyond the periphery of the RV (see Fig. J17 and p. N977), indicating that the latter had extensive marginal conchiolin fringes while the animal was alive. A special provincial stock of this genus evolved on the east and south shores of North America into the weird *C. sellae-*

Fig. J114. Ostreidae (Ostreinae) (p. N1139-N1140).

?Turkostrea

FIG. J115. Ostreidae (Ostreinae)
(p. *N1139-N1140*).

formis (CONRAD, 1832), which had a large, heavy, twisted, saddle-shaped, and auriculate shell (Fig. J70). Evolution of this provincially isolated stock has been elucidated by STENZEL (1949), compare p. *N1079*.]

Ferganea VYALOV, 1936, p. 19 [**Gryphaea sewerzowii* ROMANOVSKIY, 1883, p. 251, *nom. nud.*= *G. sewerzowii* ROMANOVSKIY, 1884, p. 54-55, pl. 12, figs. 1-3; OD]. Small to medium-sized (to 12 cm. high), mostly roughly squamate. LV high and narrow (*H* is about 120 to 200 percent of *L*), highly convex in horizontal cross section, convex in dorsoventral direction, compressed in arteroposterior direction but devoid of a radial keel or radial sulcus; beak large, pointed, prosogyral to nearly orthogyral, somewhat inrolled and with small attachment area; umbonal region of LV thickly filled with shell material so that there is no umbonal cavity. RV flat to lightly convex, exterior covered with many projecting growth squamae; outline ovate, higher than long (*H* is 123 to 132 percent of *L*). Chomata small, present only in well-preserved specimens. Some species developed a loosely spiraled orthogyral ligamental area, more rarely the spiral twist was stronger in the older, umbonal part of the shell. [These forms were described as *Exogyra ferganensis* ROMANOVSKIY, 1879, p. 153-154, text fig. 2, which was later assigned to *Amphidonta* [*recte Amphidonte*] because of its chomata (GEKKER, OSIPOVA, & BELSKAYA, 1962, v. 2, pl. 15). However, their ligamental area does not have a narrow, crestshaped posterior bourrelet, and the adductor muscle insertion is clearly reniform so that there is no doubt that *Ferganea ferganensis* (ROMANOVSKIY, 1879) is not an exogyrine oyster but an ostreine *Exogyra* homeomorph.] *Oligo.(Sumsarskiy Yarus)*, C.Asia-USSR.——FIG. J118,*1-6*; J119,*2-4*.

*F. sewerzowii (ROMANOVSKIY); J118,*1a-c*, LV with broken ventral margin, ext., int., post. views, ×1; J118,*2a,b*, RV ext., int., ×1; J118,*3a,b*, RV

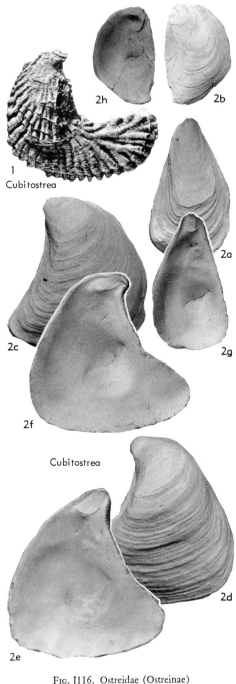

FIG. J116. Ostreidae (Ostreinae)
(p. *N1141-N1143*).

ext., int., ×1; J118,4, LV oblique view onto insertion of post. adductor muscle, ×1; J118,5a,b, RV ext., int., ×1; J118,6a,b, broken LV int., post., ×1 (Stenzel, n) (all specimens whitened for photography; specimens from Paleont. Inst. Akad. Nauk SSSR, collected by R. F. GEKKER); J119,2-4

(="*Gryphaea (Ferganea)*" VYALOV, 1936); J119, 2a,b, LV int., ant., J119,3, LV int.; J119,4a,b, RV ext., int., all ×0.75 (Romanovskiy, 1884).

Platygena ROMANOVSKIY, 1882, p. 46-47, 58-60 [**Ostrea asiatica* ROMANOVSKIY, 1879, p. 150; M] [=*Platigena* BORNEMAN, BURACHEK, & VYALOV,

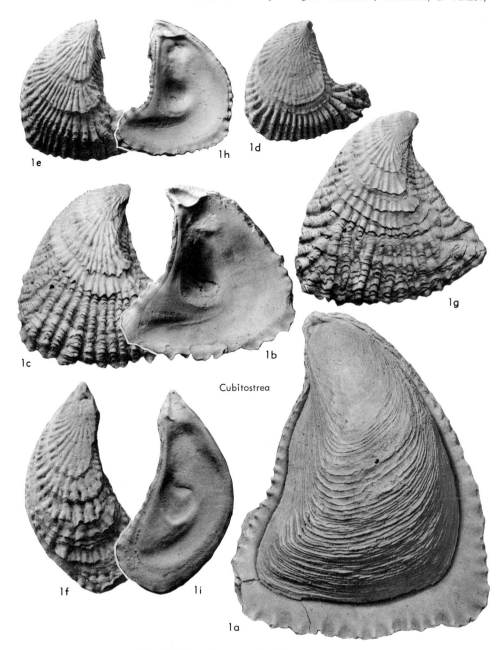

Cubitostrea

FIG. J117. Ostreidae (Ostreinae) (p. *N1141-N1143*).

1934, p. 260 *(nom. null.)*]. Medium-sized to large and flattish, width about 0.16 to 0.25 of height; outline roughly orbicular; old shells higher than long. LV gently convex, RV even less con-vex. Outline of valve cavity in LV banjo-shaped (resembling guitar-shaped valve cavity of *Sokolowia*), with slender subparallel-sided dorsal neck and suborbicular ventral banjo body. Space on

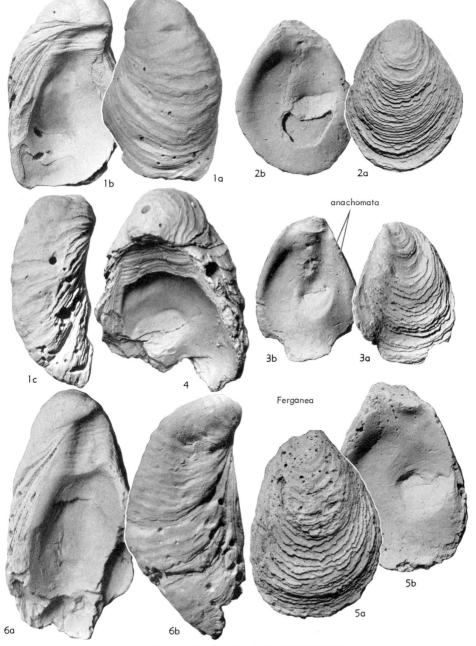

FIG. J118. Ostreidae (Ostreinae) (p. *N1143-N1144*).

either side of banjo neck filled out with former temporary anterodorsal and posterodorsal valve margins left behind during growth of shell. Later growth restricted to continued gradual shift of shell cavity in ventral direction without increase in size of cavity, which in LV is very shallow, without umbonal cavity or overhang at hinge plate. Adductor muscle imprint reniform, ·its longer axis tilted, nearly central in location. No chomata. Numerous growth squamae, many of which are frilled in harmony with nearly continuous small radial ribs. *Up.Eoc.(Rishtan.),* C. Asia-USSR-Afr.(Sudan-Libya).——Fig. J120,*1. *P.*

asiatica (Romanovskiy), USSR; *1a-e,* LV ext., int., RV int., part of ext., both valves post. view, ×0.48 (Romanovskiy, 1882).

[*Platygena* is similar to *Deltoideum,* from which it differs in central location of the reniform muscle imprint. The genera evidently are only superficially similar. *Platygena* resembles *Sokolowia,* but differs in its flatness, lack of shell twists, and in being not at all a *Gryphaea* homeomorph. Its very shallow shell cavity proves that it never had a large gonad lodged within it. Therefore, it must have been an incubatory oyster that produced few eggs. Closeness to *Ostrea* is proved by its reniform muscle imprint.]

Sokolowia J. Böhm, 1933, p. 104 [**Gryphaea buhsii* Grewingk, 1853, p. 114 (=*Gryphaea esterházyi* Pávay, 1871, p. 375, pl. 8-9); OD]

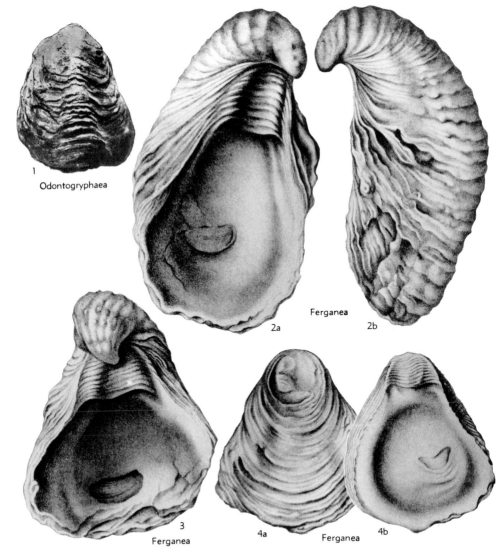

1
Odontogryphaea

Ferganea
2a   2b

3
Ferganea

4a   Ferganea   4b

Fig. J119. Ostreidae (Ostreinae) (p. *N1143-N1144, N1153-N1154*).

[=*Sokolovia* Borneman, Burachek, & Vyalov, 1934, p. 260 *(nom. van.)* (obj.); *Fatina* Vyalov, 1936, p. 19 (type, *Gryphaea esterhazyi* Pávay, *var. beldersaiensis* Gorizdro, 1915, p. 22; OD); *Kafirnigania* Gekker, Osipova, & Belskaya, 1962, v. 2, p. 115 (type, *K. orientalis;* OD)]. Small to large (to 15 cm. long and 21 cm. high), highly inequivalve, homeomorphous with *Gryphaea;* LV umbonal region extending beyond RV; outline triangular. Chomata well developed. Adductor muscle imprint reniform, centrally placed. LV with ventral margin convex, evenly curved, and marked at each end by prominent corner, anterior and posterior margins straight to concave. In all young to mature individuals length about same as height or larger (*H* 75 to 100 percent of *L*). With advancing age LV tends to quit growing in length but continues growing in ventral direction so that valve becomes much higher (*H* up to 172 percent of *L*), losing triangular outline and prominent corners. LV convex from umbo to venter and even more so from anterior to posterior; beak small, pointed, incurved, reaching over to right side. Attachment area mostly small. Surface cov-

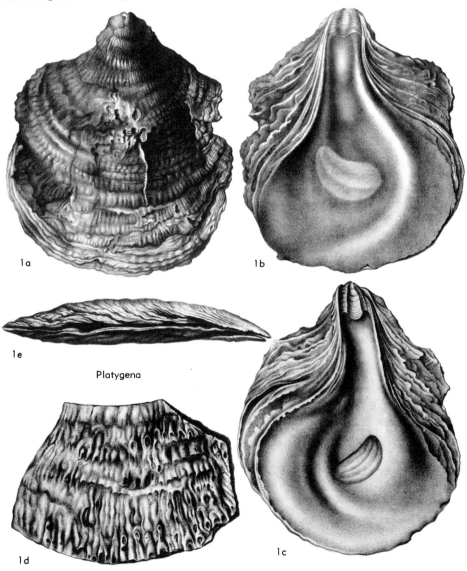

1a

1b

1e

Platygena

1d

1c

Fig. J120. Ostreidae (Ostreinae) (p. N1144-N1146).

ered partly or entirely by many subequal, parallel, even-crested, strong to obsolescent radial ribs; no free-standing growth squamae on umbonal half of valve. Valve cavity guitar-shaped in outline even in very high, old shells, because anterior and posterior valve margins are pinched in. Umbonal cavity filled in with shell deposits. RV flat to concave, with triangular outline like that of LV but truncated at umbo by ligamental area, which stands vertical to commissural plane; devoid of

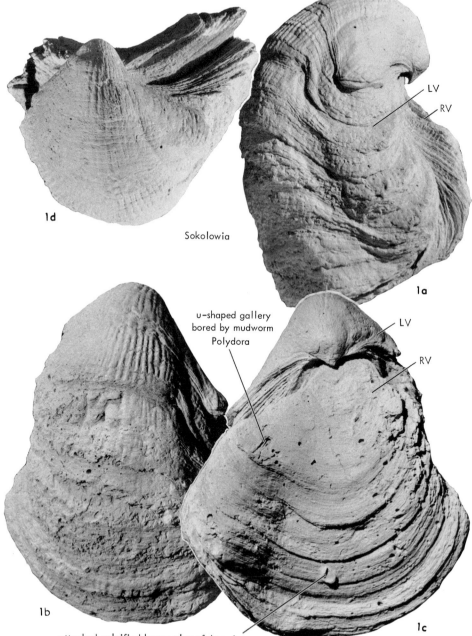

FIG. J121. Ostreidae (Ostreinae) (p. *N1146-N1150*).

ribs, but with prominent growth squamae and may have auricles. *Eoc.(Turkestan.)*(*=up.Lutet. +Auvers.*), Rumania(Transylv.)-C.Asia.——FIG. J121,1. **S. buhsii* (GREWINGK), USSR (Fergana); *1a-d,* both valves, post. view, left side, right side, umbonal view, ×0.7 (Stenzel, n). [Specimen whitened for photography; from Paleont. Inst. Akad. Nauk SSSR, collected by R. F. GEKKER.] ——FIG. J122,1. *S. beldersaiensis* (GORIZDRO,

1915) (*="Fatina"*), USSR (Fergana); *1a-d,* both valves post. view, left side, right side, umbonal view, ×1 (Stenzel, n). [Specimen whitened for photography; from Paleont. Inst. Akad. Nauk SSSR, collected by R. F. GEKKER.]——FIG. J123,1. *S. orientalis* (GEKKER, OSIPOVA, & BELSKAYA, 1962) (*="Kafirnigania"*), USSR (Baba-Tag); *1a,b,* LV ext., both valves from right side, ×1 (Gekker, Osipova, & Belskaya, 1962).

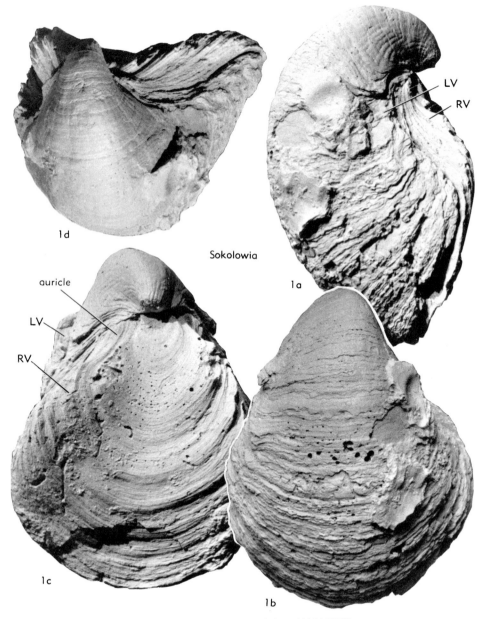

FIG. J122. Ostreidae (Ostreinae) (p. N1146-N1150).

la

Sokolowia

1b

FIG. J123. Ostreidae (Ostreinae)
(p. N1146-N1150).

[*Sokolowia* is interpreted to be a descendant of *Ostrea* (*Turkostrea*), which in turn descended from *Ostrea* (*Ostrea*). Evolution was quite rapid and produced a *Gryphaea* homeomorph (YANG KIEH, 1930; GEKKER, OSIPOVA & BELSKAYA, 1962). The stock was a provincial offshoot from *Ostrea, s.s.* and retained some generic characters of *Ostrea s.s.*, namely, reniform muscle imprints and discrepant sculpture of the valves. When VYALOV proposed *Fatina* as a new genus in 1936 he placed *Sokolowia* BöHM, 1933, under it as a mere section of *Fatina* (*Fatina*). Such a procedure is not sanctioned by the *Code*, which on grounds of priority requires that *Fatina* be placed under *Sokolowia* (*Sokolowia*). VYALOV was correct in estimating the difference between the two forms to be slight. In fact, they are so little that the type species of *Fatina* is best regarded as merely another species of *Sokolowia* and the two names treated as subjective synonyms. The same conclusion applies to *Kafirnigania*. It seems unlikely that more than one genus could evolve from *Ostrea s.s.* in the same province, in the same extensive sea basin, and in the same sedimentary environments, without separating geographic barriers. It has been generally overlooked that *Sokolowia esterhazyi* (PAVAY) was first described and figured as *Gryphaea buhsii* GREWINCK, 1853, from the vicinity of Nemekeh and Surt, in the eastern part of the Elburz Mountains in northern Iran. The species is readily identifiable from this earlier description, which is nomenclaturally available. Therefore, the prior valid synonym is the name given by GREWINCK.]

Tribe FLEMINGOSTREINI Stenzel, new tribe

Consists of the genera *Flemingostrea, Odontogryphaea, Ostreonella*, and possibly *Koķanostrea. Anulostrea* and *Quadrostrea* ultimately may find place in this tribe, but now are insufficiently known. *Odontogryphaea* is a *Gryphaea* homeomorph. The following are common features of this tribe: 1) U-shaped or terebratuloid fold on valve commissure at ventral valve margins; LV arched up toward left side to form rooflike fold; RV with semicircular or tongue-shaped ventral extension that reaches over to left side fitting into corresponding

sinus of LV. 2) Adductor muscle imprints longer than high, with straight to concave dorsal margins and imprints located rather close to ventral and posterior valve margins. 3) Shell walls exceptionally thick (up to 3.5 cm.) in umbonal half of valves. 4) Resilifer one to three times as long as each flanking bourrelet of ligamental area. 5) Ligamental area commonly rather long but low, its ends producing shoulders on shell. 6) Attachment area never large, very small to absent in *Odontogryphaea;* no tendency of one individual to grow upon another. Terebratuloid fold makes its appearance at fairly advanced stage of growth, commonly when shell reaches a height of about 2 cm. At that place growth changes its direction, and a prominent smooth hump in shell profile is the result. *U.Cret.(Cenoman.)-Mio.*

Flemingostrea VREDENBURG, 1916 [**Ostrea (Flemingostrea) flemingi* D'ARCHIAC & HAIME, 1853, p. 275; OD] [*?=Solidostrea* VYALOV, 1948a, p. 24 (type, *Ostrea hemiglobosa* ROMANOVSKIY, 1884 (in 1878-90), p. 26; OD)]. Medium-sized to large; overall shape flattish; valves subequal, neither highly convex nor compressed in anteroposterior direction. LV umbo not prominent or hook-shaped, terebratuloid fold broad and gentle, arising gradually at later growth stage than rooflike fold does in *Odontogryphaea;* no radial sulcus delimiting fold on its posterior flank which has gradual slope. Calcite prisms long and well developed in prismatic shell layers (see Fig. J20). Many species with regularly spaced concentric imbrications on RV recalling *Gryphaeostrea;* imbrications composed of prismatic shell layers and separated from each other by smooth concentric bands with sigmoidal profiles. Chomata present or absent. Ligament growing rapidly in length, but not in height in early years resulting in somewhat shouldered appearance of RV with shoulders at ends of ligamental area approaching rectangular shape. [*Flemingostrea* has several features that distinguish it from other genera of the tribe: 1) LV umbo not prominent and not beaked, 2) shell not as highly convex as in *Odontogryphaea* and not compressed anteroposteriorly at any stage, 3) shell shape less globose, tending to be flattish, 4) valves more nearly equal in size, 5) terebratuloid fold appearing later and very gradually, 6) flanks of terebratuloid fold remaining poorly delimited. These features are more primitive than corresponding ones in other genera of the tribe, because they more nearly approach those of the normal, average ostreine prototype. Therefore, *Flemingostrea* is regarded as the ancestor of the tribe. Its acme was in Late Cretaceous time]. *U.Cret.(Cenoman.)-Mio.*, S.Am.(Peru)-N.Am.(Gulf Mexico-Atl.Coast-NE.Mexico-Texas-Ark.-Miss.-Ala.-Ga.-N.Car.-N.J.-Utah)-Eu.(Belg.)-Afr.(Alg.-Egypt-Sudan-Senegal-

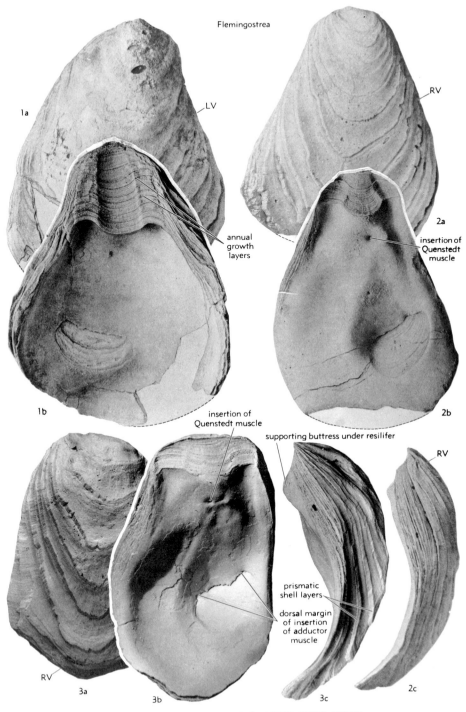

Fig. J124. Ostreidae (Ostreinae) (p. *N1150, N1152-N1153*).

Madag.) - Asia(Pak. - Afghan.) - C.Asia - USSR (Tadzik Basin-Gissar Mts.-Transalaii Mts.). — Fig.
J124,*1-3*. *F. subspatulata* (FORBES, 1845) (=*Ostrea owenana* SHUMARD, 1861), U.Cret.(Maastricht.); *1a,b*, Nacatoch Sand, Navarro Gr., near

Greenville, Hunt Co., Texas, LV ext., int., ×0.7;
2a-c, Nacatoch Sand, near Corsicana, Navarro Co.,
Texas, RV ext., int., post., ×0.7; *3a-c*, Ripley F.,
near Braggs, Lowndes Co., Ala., RV ext., int.,
post., all × 0.7 (*1,2*, Stephenson, 1941;*3*, Stenzel,

Flemingostrea ?

Fig. J125. Ostreidae (Ostreinae) (p. *N*1150, *N*1152-*N*1153).

n). [All specimens whitened for photography.]
—FIG. J125,1,2. F.? *hemiglobosa* (ROMANOVSKIY, 1884) (=*Ostrea gigantica* Cox, 1938 (*non* SOLANDER, 1766), =*Ostrea (Solidostrea)* (VYALOV, 1948c), Eoc. (Suzakian Stage); *1a,b,* Tash Kurghan, N. Afghan., LV ant., ext., ×0.5 (Cox, 1938); *2a-d,* Tadzik Basin, USSR, LV ext., int., RV ext., int., ×0.5 (Vyalov, 1948c). [*See* also Fig. J20.]

[*Solidostrea* was described as new a second time, by VYALOV, 1948b, p. 13-19, pl. 3-7. The type species has some superficial resemblance to "*Gigantostrea*" *gigantica* SOLANDER in BRANDER, 1766 (*recte Pycndonte (Pycnodonte) gigantica*) and Cox in CIZANCOURT & COX (1938, p. 39-42, pl. 5, fig. 5a,b) described specimens of *Flemingostrea? hemiglobosa* from Afghanistan under the former name in spite of his observations that they lacked vesicular shell structure and had a reniform adductor-muscle imprint. The two differ greatly, because *Pycnodonte* and its junior synonym *Gigantostrea* have orbicular muscle imprints and vesicular shell structure. *Solidostrea* is presumably a junior synonym of *Flemingostrea*, because the type species of both have the characteristic well-spaced concentric undulations of sigmoidal cross section, each one ending in a growth squama, a very thick shell wall of the LV, and a humped profile of the LV, noted and well-figured by Cox.]

Kokanostrea VYALOV, 1936, p. 19 [**Ostrea kokanensis* SOKOLOV, 1910, p. 73; OD]. Small (up to 3.5 cm. high), highly inequivalve; outline oval with umbones at small end. LV highly convex, profile hook-shaped, with prominent hump about 15 mm. from umbo which is acute or bluntly cut off, not prominent and not *Gryphaea*-like, but prosogyral; tendency to form wing at anterior valve margin. RV operculiform, flat to concave. Chomata present. Adductor muscle imprint semicircular, longer than high, located closer to posterior and ventral valve margins. Both valves devoid of radial ribs, both with uniformly spaced smooth concentric growth squamae. *Eoc.(Turkestan.) (up.Lutet. + Auvers.),* C.Asia. —— FIG. J126,1. **K. kokanensis* (SOKOLOV); *1a-c,* both valves, oblique post. view, LV int., RV int., ×1 (Sokolov, 1910).

[Information on this genus is scanty and its taxonomic position is open to question. VYALOV (1936, p. 19) gave the following definition for it: Beak well developed, gryphoidally inrolled; ligamental area inclined toward anterior; posterior slope of shell steeper than the anterior one. He placed it as a subgenus of *Liostrea,* as did GEKKER, OSIPOVA, & BELSKAYA (1962). However, because of differences in outline and position of the adductor muscle imprints, the two taxa evidently are not closely related. *Kokanostrea* is possibly a member of the Flemingostreini; indications for this placement are the humped and hooked profile of the LV, recalling *Odontogryphaea,* and the outline and position of the adductor muscle imprint.]

Odontogryphaea IHERING, 1903, p. 194 [**Gryphaea consors* var. *rostrigera* IHERING, 1902, p. 113 (= *G. (Odontogryphaea) rostrigera* IHERING, 1903, p. 212; OD] [=*Sinustrea* VYALOV, 1936, p. 18 (type, *Ostrea (Flemingostrea) morgani* VREDENBURG, 1916, p. 197; OD); *Sinostrea* HAAS, 1938, p. 294 (*nom. van.*)]. Small to medium-sized (up to 13 cm.), composed of lamellar and prismatic layers; vesicular and chalky layers absent. Outline orbicular or oval to triangular, umbo at small end. No auricles near hinge. Shell subequivalve to inequivalve, but in most species not highly inequivalve. Valve commissure twisted, with terebratu-

1a

1b

Kokanostrea

1c

FIG. J126. Ostreidae (Ostreinae) (p. *N1153*).

loid fold at venter. LV convex and capacious, with prominent beak mostly somewhat opisthogyral, rarely prosogyral, incurved unevenly in hook-shape. In nearly all species beak does not reach through plane of valve commissure to right side of shell. LV compressed in anteroposterior direction in most species in early growth stage; developing rather abruptly rounded rooflike radial fold at beginning of later growth stage; fold delimited on its posterior flank by radial sulcus. RV never really operculiform, simply convex in early growth stage, convex along line from umbo to mid-ventral margin except at tonguelike ventral end, which may be slightly concave and fits into rooflike fold of LV. RV slightly convex to deeply concave in transverse (anteroposterior) direction. Umbonal cavity of LV filled with solid shell material, and LV quite thick-walled from tip of umbo to adductor muscle. Ligamental area usually longer than high; resilifer is 1.5 to 2 times as long as either flanking bourrelet. Chomata to both sides of ligament, very extensive in type species, placed so close to valve margins of RV that their growth tracks are visible on the outside. Posterior adductor muscle imprint never orbicular; placed close to venter. Imprint outline different in each valve, one on left side longer and more reniform to ribbon-shaped. Surface features of both valves similar, consisting of foliaceous growth squamae irregularly spaced, but farther apart in early growth stage. Radial costae, either absent or small, developed only on rooflike fold of LV as discontinuous fine frills of crowded foliaceous growth squamae. *U.Cret.(Maastricht.)-Eoc.,* France (Paris Basin, *Thanet.;* Corbières, *Sparnac.-Lutet.*)-S. Asia(Baluch.)-S. Am.(Patag.)-N. Gulf Mexico(Ga.-Ala.-La.-NE.Mexico). —— FIG. J127,-1-3. **O. rostrigera* (IHERING), Eoc., Arg. (Patagonia, Chubut Terr.); *1,* LV int.; *2,* RV int.; *3a-c,* both valves, oblique left ventral view, oblique right ant. view, ant. view, all ×0.75 (Ihering, 1903).——FIG. J119,1. *O. morgani* (VREDENBURG, 1916) (=*"Liostrea (Liostrea)* section *Sinustrea"*

VYALOV, 1936), U.Cret.(Maastricht.), NW.Pak. (Des Valley, Baluch.); both valves, left side, ×0.75 (Vredenburg, 1916). [*See* also Fig. J9, J34, J64.]

[VREDENBURG was unaware of *Odontogryphaea* and did not separate this genus from *Flemingostrea* which he regarded as a subgenus of *Ostrea,* showing thereby that he was not impressed by the homeomorphous similarities with *Gryphaea.* VYALOV (1936, 1948a) was unaware of *Odontogryphaea.* Therefore, he proposed a new taxon, *Sinustrea,* as a section of *Liostrea* (*Liostrea*) where he also placed *Ostreonella* as a section. Thus both were somewhat removed from *Flemingostrea,* which he left standing as a subgenus of *Ostrea,* following VREDENBURG. All three were placed in the Ostreinae in spite of the *Gryphaea*-like beaks in some of the species. His apparent reason, not expressed in print, for not placing *Sinustrea* in the Gryphaeinae was that the Gryphaeinae, according to him, must have a flat or concave upper valve, an astute observation in this case. Indeed, *Odontogryphaea* has a RV that is convex in its early adult growth stage, presumably because it is descended from *Flemingostrea,* which has a similarly convex RV. Both are clearly members of the Ostreinae, although *Odontogryphaea* is a homeomorph of *Gryphaea.*——Some species, notably *Odontogryphaea thirsae* (GABB, 1861, p. 329) from the Nanafalia Formation (Thanet.) of the Wil-

cox Group in Alabama, lack imprints of former attachment at their left umbones. They must have grown from larval to adult stage without becoming attached to a firm substrate. The feature is probably the result of their adaptation to a substrate lacking places of firm attachment. (See Fig. J34 and J64.)]

Ostreonella ROMANOVSKIY, 1890, p. 101[**O. prima;* OD] [=*Ostronella* NIKITIN, 1894, p. 171 (*nom. null.*)]. Small (to 5 cm. high), overall shell shape biconvex lenticular when young to globose when old. Valves almost equal, highly inflated, highly convex, somewhat compressed anteroposteriorly so that they may become wider than long (length is 75 percent, width 95 percent of height). LV umbo not prominent, not hook-shaped, with small attachment area. Terebratuloid fold obscure in young; poorly delimited at its flanks, rising rather slowly, gradually, and fairly late during growth; ending in U-shaped semicircular hollow at valve commissure. Concentric growth squamae on both valves evenly and regularly

FIG. J127. Ostreidae (Ostreinae) (p. *N1153-N1154*).

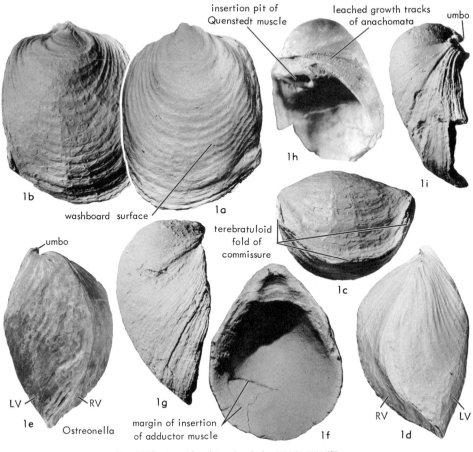

insertion pit of Quenstedt muscle

leached growth tracks of anachomata

umbo

1h

1i

1b

washboard surface

1a

umbo

terebratuloid fold of commissure

1c

LV RV

1e   Ostreonella

1g

margin of insertion of adductor muscle

RV LV

1f

1d

FIG. J128. Ostreidae (Ostreinae) (p. *N1154-N1157*).

spaced, separated from each other by smooth concentric bands that have sigmoidal profiles; growth squamae so delicate that they are mostly broken off. As a result the successive smooth concentric bands appear to merge forming a wavy "wash-board" surface. Chomata present. In old individuals the shell wall is very thick from the umbo to the adductor muscle. *U.Cret.(Campan.),* USSR (Gissar Mt. Range, near Shirabad town, Uzbek.)-C.Asia.——FIG. 128,*1.* **O. prima,* Gissar Mt. Range, USSR; *1a-e,* both valves, left and right sides, ventral, ant., post. views; *1f,g,* LV int., post.; *1h,i,* RV oblique post. and ant. views of an extremely humped and thick-walled individual, all ×1.3 (Stenzel, n). [All figures except *1e* whitened for photography; specimens by courtesy of KH. KH. MIRKAMALOV, Tashkent, and D. P. NAIDIN, Moscow State Univ., SSSR.]

[The genus is monotypic and so unusual that ROMANOV-SKIY remarked that it does not have much resemblance to *Ostrea.* He is to be commended for recognizing that it is an ostreid genus. BORNEMAN, BURACHEK, & VYALOV (1934, p. 255) noticed that the convexity of the right valve,

which ROMANOVSKIY had regarded as a generic character setting the genus apart, was fairly variable and ranged from very strong to weak. Therefore, they felt justified in doubting its value as a generic character setting *Ostreonella* apart from other smooth-shelled oysters. VYALOV in his elaborate scheme of classification of the Ostreidae (1936, p. 18; 1948a, p. 35) lowered the rank of *Ostreonella* to that of a section and placed it under *Liostrea* (*Liostrea*) DOU-VILLÉ (1904a, p. 273) in the subfamily Ostreinae VYALOV, 1936. This arrangement has been followed by other authors in the USSR but is nonetheless faulty both as to nomenclature and phylogenetic systematics. First, it conflicts with *Code* Art. 23(e)(i). A subgenus or section of a genus dating from 1890 cannot legally be placed under a genus dating from 1904. If the two taxa are truly so closely related that one must be placed under the other, *Ostreonella,* having nomenclatural priority, would have to stay as a genus and *Liostrea* would have to be subordinated as a subgenus or possibly as a section. Second, *Liostrea* and *Ostreonella* are not at all closely related, notwithstanding the opinions and arguments of VYALOV (1936; 1948a) and BOBKOVA (1961, p. 42). Aside from the convexity of the RV, *Ostreonella* has several other generic features separating it from *Liostrea.*——Comparison of the type species of the two shows the following definitive differences: *Liostrea* has orbicular muscle imprints situated halfway between hinge and ventral valve margins, or even nearer to the hinge; although the shell and its valve commissure are variable and somewhat irregular, in average the valve commissure is a flat plane and chomata are absent on the commissural shelves. It is one of the Gryphaeinae as redefined by STENZEL (1959, p. 16). In contrast, *Ostreo-*

nella has reniform muscle imprints, longer than high in outline; they are placed rather close to the ventral valve margins. Its valve commissure, particularly in older, more inflated individuals, is heteroclite, that is, no longer a flat plane. For this reason, the valve commissure describes an "S" curve (BOBKOVA, 1961, p. 40), when seen from the posterior side. A broad semicircular terebratuloid fold is well shown on older, more inflated shells. The differences between *Liostrea* and *Ostreonella* in regard to outline and position of the adductor muscle imprints are proof that the internal soft anatomies of the two were quite different. Furthermore, *Ostreonella* has valves approaching equal convexities so that the general shell shape in younger individuals is like a biconvex lens and in older, more inflated individuals it approaches globose. The umbonal half of the shell, extending from umbo to region of the adductor muscle, has greatly thickened (15 mm.) shell walls, whereas the other, ventral, half of the shell has walls of normal thickness (2 mm.). Without doubt *Ostreonella* is one of the Ostreinae as redefined (STENZEL, 1959, p. 16) and a member of the *Flemingostrea* stock.——VYALOV's scheme of classification fails to take notice and make use of any

of internal features of the shell, such as structure of shell walls, outline and position of the adductor muscle imprints, and morphology of the ligamental areas. Rather, it is based on external morphology alone. For this reason it places side by side genera having superficial external similarities but not really related. *Liostrea* and *Ostreonella* are examples. The position VYALOV gave to *Ostreonella* served only to obscure its true relationships. Recognition of these had to wait until a better, objective, detailed redescription of its type species became available. In providing this, however, BOBKOVA (1961, p. 39-43, pl. 3, fig. 1-5, pl. 4, fig. 1-5) chose to follow VYALOV's classification and again called it *"Liostrea" prima* (ROMANOVSKIY), but she pointed out that two North American species, *"Liostrea" thirsae* (GABB, 1861) (see HEILPRIN, 1884, p. 311, pl. 63, fig. 3-6; STENZEL, 1959, fig. 9, 15-16, 18-19; see Fig. J34 and J64) and *"L." oleana* (STEPHENSON, 1945, p. 72-74, fig. 1-7), are quite similar to it. Although BOBKOVA relegated these two species to *Liostrea* they furnish important clues to true affinities of *Ostreonella*. BOBKOVA may be credited as the first to recognize close relatives of *Ostreonella prima* and to open the way for the present

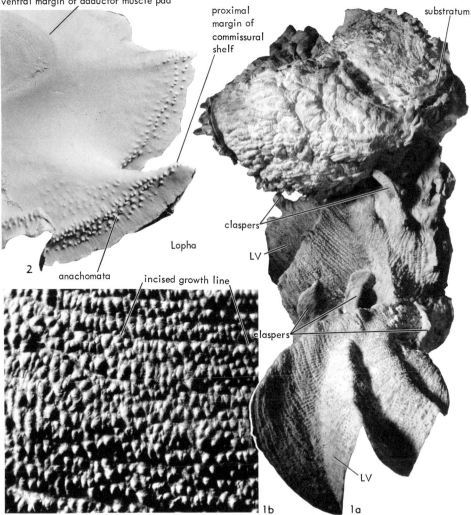

Fig. J129. Ostreidae (Lophinae) (p. *N1157-N1158*).

evaluation of this baffling genus. Fortunately, several specimens were kindly furnished to me by Prof. KH. KH. MIRKAMALOV of Tashkent.——The two North American species have been critically re-examined (STENZEL, 1945a,b; 1959, p. 32-33, fig. 3-19). The former was found to be a species of *Odontogryphaea* IHERING, 1903, and is rather similar to its type species; the latter was found to fit in the genus *Flemingostrea* VREDENBURG, 1916.]

Subfamily LOPHINAE Vyalov, 1936
[Lophinae VYALOV, 1936, p. 19]

Incubatory. Attachment area medium-sized to large; some genera provided with shelly claspers that grow out at intervals to lock onto substrate. Valves subequal in size and convexity, and with similar rib patterns consisting of regular, subequal, fairly sharp-crested plicae which produce a regularly plicate valve commissure; pattern not greatly variable in individuals of the same species. Innumerable very small, slightly elongate tubercles are scattered over interior faces of valves, especially near their margins (Fig. J129,2). *U.Trias.-Rec.*

Lopha RÖDING, 1798, p. 168 [**Mytilus cristagalli* LINNÉ, 1758, p. 704; SD DALL, 1898, p. 672] [=*Alectryonia* FISCHER DE WALDHEIM, 1807, p. 269 (type, *A. rara*, =*Mytilus cristagalli*; SD GRAY, 1847, p. 201); *Alectrionia* FISCHER DE WALDHEIM, 1808, Tab. 5 *(nom. van.);* *Dendostrea* SWAINSON, 1835, p. 39 (type, *Ostrea folium* LINNÉ, 1758, p. 699; SD HERRMANNSEN, 1847, v. 1, p. 378); *Dendostraea* SOWERBY, 1839, p. 38 *(nom. van.);* *Dendrostraea* SWAINSON, 1840, p. 389 *(nom. van.);* *Dendrostrea* AGASSIZ, 1846, p. 118 *(nom. van.);* *Alectronia* LOGAN, 1898, p. 485 *(nom. null.);* *Actryonia* DOLLFUS, 1903, p. 271 *(nom. null.);* *Alcetryonia* BÖSE, 1910, p. 105 *(nom. null.);* *Alectryouia* STRAUSZ, 1928, p. 277 *(nom. null.);* *Ostrea (Pretostrea)* IREDALE, 1939, p. 397 (type, *O. (P.) bresia*, =*Lopha cristagalli* (LINNÉ)); *Alektryonia* VYALOV, 1948a, p. 29 *(nom. null.);* *Alectryossia* SALISBURY & EDWARDS, 1959, p. 128 *(nom. null.)*]. Small to medium-sized (to about 11 cm. long); both valves convex, subequivalve, with 6 to more than 50 sharp radial plicae the pattern of which is not greatly variable in each subgenus. *Trias.-Rec.*, mostly trop. and partly subtrop., worldwide.

L. (Lopha). The subgenus has two distinct ecomorphs. Ecomorph 1 [commonly called *Lopha cristagalli* (LINNÉ) (=*Lopha s.s.* AUCTT.)] grows on mangrove or on each other or on other bivalves, for example, *Chama*. Shell small to medium-sized (to 11 cm. long); both valves convex, equivalve; shape roughly globular to irregular with 6 to 9 or even 12 deep sharp radial plicae, angles of which decrease from near hinge to point farthest away from it according to RUDWICK's rule (see RUDWICK, 1964, and p. N1025-N1026). Surface of both valves roughened by countless small,

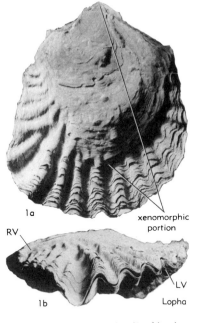

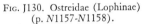

FIG. J130. Ostreidae (Lophinae)
(p. N1157-N1158).

low, rounded, almost equal-sized protuberances, which are elongate and arranged in obscure parallel radial rows disposed at right angles to incised growth lines (Fig. J129,1). No imbricating growth squamae. Hollow recurved compressed hyote spines arise intermittently from crests of some plicae; some spines recurved as claspers for support of LV on substrate. Ecomorph 2 [called *Dendostrea folium* (LINNÉ), 1758, p. 699, by some authors] grows on stems of Gorgonacea or sea fans (Fig. J47). Small (to 8.5 cm. high); both valves convex, equivalve; shape lanceolate to ovate, with many sharp radial plicae. Surface without any protuberances, but with many fine, closely set growth squamae; hollow recurved hyote spines arise intermittently from crests of plicae, several of them serving as claspers for LV embracing stem of gorgonacean. Many individuals with xenomorphic sculpture consisting of rounded longitudinal ridge extending from umbo to opposite end of RV, corresponding to gorgonacean substrate. *Mio.-Rec.*, trop. and partly subtrop., worldwide.——FIG. J129,1,2. *L. (L.) folium* (LINNÉ, 1758), ecomorph 1, called *L. cristagalli* (LINNÉ, 1758), living, Indo-Pac.; *1a,b*, specimens from Ponape I., Carolines, growth group on *Chama* shell *(1a)*, surface sculpture on ext. *(1b)*, ×0.7, ×3; *2*, specimen from Guam showing band of small tubercle-shaped anachomata along margin of commissural shelf of RV, ×1.3 (Stenzel, n). [Specimen by courtesy of

H. A. LOWENSTAM, California Inst. Technology, Pasadena, Calif.]——FIG. J130,*1. L. (L.) folium* (LINNÉ, 1758), ecomorph called *Ostrea (Pretostrea) brezia* by IREDALE, 1939, living, Australia; *1a,b,* right side, posteroventral view, ×0.8 (Carter, 1968). [*See* also Fig. J47.]

[Many authors, SWAINSON included, mistook the gorgonacean stems on which ecomorphs 2 grow for mangrove roots, hence the name tree oyster or *Dendostrea* SWAINSON, 1835 (an obvious error *pro Dendrostrea*). Indeed, when kept in a museum and dried out, the stems are blackish and wrinkled, resembling tree stems or roots with wrinkled blackish bark. THOMSON, 1954 (p. 146-149, and letter of 14 Dec. 1966) was the first to discover that *Ostrea folium* LINNÉ, *Mytilus cristagalli* LINNÉ, and *Ostrea (Pretostrea) bresia* IREDALE, 1939, are names for morphs of the same polymorphic species. Although his idea is in opposition to those expressed by many other workers (LAMY, 1929, p. 244-246, 254-256; DODGE, 1952, p. 190-191, 205), it is accepted here, because his work is based on extensive observations made on live oysters. According to THOMSON the morphs generally identified as *Lopha cristagalli* live in intertidal situations and those identified as *Ostrea folium* live subtidally on Gorgonacea (Fig. J47). The former are found in the tropical Indo-Pacific; the latter are found, in addition, in the tropical and subtropical W. Atlantic, Caribbean, and Gulf of Mexico. If THOMSON's interpretation and synonymization are accepted, the three above-given species names become subjective synonyms. THOMSON as the first reviser of this situation selected *Lopha folium* (LINNÉ) as the name designated for the biologic species, because it has priority and page precedence over the other two names. Thus the three genus-group names are subjective synonyms from a nomenclatural point of view. Present-day rules (*Code*, 1964) do not support the conclusion of STENZEL (1947, p. 177) that *Lopha* is unavailable as of 1798 and that *Alectryonia* FISCHER DE WALDHEIM, 1807, must be used in its stead. Until RANSON (1941) corrected the generic assignment, *Hyotissa hyotis* (LINNÉ) had been regarded as a *Lopha* (see HIRASE, 1930, p. 23). However, there is no question that *Hyotissa* is not closely related.]

L. (**Abruptolopha**) VYALOV, 1936, p. 20 [*Ostrea abrupta* D'ORBIGNY, 1842, p. 59; OD]. Medium-sized (to about 10 cm. high), outline pyriform, both anterior and posterior valve margins concave, but each surmounted by short, subequal auricle; valves subequal and with same pattern of plicae, consisting of many (more than 50) narrow angular even-crested radial plicae which diverge from umbones and are interrupted at 3 to 5 places by abrupt angulations of shell profile where valves stopped growing in height but continued to accumulate wall thickness. *Cret.,* S.Am. (Colom.).——FIG. J131,*1.* *L. (A.) abrupta* (D'ORBIGNY, 1842), Maastricht.; *1a-c,* types of D'ORBIGNY, RV ext., profile, profile, ×0.7 (D'ORBIGNY, 1842).

L. (**Actinostreon**) BAYLE, 1878, explanation to pls. 132, 143 [*Ostrea solitaria* J. SOWERBY, 1824, v. 5, p. 105, pl. 468, fig. 1; SD DOUVILLÉ, 1879]. Shell size small to medium, about 6-10 cm. high. Shell close to equivalve, outline mostly elongately oval and slightly crescentically curved, a slight posterior auricle is common; adult shells have about 25 narrow, dichotomous slightly curved plicae with narrow but not angular crests. *Jur.-Cret.,* W.Eu.-N.Am.——FIG. J132,*1.* *L. (A.) solitaria* (J. SOWERBY), U.Jur.(Sequan., *Trigonia clavellata* Beds, Glos Oolite Series), Eng. (Weymouth, Dorsetshire); *1a-h,* topotypes, RV, LV,

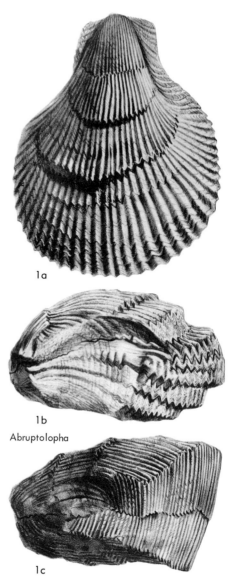

1a

1b

Abruptolopha

1c

FIG. J131. Ostreidae (Lophinae) (p. N1158).

ant. view of both valves, LV, RV, post. view of both valves, LV, RV, ×0.5 (Arkell, 1933).

Agerostrea VYALOV, 1936, p. 20 [*Ostracites ungulatus* VON SCHLOTHEIM, 1813, p. 112 (=*Ostrea larva* LAMARCK, 1819, p. 216); OD]. Small to medium-sized (to 12 cm.), outline falcately curved, flanks tapering gently from umbo to opposite end; anterior flank convex, posterior one concave, both nearly vertical to generalized commissural plane; auricles at either end of hinge present or absent. Commissure folded into high and narrow plicae (up to 4 cm. high), ending in acute or rounded points (up to 20). Central field of valves flat, lacking costae or plicae. Many small rounded tubercles on commissural shelf. Adductor muscle imprint comma-shaped, situated close to hinge. *U.Cret.(Campan.-Maastricht.)*, worldwide.——Fɪɢ. J133,*1.* *A. ungulata* (ᴠᴏɴ SCHLOTHEIM), Maastricht., Neth.(St. Pietersberg near Maastricht); *1a-e,* LV ext., LV int., ant. side, RV int., LV int., all ×1.5 (adductor muscle insertion is outlined by pencil line on *1b*) (Stenzel, n).

[*Agerostrea* seems to be connected by transition species with *Arctostrea,* which is probably its ancestor. It differs

Actinostreon

Fɪɢ. J132. Ostreidae (Lophinae) (p. *N1158*).

from the latter in lacking ribs or plications on the central fields of the valves, in lacking hyote spines arising from crests of the plicae, and in being less compressed from one flank to the other. The lineage appears to have died out with the end of the Maastrichtian.]

Alectryonella SACCO in BELLARDI & SACCO, 1897a, p. 99 [*Ostrea plicatula* GMELIN, 1791, p. 3336, no. 111; OD]. Small (up to 9 cm. long), outline

semicircular or crescentic so that outside curve, from hinge to branchitellum, approximates semicircle or well-curved loop, and distance on posterior side from hinge to branchitellum is very short. Highly inequivalve; LV deep, RV flat or rarely slightly convex; attachment area large to very large, in many individuals showing imprints

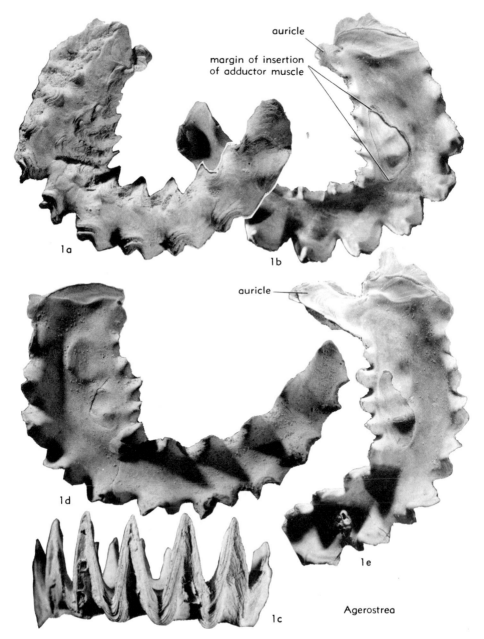

FIG. J133. Ostreidae (Lophinae) (p. *N*1158-*N*1160).

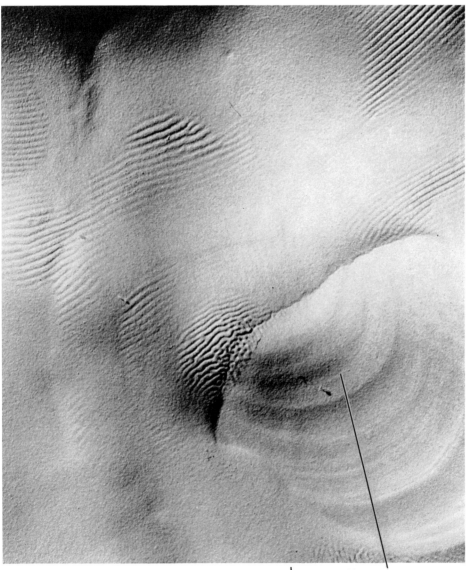

TO VENTRAL MARGIN ↑   adductor muscle pad

Fig. J134. Surface configuration on patches of fingerprint shell structure on RV of *Alectryonella plicatula* (same as Fig. J135,*1d*), ×5.5 (Stenzel, n). [Specimen whitened for photography.]

of compound corals. Both valves have about 14 to 22 continuous, commonly well-rounded, very nearly equal radial plications separated by interspaces of same size and equally well rounded, most plications not anastomosing, but widening toward the valve margins, and accompanied by a very few intercalated plications; growth squamae very few but irregular concentric growth wrinkles numerous. Adductor muscle imprint somewhat higher than long, obliquely distorted, and slightly concave at its dorsal margin, both of its horns well rounded and not prominent. RV margins bearing slightly elongate pustules in band 3 to 5 mm. wide; LV margins without pits except near hinge. Many individuals have an umbonal cavity in LV. Quenstedt muscle imprint in pit placed less than 1 mm. from ventral margin of ligament. Fingerprint shell structure often visible on internal

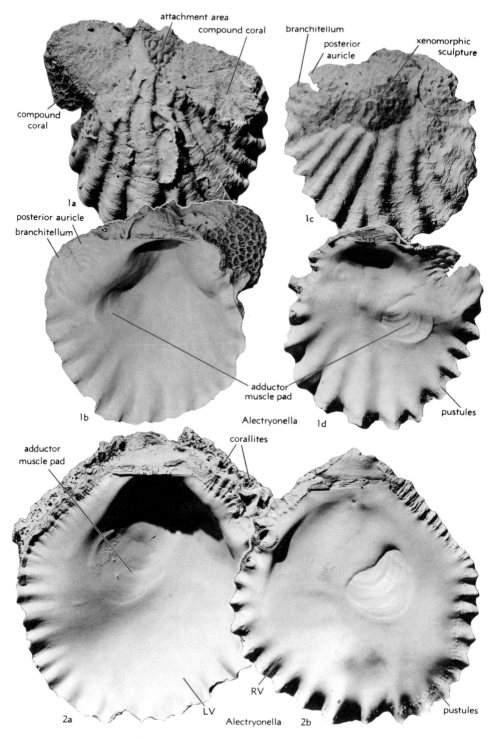

Fɪɢ. J135. Ostreidae (Lophinae) (p. *N*1160-*N*1161, *N*1163-*N*1164).

surfaces of the valves (Fig. J134). ?*Mio., Plio.-Rec.,* tropical SW.Pac.——Fig. J135,*1,2.* **A. plicatula* (GMELIN), growing on compound hexacoral, Mad-ag.(Nosi-Bé); *1a-d,* LV ext., int., RV ext., int., ×0.75; *2a,b,* LV int., RV int., ×0.75 (Stenzel, n). [Specimens by courtesy of R. TUCKER AB-BOTT, formerly of Acad. Nat. Sci. Philadelphia; all whitened for photography.] [*See* also Fig. J29.]

[Interpretation of this genus depends entirely on under-standing of the type species, which is attainable in manner explained as follows. When SACCO (1897a) selected *"Alec-tryonella plicatula* (Gm. Lk.)" as the type species of his new subgenus *Alectryonella,* he made it quite clear through the authorships indicated that the particular species he had in mind was LAMARCK's interpretation of *Ostrea plicatula* GMELIN (1791, p. 3336, no. 111). LAMARCK's concept of this species can be analyzed, because fortunately the speci-mens studied by him are preserved at the Laboratoire de Malacologie of the Muséum National d'Histoire Naturelle in Paris. There they were studied by me in September 1962 with the generous aid of GILBERT RANSON. The six type lots of *Ostrea plicatula* LAMARCK (1918, p. 211) con-tain several different species. Which one of the species should retain the name *plicatula* LAMARCK can be solved only by a judicious selection of a lectoholotype from among his specimens. LAMARCK himself recognized that more than one so-called variety were represented among his types, for he closed his description with the words: "It . . . offers a quantity of varieties which to distinguish would be more detrimental than useful to science." Some of the type lots are labeled var. [b], var. [c], and var. [d], and these designations apparently go back to LAMARCK. If a lectoholotype is to be selected, it is evident that all the type lots which are labeled var. [b], var. [c], and var. [d] were regarded by LAMARCK as atypical and should be ex-cluded from consideration at the start. This restriction narrows the selection down to one stiff cardboard to which are glued two well-preserved bivalved specimens. This same cardboard was listed by LAMY (1929-30, p. 82-89) as the "1er carton" and bears two labels *"ostrea plicatula"* and *"huitre plicatule/ostrea plicatula"* in LAMARCK's hand-writing attached to its back; in addition, the front of the cardboard is labeled *"Ostrea/Ostrea plicatula* Lamk/Nelle Hollande (individus nommés par LAMARCK)." The two specimens are much alike and certainly are of the same species. Also, they correspond to the figure given by CHEMNITZ (1780-95, v. 8, pl. 73, fig. 674), as LAMY has pointed out. This fact is rather important, because GMELIN had not illustrated his species, but had defined it by a short Latin description and two references to illustrations in older publications, one of which was this figure in CHEMNITZ. By restricting LAMARCK's specific name to the two types of the "1er carton" of LAMY, one succeeds in establishing the identity of "Die faltenvolle Auster" of CHEMNITZ with *Ostrea plicatula* GMELIN and with *O. pli-catula* LAMARCK. At the same time one takes into account LAMARCK's own revision of his type materials as implied by the way in which he had separated the materials into several so-called varieties. The larger one of the two type specimens of the "1er carton" is herewith selected as the lectoholotype of *Ostrea plicatula* LAMARCK, 1819. Its di-mensions are: largest dimension 7.66 cm., dimension at right angles to the largest 6.11 cm., and width 2.66 cm. These dimensions, measured in 1962, are somewhat differ-ent than those given by LAMY, who probably did not use calipers. The peculiar finger-print shell structure is visible on the inside of both valves in both specimens. The struc-ture is believed to be a definitive feature of this species, even though it was not noticed, or at least not mentioned by CHEMNITZ, GMELIN, LAMARCK, LAMY, and several other authors. The same structure was seen in the following additional museum specimens, which are regarded as rep-resenting *Ostrea plicatula* GMELIN, 1791: (1) *Ostrea lactea* G. B. SOWERBY (1871, pl. 21, figs. 48 a-b), 1 specimen, Acad. Nat. Sci. Philadelphia. (2) *O. cumingiana* DUNKER (DUNKER in PHILIPPI, 1845-47, *Ostrea* issue, p. 81-82, and pl.), 10 specimens from Indochina, Mus. Natl. d'Histoire Nat. (Paris). DUNKER's own description mentions this peculiar structure, and it is obvious that the 10 specimens are correctly identified as DUNKER's species. The Phila-delphia specimen is most likely correctly identified and may have been identified originally by SOWERBY himself, because some of the older specimens at Philadelphia were obtained from SOWERBY or REEVE. GILBERT RANSON agreed

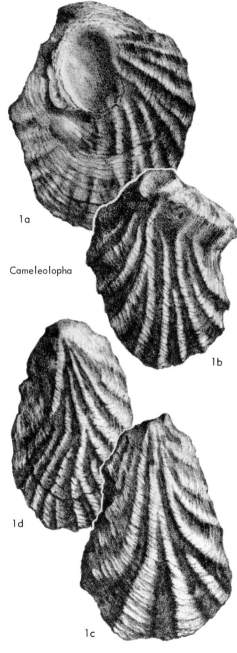

1a

Cameleolopha

1b

1d

1c

FIG. J136. Ostreidae (Lophinae) (p. *N*1164).

in September 1962 that the 10 specimens labeled *O. cumin-giana* are the same species as the two types of the "1er carton" of *O. plicatula* LAMARCK, 1819. The type species is incubatory (AMEMIYA, 1929).] [Circumstances relating to designation and identification of the type species of *Alectryonella* GMELIN as recorded and interpreted by STEN-

ZEL have been reported clearly by him in foregoing text now published at his insistence. STENZEL, thinking that his discussion is adequate, has not accepted suggestion of the Editor that the systematic fixation of *Alectryonella* can be most simply and firmly established by designating now a neotype specimen—say from one of the lots of LAMARCK's *Ostrea plicatula* (1819)—to serve as holotype of *O. plicatula* GMELIN, 1791, type species of *Alectryonella* by original designation.—R. C. MOORE.]

Cameleolopha VYALOV, 1936, p. 20 [**Ostrea cameleo* COQUAND, 1869, p. 149; OD]. This taxon proposed as a section of *Lopha s.s.* includes small shells (to 6 cm. high) with outlines orbicular or oval to spatulate and rounded; auricles generally absent, never large. Both valves have 12 to 20

angular, narrow-crested, dichotomous and intercalating radial ribs that continue to commissure. RV flat, LV convex to gibbous. *U.Cret.(Cenoman.)*, N.Afr.(Alg.).——FIG. J136,1. **C. cameleo* (COQUAND), Bou Sâada, Subdivision de Sétif; *1a-d*, outlines of 4 individuals, ×1 (Coquand, 1869).

Nicaisolopha VYALOV, 1936, p. 20 [**Ostrea nicaisi* VYALOV, 1936, *nom. null.* (=*O. nicasei* COQUAND, 1862=*O. elegans* BAYLE in FOURNEL, 1849, p. 366, *non* DESHAYES, 1832, v. 1, p. 361); OD]. Medium-sized (to 10 cm. long and 11 cm. high),

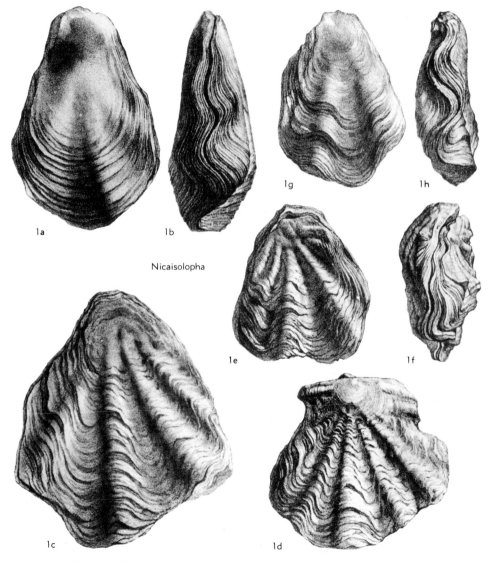

1g

1h

1a   1b

Nicaisolopha

1e   1f

1c   1d

FIG. J137. Ostreidae (Lophinae) (p. *N1164-N1165*).

outline orbicular or ovate to spatulate, with greatest length very near ventral margin and height
about 110 to 130 percent of length; flattish, width
25 to 40 percent of height, and subequivalve; LV
very similar to RV but slightly more convex and
capacious. Attachment area variable, very small
to large; neanic part of shell smooth, orbicular,
and only slightly convex at diameter of 2 to 4
cm.; marginal commissural folds beginning to
develop at end of this growth stage, such folds
few (4 to 7 large ones at ventral margin, small
ones elsewhere), well rounded at their crests.
Radial folds similar on both valves, not dichotomous, undulatory, round-crested and separated by
equal, rounded interspaces, ending at commissural
folds. Growth squamae slightly raised in later
growth stages. Adductor muscle imprint reniform, deeply concave at its dorsal margin, about
twice as long as high, placed slightly ventral of
valve center. Ligamental area low and long, its
height is about 40 percent of length; LV resilifer
shallowly excavated and slightly longer than bourrelets; RV resilifer flat. No chomata. [*Nicaisolopha* was defined by VYALOV as a section of
Lopha s.s. as follows: "Sculpture consists of
vague folds."] *U.Cret.(Turon.-Maastricht.),* Eu.
(Port.) - N.Afr.(Alg.-Tunisia-Egypt)-Mexico-S.Am.
(Peru).——FIG. J137,*1.* **N. nicasei* (COQUAND,
1869), Campan., Alg. (M'zâb-el-Messai and other
local.); *1a-h,* 5 individuals, ×0.7 (Coquand,
1869).

Rastellum FAUJAS-SAINT-FOND, 1799 [?1802], p.
167 [**Ostrea macroptera* J. DE C. SOWERBY, 1824,
v. 5, p. 105; SD WINKLER, 1863-67, p. 251]
[=*Rostellum* PERVINQUIÈRE, 1910a, p. 119 *(nom.
null.)* *(non* MÖRCH, 1850, p. 26)]. Small to large
(largest diameter up to 24 cm.), almost equivalve,
sculpture same on both valves, outlines crescentic
to triangular-crescentic, latter accompanied by large
triangular auricles anterior and posterior to ligament producing long straight hinge line; posterior
auricle commonly larger one. Shells with crescentic outline long-tapering, curved like comma or
hook, semicircular or almost full circle, with convex anterior and concave posterior flanks that
tend to be vertical to commissural plane or nearly
so, resulting in narrow roof along mid-axis of
valves. Attachment area mostly large. Adductor muscle large and subtriangular to comma-
shaped in cross section, located rather close to
hinge, i.e., less than half of shell height from
hinge. Roofs of valves with many branching dichotomous costae, subdued to well-developed;
costae turn into sharp plicae on reaching steep
flanks and descend straight down flanks without
any dichotomy. Crests of plicae either smooth or
raised at intervals into narrow hyote spines. Valve
commissure has many (100 or more) interlocking
zigzags, the tips of which are acute-angled;
grooves between plicae end in projecting acute
tips and plicae end in corresponding recesses.

1b

Rastellum

1a

FIG. J138. Ostreidae (Lophinae) (p. N1165).

M.Jur.(Callov.)-U.Cret.(Maastricht.), worldwide.

[The confused nomenclatural history of this generic name
has been discussed by PERVINQUIÈRE, 1910a, 1911, and
STENZEL, 1947. (See also discussion herein, p. N1201.)]

R. (Rastellum). Shell outline tending more to
triangular crescentic than to crescentic with
parallel flanks. Plicae mostly without hyote
spines.   *M.Jur.(Callov.)-U.Cret.(Maastricht.),*
worldwide.——FIG. J138,*1.* **R. (R.) macropterum* (J. DE C. SOWERBY), Maastricht. (Tuffeau
de Maestricht), Neth. (St. Pietersberg, S. of Maa-

stricht.); *1a,b,* LV int., LV oblique ant. view, on rock matrix of Maastricht grainstone, ×1 (Stenzel, n). [Specimen loaned by courtesy of C. O. VAN REGTEREN ALTENA, Teyler's Museum, Haarlem, Neth., no. 11046, type of FAUJAS-SAINT-FOND [?1802], pl. 28, fig. 7, as *Rastellum. See* also Fig. J153,*4-5.*]

R. **(Arctostrea)** PERVINQUIÈRE, 1910a, p. 119 [**Lopha (Arcostrea) carinata* LAMARCK, 1806a, p. 166; OD] [=*Arctostraea* JOURDY, 1924, p. 17 *(nom. van.); Artostrea* VYALOV, 1936, p. 20 *(nom. null.); Arcostrea* CHARLES & MAUBEUGE,

1951, p. 114 *(nom. null.)*]. Shell outline tending to crescentic with both flanks parallel and very steep; outlines long-tapering and curved like comma, hook or semicircle to almost full circle; steep flanks (up to 5 cm. tall) on each valve. Plicae rise at their crests at intervals to form slender hyote spines, which in some grow into long tubules. Commissure very serrate; serrae as many as 3 per cm. of commissure length. Tips of zigzags sharp-pointed, long, and narrow. [Distinction between *R. (Rastellum)* and *R. (Arctostrea)* is uncertain.] *Cret.(U.Alb.-L.Cenoman.),*

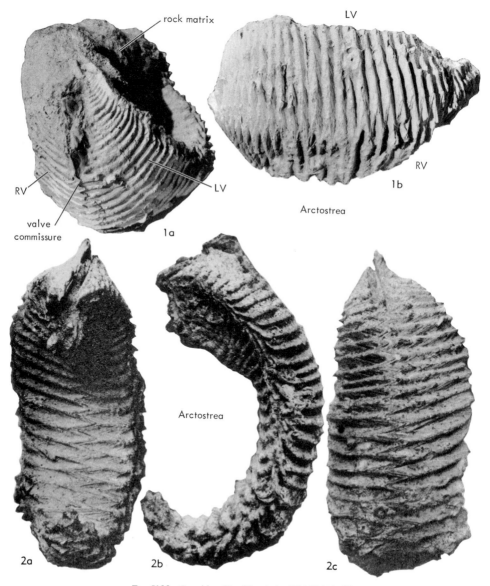

FIG. J139. Ostreidae (Lophinae) (p. *N1166-N1167*).

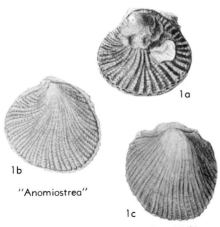

FIG. J140. Genera Uncertain (p. N1167).

worldwide.——FIG. J139,*1,2.* **R. (A.). carinatum* (LAMARCK), Cenoman., France; *1a-b,* oblique umbonal view, ant., ×1; *2a-c,* LeMans, both valves post., right side, ant., ×1 (holotype) (*1,* Stenzel, n; *2,* Pervinquière, 1910b). [Specimen loaned by courtesy of Naturhist. Museum, Basel, Switz.]

[An interesting study of the functional morphology of *Arctostrea* was made by CARTER, 1968.]

NOMINA DUBIA AND GENERA OF DUBIOUS TAXONOMIC VALUE OR POSITION

Listed below are generic names published for Ostreidae AUCTORUM which were obscurely described and are difficult to recognize or interpret. Their type species were not available for this study.

Anomiostrea HABE & KOSUGE, 1966, p. 323 [*Ostrea pyxidata ADAMS & REEVE, 1848, p. 72 (*non* BROCCHI, 1814, p. 579, =*O. pyxidata* BORN, 1778, p. 93); OD]. Described as follows: "The shell is small, thin, orbicular, dark green to pale yellow, inequivalve. The upper valve is flat, divergently ribbed and has the ovate white muscular scar situated near the center of its inner surface. The lower valve attached to the coral is deeply concaved, forming a cup-shape and crenulated at the margin by the ribs on the surface and its muscular scar is distinctly elevated by the white callus. The ligament is small and short. This is a unique *anomia*-like oyster found on the branches of corals in the Philippines and rather commonly collected at Samboanga, Mindanao by Mr. Ichiro Yamamura. No related species has been reported until today." Figure and description given are not sufficiently informative.——FIG. J140,*1.* **A. pyxidata* (ADAMS & REEVE, 1850), Philip.I.; *1a-c,* both valves, seen from right(?) side, ×1 (*1a,* Adams & Reeve, 1850, pl. 21, fig. 19; *1b,c,* G. B. Sowerby, Jr., 1870-71, v. 18, *Ostraea* sp. 16). [All

figures from copies furnished by courtesy of RUTH TURNER, Museum of Comparative Zoology, Harvard Univ., Cambridge, Mass.]

Anulostrea VYALOV, 1936, p. 19 [**Ostrea bourguignati* COQUAND, 1869, p. 86; OD]. Proposed as subgenus of *Liostrea.* All specimens figured by COQUAND have valves closed, so that nothing is known of the interior, and the affinities of this taxon remain unknown. General outline of the shell, approaching rectangular, well-developed

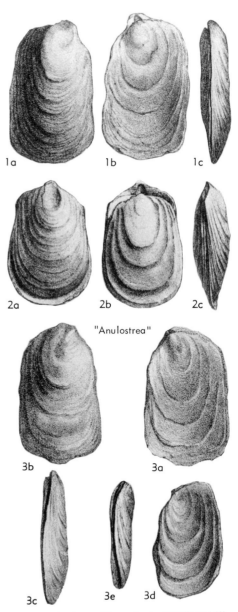

FIG. J141. Genera Uncertain (p. N1167-N1168).

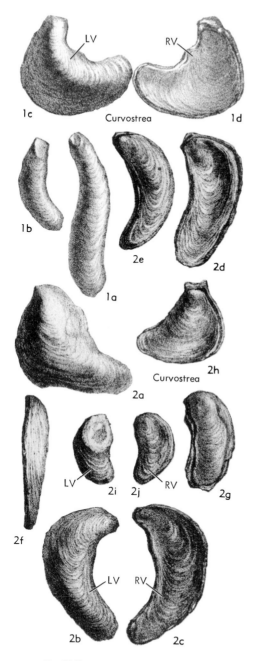

FIG. J142. Genera Uncertain (p. *N*1168).

U.Cret.(*Coniac.-Santon.*),   N. Afr. (Tunisia) - Eu.
(France).——FIG. J141,*1-3.* **A. bourguignati* (Co-
QUAND), Coniac.-Santon.; *1a-c,* Alg.(Sétif), both
valves, left side, right side, ant. view, ×1; *2a-c,*
France (Saint-Paterne), both valves, left side,
right side, ant. view, ×1; *3a-e,* Tunisia (R'fana
near Tebessa), both valves right side, left side,
ant. view, both valves right side, post. view, ×1
(COQUAND, 1869).

Curvostrea VYALOV, 1936, p. 18 [*Ostrea rediviva
COQUAND, 1869, p. 154; OD]. All figured specimens of this small species have valves closed,
so nothing is known concerning muscle imprints
and other internal features. Thus the taxon remains largely unknown and cannot be classified.
——FIG. J142,*1,2.* **C. rediviva* (COQUAND), U.
Cret.(Cenoman.); *1a-d,* France (Saint-André near
Goudargues, Département du Gard), LV ext.,
LV ext., both valves left side, right side, ×1;
2a-j, Alg. (Bou-Sâada), LV ext., both valves left
side, right side, both valves right side, both valves
right side, ant. view, both valves right side, both
valves right side, LV, RV, ×1 (Coquand, 1869).

Margostrea VYALOV, 1936, p. 20 [*Ostrea merceyi
COQUAND, 1869, p. 93; OD]. Proposed as section
of *Lopha s.s.* with the following definition: "Shell
not incurved, with a very large median smooth
surface; beyond, the sculpture presents only one
undulation or crenelations at the shell margins."
U.Cret.(*Santon.*), Eu.(France).——FIG. J143,*1,2.*
**M. merceyi* (COQUAND); *1,* Parnes, RV ext., ×1;
2a-g, Tartigny, Département Oise, LV ext., int.,
RV ant., RV int., LV int., ext., RV int., ×1 (Coquand, 1869).

[Only one unlabeled specimen possibly referable to the
type species was found in collections of the École des Mines
in Paris. The central unribbed area of RV evidently is
xenomorphic in this bivalved specimen, because its LV has
a correspondingly large attachment area. It is also evident
from figures published by COQUAND that all specimens of
the type species are young and grew on an elongate and
convex, that is, more or less cylindrical, substrate. In
these figures the RVs have an unribbed convex central area
and the LVs have the corresponding concave attachment
area. Therefore, the central unribbed areas are xenomorphic and cannot be diagnostic for supraspecific taxa.
For this reason the species too remains unrecognizable; it
is perhaps a lophine oyster. However, even that guess is
dubious, because WOODS, 1913, p. 381, listed *Ostrea merceyi*
COQUAND in the synonymy of *Ostrea semiplana* J. DE C.
SOWERBY, 1825, =*Hyotissa semiplana* (J. DE C. SOWERBY,
1825). Thus *Margostrea* must be set aside as indeterminate.]

Notostrea FINLAY in MARWICK, 1928, p. 432
[**Ostrea subdentata* HUTTON, 1873, p. 34; OD].
Only one valve, best illustrated by BOREHAM
(1965, pl. 13, fig. 2-3, p. 48), has been figured.
Contrary to statements by FINLAY (1928b, p. 266)
this lectoholotype is a LV, not RV. Descriptions
given and material available so far are insufficient.
BOREHAM's illustrations seem to show vermiculate
chomata. If it turns out that the shell wall has
vesicular shell structure, *Notostrea* would fall into
the Pycnodonteinae as a rather unusual monotypic
genus. *Low.Oligo.(Duntroon.),* N.Z. —— FIG.
J144,*1.* **N. subdentata* (HUTTON), Broken River,
Trelissick Basin; *1a,b,* holotype, LV ext., int., ×2
(Stenzel, n). [Plaster cast by courtesy of C. A.

shoulders at ends of the ligament, and regularly
and widely spaced concentric growth squamae separated by smooth intervals are all reminiscent of
Flemingostrea, but the specimens figured are too
small to show the fold of the commissure at the
venter of the shell, which would be diagnostic.

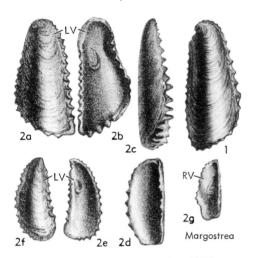

FIG. J143. Genera Uncertain (p. *N1168*).

FLEMING and Mrs. A. U. E. SCOTT, New Zealand Geol. Survey, Lower Hutt, N.Z.]

Pulvinostrea VYALOV, 1936, p. 17 [**Ostrea (Cymbulostrea) fluctuosa* MAYER-EYMAR, 1889, p. 3; OD]. Described as a section of the subgenus *Ostrea (Cymbulostrea):* "Beak pointed, shell high, the sculpture consists of some distant, interrupted costae." The type species was described but not figured by MAYER-EYMAR, who placed it in the "series of *Ostreae Polyphemi.*" It was redescribed by OPPENHEIM (1903, p. 32-33, pl. 1, fig. 12-12b) from a single LV which had been identified by

MAYER-EYMAR and deposited at Munich. Although this is believed to be the type, or one of the types, some serious discrepancies between it and the description given by MAYER-EYMAR were noted by OPPENHEIM (p. 33), so that a mixup of labels is not improbable. Although MAYER-EYMAR described both valves, no RV was available to OPPENHEIM. *Eoc.,* Afr.(Egypt).——FIG. J145,*1.* *?P. fluctuosa?* (MAYER-EYMAR) (=*Ostrea (Cymbulostrea)* section *Pulvinostrea* VYALOV, 1936), Libyan, Nobka; *1a-c,* LV ext., int., post., ×1 (Oppenheim, 1903). [Labeled as a type specimen by MAYER-EYMAR, 1889, but possibly erroneously.]

[Because MAYER-EYMAR stated that the upper valve is devoid of ribs and the lower valve has 9 radial ribs, it is likely that this species falls into the general group of genera and

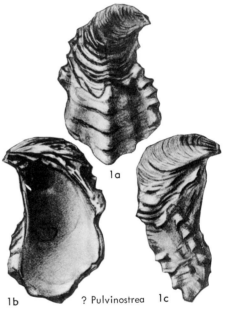

FIG. J145. Genera Uncertain (p. *N1169*).

subgenera that cluster around *Ostrea s.s.* For the rest, the material available to MAYER-EYMAR, OPPENHEIM, and VYALOV is too inadequate to permit one to establish a species firmly, let alone a supraspecific taxon.]

Quadrostrea VYALOV, 1936, p. 18 [**O. tetragona* BAYLE in FOURNEL, 1849, v. 1, p. 367; OD]. Described as a section of *Liostrea (Liostrea):* "Beak slightly prominent, cardinal margin straight, shell angular." The holotype of the type species at the École des Mines in Paris is a bivalved specimen with valves closed and interior filled with matrix and inaccessible. Its systematic position remains unknown. *U.Cret.(Senon.),* Afr.(Alg.). —— FIG. J146,*1.* **Q. tetragona* (BAYLE), Campan., M'zab-el-Messaï and other local.; *1a-c,* two individuals, ×1 (COQUAND, 1869).

FIG. J144. Genera Uncertain (p. *N1168*).

"Quadrostrea"   1b

1c

Fig. J146. Genera Uncertain (p. N1169).

GENERA ERRONEOUSLY ASSIGNED TO OSTREIDAE

The following genera originally were described as or suspected to be Ostreidae AUCTT., but found to belong to other families.

Heterostrea JAWORSKI, 1913, p. 192-195, pl. 6-7 [*Crassostrea (Heterostrea) steinmanni;* M]. Heterostrea was originally believed to be a primitive, dimyarian, ancestral stage (Stadium in German) of *Crassostrea.* It was found in the Jur. (M. Bajoc.) of Chunumayo, Peru. JAWORSKI (1951) later removed it from the Ostreidae and placed it in *Myoconcha* SOWERBY, 1824, p. 5, 103, near *M. unguis* WHIDBORNE (1883, p. 530, pl. 18, fig. 21), family Pleurophoridae, subfamily Myoconchinae NEWELL, 1957 (p. N547).——Fig. J147,1. *Myoconcha steinmanni* (JAWORSKI) [="Crassostrea (Heterostrea)"]; 1a-c, RV int., LV int., ext., ×0.4 (Jaworski, 1913).

Holocraspedum CRAGIN (1893, p. 190-191) [*Ostrea anomiaeformis* ROEMER, 1849, p. 394; OD]. CRAGIN correctly placed this genus in the Anomiidae, although the type species had been de-

scribed as an *Ostrea.* All the specimens of the type species that were available to CRAGIN are preserved at the University of Texas in Austin, Texas; they are all anomias and *Holocraspedum* is here questionably regarded by Cox as a junior synonym of *Placunopsis* (p. N380).

Lithiotis GÜMBEL, 1871, p. 48-51 [*L. problematica;* M]. Since GÜMBEL described it, believing it to be a lime-secreting alga, many authors have discussed its systematic position and some believed it to be an oyster close to *Crassostrea.* It is now placed in the Lithiotidae REIS, 1903, near the Spondylidae (*see* p. N1200).

Ostreinella COSSMANN in COSSMANN & PEYROT, 1914, p. 398-400 [*Liostrea (Ostreinella) neglecta* (MICHELOTTI), 1847, p. 81, pl. 3, fig. 6, not fig. 3; OD]. Originally described as a subgenus of *Liostrea* it was recognized by GLIBERT & VAN DE POEL, 1965, p. 6, as a member of the Vulsellidae [Malleidae] (*see* p. N331).——Fig. J148,*1.* *O. neglecta* (MICHELOTTI), Mio. (Burdigal.), SW. France (Aquitaine); *1a-d,* loose valves, ×2 (Cossman & Peyrot, 1913-14).

Palaeostrea GRABAU, 1936, p. 284-286, pl. 28, fig. 1 [*P. sinica;* OD]. This incompletely described fossil is wholly unidentifiable and probably not an oyster, see p. N1051. [*See* also Fig. J59,4.]

Praeostrea BARRANDE, 1881a, p. 147, pl. 3, fig. 1-2 and 3-4 [*P. bohemica;* M]. The taxon is now placed as a junior synonym of *Vevoda* BARRANDE, 1881, in the family Antipleuridae NEUMAYR, 1891, Superfamily Praecardiacea HÖRNES, 1884 (*see* p. N247, p. N1051). It has been restudied by KŘÍŽ, 1966. [*See* Fig. J59,5.]

UNAVAILABLE GENERIC NAMES

The generic names listed below have been found to be unavailable according to the present *Code.*

crist. VON SCHLOTHEIM (1820, p. 240-245). This abbreviation written in lower-case letters was used in connection with 9 species names of the genus *Ostracites.* The list was headed by the vernacular "D. Cristaciten. (Hahnenkämme)"; the abbreviation was not explained. Later VON SCHLOTHEIM (1823, p. 82) used "*cristacit.*" instead in connection with two of these species. STENZEL (p. N1210) has shown that these abbreviations stand for *Cristacites* and were meant to be a subgeneric name. *Cristacites* VON SCHLOTHEIM is a *nomen oblitum,* because until STENZEL'S investigation no one had recognized it as a subgenus or spelled it out correctly. Because STENZEL is the first reviser, his conclusion that *crist.* and *cristacit.* are unavailable is to be followed notwithstanding that VOKES (1967), p. 193, listed *Cristacites* VON SCHLOTHEIM as valid nomenclaturally.

Acuminata ARKELL (1934, p. 64).

Bilobata ARKELL (1934, p. 64).

Corrugata ARKELL & MOY-THOMAS (1940, p. 404).

Cretagryphaea ARKELL (1934, p. 62).

Dilatata ARKELL (1934, p. 64).
Incurva ARKELL (1934, p. 64).
Jurogryphaea ARKELL (1934, p. 62).
Knorrii ARKELL (1934, p. 64).
Marcoui ARKELL (1934, p. 64).
Virgula ARKELL (1934, p. 64).

These ten names were proposed as "rationalised names" admittedly under disregard of the zoological Rules. They were called "name of the lineage," that is, they were meant to designate monophyletic groupings of some sort. However, each lacked definition or indication as required by the *Code* and remained unavailable.

Neogryphaea VYALOV & SOLUN (1957, p. 197).
Pseudogryphaea VYALOV & SOLUN (1957, p. 197).

These two names were proposed as hypothetical names for as yet undescribed genera. They lack species and definitions. Contrary to the listing by VOKES (1967, p. 196-197), they are unavailable.

Neogyra CHELTSOVA (1969, p. 9).

This name is a *nom. nudum* in CHELTSOVA. It

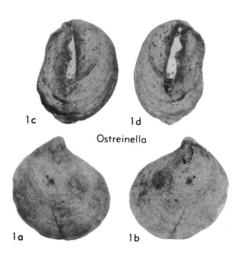

Ostreinella

FIG. J148. Malleidae (*Ostreinella*) (p. N1170).

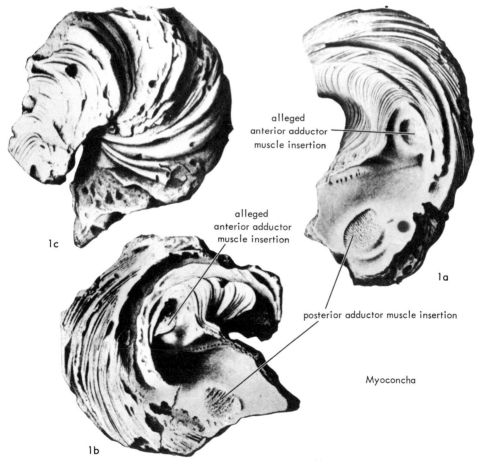

FIG. J147. Permophoridae (*Myoconcha*) (p. N1170).

was attributed by her to Vyalov & Solun (1957), but no such name is to be found there.

Sdikia de Gregorio, 1884, p. 48 [*Ostrea (Sdikia?) bonfornellensis*; OD]. *Sdikia* was proposed as a subgenus of *Ostrea*; the only species was described but not illustrated and was listed with a question mark as shown above. Species and genus were based on a single valve only and remain unrecognizable. In view of the question mark the species listed cannot be accepted as the type species (*Code*, Art. 67h). Thus *Sdikia* remains a nominal genus without included species, besides being unrecognizable. *?U.Mio.*, Italy (Buonfornello).

REFERENCES

Adams, Arthur, & Reeve, L.
(*1) 1848, *Mollusca:* in Arthur Adams (ed.), 1848-50, The zoology of the voyage of H. M. S. *Samarang* under command of Captain Sir E. Belcher . . . during . . . 1843-46, 4 pts., x+87 p., 24 pl., Reeve & Benham (London).

Agassiz, Louis
(*2) 1842-46, *Nomenclatoris zoologici index universalis continens nomina systematica classium, ordinum, familiarum et generum animalium omnium tam viventium quam fossilium*, etc.: Fasc. 9-10 (Nomina systematica generum molluscorum), viii+393 p., Jent & Gassmann (Soloturn, Switz.). [Paged separately for each chapter.]

Albrecht, J. C. H., & Valk, W.
(*3) 1943, *Oligocäne Invertebraten von Süd-Limburg:* Geol. Stichting Mededeel. (Maastricht), ser. C-IV-1, no. 3, 163 p., 27 pl.

Amemyia, Ikasaku
(*4) 1929, *Another species of monoecious oyster, Ostrea plicata Chemnitz:* Nature, v. 123, no. 3110, p. 874 (June 8).

Anderson, A. E., Jonas, E. C., & Odum, H. T.
(*5) 1958, *Alteration of clay minerals by digestive processes of marine organisms:* Science, v. 127, no. 3291, p. 190-191 (Jan. 24).

Anonymous [=d'Argenville, Desallier]
(*6) 1742, *L'histoire naturelle éclaircie dans deux de ses parties principales, la lithologie et la conchyliologie, dont l'une traité des pierres et l'autre des coquillages, ouvrage dans lequel on trouve une nouvelle méthode & une notice critique des principaux auteurs qui ont écrit sur ces matières:* vi+492 p., frontispiece, 33 pl. [1st edition], de Bure l'aîné (Paris).

Anonymous [H. G. Bronn probably]
(*7) 1831, *Review of G. Fischer's neue fossile Konchylien-Geschlechter und Arten in Russland* (Bull. Soc. Nat. Moscou, Mosc. 1829. I. 12 cah. 8vo*>Bibl. univers. 1830. Août.- Science. Arts. XLIV. 412-416): Jahrb. Mineralogie, Geognosie, Geologie u. Petre-faktenkunde, K. C. von Leonhard & H. G. Bronn (ed.), v. 2 (1831), p. 335-336.

Archiac, Adolphe d'
(*7a) 1850, *Description des fossiles du Groupe Nummulitique recueillis par M. S.-P. Pratt et M. J. Delbos aux environs de Bayonne et de Dax:* Soc. Géol. France, Mém., sér. 2, v. 3, p. 2, mém. 6, p. 397-456, pl. 8-13.

——, **& Haime, Jules**
(*8) 1853, *Description des animaux fossiles du Groupe Nummulitique de l'Inde précédée d'un résumé géologique et d'une monographie des nummulites:* vii+373+iii p., unnumbered fig.; atlas, 36 pl., Gide & J. Baudry (Paris).

Arkell, W. J.
(*9) 1932-36, *A monograph of British Corallian Lamellibranchia:* Palaeontograph. Soc., v. 84 (1930) [Dec. 1932], pt. 4, p. 133-180, fig. 22-48, pl. 13-20; v. 88 (1934) [Dec. 1935], pt. 8, xvi+325-350, pl. 45-49; v. 89 (1935) [Nov. 1936], pt. 9, xvii-xxii+p. 351-376, pl. 50-56.

(*10) 1933, *The Jurassic System in Great Britain:* xii+681 p., 97 fig., 41 pl., frontispiece, Oxford Univ. Press (Oxford).

(*11) 1934, *The oysters of the Fuller's Earth; and on the evolution and nomenclature of the Upper Jurassic Catinulas and Gryphaeas:* Cotteswold Naturalists' Field Club, Proc., v. 25 (1933-1935), pt. 1, p. 21-68, 5 fig., pl. 1-6, table.

——, **& Moy-Thomas, J. A.**
(*12) 1940, *Paleontology and the taxonomic problem:* p. 395-410, 1 fig., in Julian Huxley (ed.), The new systematics, viii+583 p., illus., Clarendon Press (Oxford).

Asari, Tamiya
(*13) 1950, *Geochemical distribution of strontium. VII. Strontium content of shells:* Chem. Soc. Japan, Jour., v. 71, p. 156-158. [In Japanese.]

Astre, Gaston
(*14) 1922, *Recherches critiques sur l'Ostrea, dite stricticostata, des terrains nummulitiques de la Montagne-Noire (=Ostrea moussoulensis nov. spec.):* Soc. Histoire Nat. Toulouse, Bull., v. 50, p. 141-204, pl. 1-6.

Awati, P. R., & Rai, H. S.
(*15) 1931, *Ostrea cucullata (The Bombay oyster):* Indian Zool. Mem. on Indian animal types, no. 3, 107 p., 57 fig. (Jan.), Methodist Publ. House (Lucknow).

Azzaroli, Augusto
(*16) 1958, *L'Oligocene e il Miocene della Somalia, Stratigrafia, Tettonica, Paleontologia (Microforaminiferi, Coralli, Molluschi):* Palaeontographia Italica, v. 52, new ser. v. 22 (1958), 143 p., 34 fig., 36 pl. (Nov. 10).

Barrande, Joachim
(*17) 1881a, *Système silurien du centre de la Bohême, pt. 1, Recherches Paléontologiques, v. 6, Classe des Mollusques, Ordre des Acéphalés (section 1 of v. 6):* xxiv+342 p., pl. 1-48, folio edit. chez l'auteur et éditeur (Prague and Paris).
(*18) 1881b, *Acéphalés. Études locales et comparatives. Extraits du Système Silurien du centre de la Bohême, v. 6, Acéphalés:* xxxii+536 p., 10 pl. (irregular numbers), octavo edit. reprint of folio edition with different pagination, chez l'auteur et éditeur (Prague and Paris).

Baughman, J. L.
(*19) 1948, *An annotated bibliography of oysters, with pertinent material on mussels and other shellfish and an appendix on pollution:* Agr. & Mech. Coll. of Texas, 794 p. (May 24).

Bayle, Émile
(*19a) 1849, *Sur quelques fossiles de la Province de Constantine:* Note A, p. 359-379, pl. 17-18, in Henri Fournel, 1849 (see *122b).
(*20) 1878, *Fossiles principaux des terrains:* Explic. Carte Géol. France, v. 4, atlas, pt. 1, pl. 1-158 and pl. expl. [France, Service carte géol.] [Text was never published.]

Benett, Etheldred
(*20a) 1831a, *A catalogue of Wiltshire fossils:* in Sir R. C. Hoare, A history of modern Wiltshire (1822-44, 14 pt. in 6 vol.), v. 3, pt. 2, p. 117-126, J. Nichols & Son (London).
(*20b) 1831b, *A catalogue of the organic remains of the County of Wilts.:* iv+9 p., 18 pl., J. L. Vardy (Warminster).

Bernard, Félix
(*21) 1896, *Troisième note sur le développement et la morphologie de la coquille chez les lamellibranches (Anisomyaires):* Soc. Géol. France, Bull., ser. 3, v. 24, pt. 6, p. 412-449, 15 fig. (Aug. 22).

Beurlen, Karl
(*22) 1958, *Die Exogyren. Ein Beitrag zur phyletischen Morphogenese der Austern:* Neues Jahrb. Geologie Paläontologie, Monatsh., no. 5, p. 197-217, 3 fig. (May).

Beyrich, H. E.
(*23) 1852, *Bericht über die von Overweg auf der Reise von Tripoli nach Murzuk und von Murzuk nach Ghat gefundenen Versteinerungen:* Deutsche Geol. Gesell., Zeitschr., v. 4, p. 143-161, pl. 4-6 (Jan.); also same author, date, and title: Gesell. für Erdkunde zu Berlin Monatsber. über die Verh., new ser., v. 9 (1852), p. 154 etc.
(*24) 1862, *Zwei aus dem deutschen Muschelkalk noch nicht bekannte Avicula-artige Muscheln:* Same, Zeitschr., v. 14, p. 9-10.

Bittner, Alexander
(*25) 1895, *Lamellibranchiaten der alpinen Trias, pt. I: Revision der Lamellibranchiaten von St. Cassian:* [K.K.] Geol. Reichsanst., Abh., v. 18, no. 1, 236 p., 24 pl. (Oct.).
(*26) 1901-12, *Lamellibranchiaten aus der Trias des Bakonyer Waldes:* Ungar. Geog. Gesell. Balaton Ausschuss, Result. wiss. Erforschung des Balatonsees, v. 1, pt. 1 (1901), Anhang—Paläontologie der Umgebung des Balatonsees, v. 2, pt. 3 (1912), 107 p., 9 pl.
(*27) 1902, *Lamellibranchiaten aus der Trias von Hudiklanec nächst Loitsch in Krain:* [K.-K.] Geol. Bundesanst., Jahrb., v. 51 (1901), p. 225-234, pl. 7 (Jan. 20).

Bjerkan, Paul
(*28) 1918, *Gamle østers: Norsk Fiskeritidende* [Bergen, Norway], v. 37, no. 2, p. 42-46, text fig. (Feb.).

Blainville, H. M. D. de
(*28a) 1821, GRYPHÉE, *Griphoea (Conchyl.):* in Dictionnaire des Sciences Naturelles [edit. 2, v. 19, 540 p.], p. 533, F. G. Levrault (Paris).
(*28b) 1825, *Manuel de malacologie et de conchyliologie, . . . :* text vol., viii+647 p., 87 pl., table synoptique, F. G. Levrault (Paris).

Bobkova, N. N.
(*28c) 1949, *Atlas rukovodyashchikh form iskopaemykh faun SSSR,* v. 11, Verkhniy mel gosgeolnzdat [Atlas of guide forms of the fossil fauna of the USSR, v. 11, Upper Cretaceous] [not seen].
(*29) 1961, *Pozdnemelovye ustritsy Tadzhikskoi depressii* [Late Cretaceous oysters of the Tadzhik Basin]: Problema neftegazonosnosti Srednei Azii, v. 7, Russia, Vses. Nauchno-Issledov. geol. Inst. (VSEGEI), Trudy, new ser., v. 50, 140 p.+1 index p., 32 pl.

Bøggild, O. B.
(*30) 1930, *The shell structure of the mollusks:* K. Danske Vidensk. Selsk., Skr., Naturv. og mat. Afd., 9 Raekke, v. 2, art. 2, p. 231-326, 10 fig., pl. 1-15.

Böhm, Georg

(*31) 1892, *Lithiotis problematica, Gümbel:* Naturf. Gesell. zu Freiburg im Breisgau, Ber., v. 6, no. 3, p. 55-80, pl. 2-4.

(*32) 1906a, *Zur Stellung von Lithiotis:* Centralbl. Mineralogie, Geologie, Paläontologie, p. 161-167, 2 fig.

(*33) 1906b, *Apicalhöhlung bei Ostrea und Lage des Muskeleindrucks bei Lithiotis:* Same, p. 458-461, fig.

Böhm, Johannes

(*34) 1904, *Über die obertriadische Fauna der Bäreninsel:* Svenska Vetensk.-Akad. Handlingar, new ser., v. 37, no. 3, 76 p., 7 pl., 10 fig. (Dec. 10).

(*35) 1911,*Ueber cretaceische und eocäne Versteinerungen aus Fergana:* in K. Futterer, Durch Asien; Erfahrungen, Forschungen und Sammlungen während der von Amtmann Dr. Holderer unternommenen Reise, v. 3, Naturwissenschaftliche und meteorologische Ergebnisse, pt. 3, Palaeontologie, no. 1, p. 93-111, 5 fig., 1 pl., Dietrich Reimer (Berlin).

(*36) 1933, *Die palaeogene Fauna Ost-Turkestans:* Deutsche Geol. Gesell., Zeitschr., v. 85 (1933), no. 2, p. 99-118, pl. 9-11 (March 25).

Böse, Emilio

(*37) 1906, *La fauna de moluscos del Senoniano de Cárdenas, San Luis Potosí:* Inst. Geol. Mexico, Bol. 24, 95 p., 18 pl.

(*38) 1910, *Monografía geológica y paleontológica del Cerro de Muleros cerca de Ciudad Juarez, Estado de Chihuahua, y descripcion de la fauna cretácea de la Encantada, Placer de Guadalupe, Estado de Chihuahua:* Same, Bol. 25, vi+193 p.; atlas, 2 maps+48 pl.

(*39) 1919, *On a new Exogyra from the ·Del Rio Clay and some observations on the evolution of Exogyra in the Texas Cretaceous:* Univ. Texas, Bull. 1902, 22 p., 1 fig., 5 pl. (Sept.).

Bonarelli, Guido, & Nágera, J. J.

(*39a) 1921, *Observaciones geologicas en las inmediaciones del Lago San Martin (Territorio de Santa Cruz):* Minister. de Agric., Argentine Repub., Direcc. Gen. Minas, Geologia e Hidrologia, Ser. B (Geol.), Bol. 27, 39+index p., 6 text fig., 6 pl.

Boreham, A. U. E. [Mrs. G. H. Scott]

(*40) 1965, *A revision of F. W. Hutton's pelecypod species described in the catalogue of Tertiary Mollusca and Echinodermata (1873):* New Zealand Geol. Survey, Paleont. Bull. 37, 84 p., 6 fig., 20 pl. (May).

Born, Ignatius von

(*41) 1778, *Index rerum naturalium musei Cae-*

sarei Vindobonensis, Pars I.ma, Testacea. Verzeichniss der natürlichen Seltenheiten des K. K. Naturalien Kabinets zu Wien, Erster Theil, Schalthiere: 458+several unnumbered p., Officina Krausiana (Vienna).

(*42) 1780, *Testacea musei Caesarei vindobonensis quae jussu Mariae Theresiae Augustae disposuit et descripsit:* Joannus Paulus Kraus (Vindobonae [Vienna]).

Borneman, V. A., Burachek, A. R., & Vyalov, O. S.

(*43) 1934, *K voprosu o rasprostranenii tretichnykh i melovykh ustrits v Sredney Azii* [Problem of distribution of Tertiary and Cretaceous oysters in central Asia]: Soc. Nat. Impér. Moscou, Bull., new ser., v. 42, sec. géol., v. 12 (2), p. 251-261.

Bosc, L. A. G.

(*44) 1802, *Histoire naturelle des coquilles,* etc.: 5 v. [edit. 1], Deterville (Paris).

Boule, Marcellin

(*45) 1913, *Types du prodrome de paléontologie stratigraphique universelle (suite):* Annales Paléontologie, v. 8, pt. 2, p. 145-176, pl. 27-36 (June).

Brander, Gustav

(*46) 1766, *Fossilia Hantoniensia collecta, et in Musaeo Britannico deposita:* vi+43 p., 9 pl. (London).

Brauer, Friedrich

(*47) 1878, *Bemerkungen über die im kaiserlich zoologischen Museum aufgefundeten Original-Exemplare zu Ign. von Born's Testaceis Musei Caesarei Vindobonensis:* K. Akad. Wiss. (Wien), Sitzungsber., Math-Naturw. Cl., v. 77, Abt. 1, no. 2, p. 117-192.

Breuer, J. P.

(*47a) 1962, *An ecological survey of the Lower Laguna Madre of Texas, 1953-1959:* Univ. Texas, Inst. Marine Sci. Pub., v. 8, p. 153-183, 4 text fig. (Nov.).

Brocchi, G. B.

(*47b) 1814, *Conchiologia fossile subapennina con osservazioni geologiche sugli Apennini e sul suolo adiacente:* v. 2, p. 241-712, pl. 1-16, G. Silvestri, Stamperia Reale (Milano).

Broili, Ferdinand

(*48) 1904, *Die Fauna der Pachycardientuffe der Seiser Alp:* Palaeontographica (K. A. von Zittel, ed.), v. 50, pt. 4, 5, p. 145-227, pl. 17-27 (Jan.).

Brongniart, Alexandre

(*48a) 1823, *Mémoire sur les terrains de sédiment supérieurs calcaréo-trappéens du Vincentin,* . . .: 86 p., 6 pl., F. G. Levrault (Paris).

Bronn, H. G.

(*48b) 1834-37, *Lethaea geognostica oder Abbil-*

dungen und Beschreibungen der für die Gebirgs-Formationen bezeichnendsten Versteinerungen . . .: edit. 1, text vol. and atlas, E. Schweizerbart (Stuttgart). [p. 193-480 are Lief. 3-5 and dated 1836.]

(*49) 1862, *Die Klassen und Ordnungen der Weichtiere (Malacozoa), wissenschaftlich dargestellt in Wort und Bild:* v. 3, pt. 1, Kopflose Weichtiere (Malacozoa Acephala), 518 p., 34 fig., 44 pl., C. F. Winter (Leipzig and Heidelberg).

Bruguière, J. G.

(*50) 1791-92, *Tableau encyclopédique et méthodique des trois règnes de la nature contenant l'helminthologie, ou les vers infusoires, les vers intestins, les vers mollusques,* etc.: livr. 7, pt. 1 (1791), i-viii+1-83 p., pl. 1-95; pt. 2 (1792), p. 85-132, pl. 96-189 (including pl. 107A,B,C), Panckouke (Paris).

Buch, Léopold von

(*51) 1835, *Note sur les huîtres, les gryphées et les exogyres:* Annales Sci. Nat., ser. 2, v. 3, Zoologie, p. 296-299.

Burnaby, T. P.

(*52) 1965, *Reversed coiling trend in Gryphaea arcuata:* Liverpool Geol. Soc. & Manchester Geol. Assoc., Geol. Jour., v. 4, pt. 2, p. 257-278, 1 fig. (March 18).

Cahn, A. R.

(*53) 1950, *Oyster culture in Japan:* Gen. Headquarters, Supreme Commander for the Allied Powers, Nat. Resources Sec., Rept. 134 (Tokyo), 83 p., 40 fig. (Sept.).

Carter, R. M.

(*53a) 1968, *Functional studies on the Cretaceous oyster Arctostrea:* Palaeontology, v. 11, pt. 3, p. 458-485, 11 text fig., pl. 85-90 (Aug.).

Caspers, Hubert

(*54) 1950, *Die Lebensgemeinschaft der Helgoländer Austernbank:* Biol. Anst. Helgoland, Helgoländer Wiss. Meeresunters. List (Sylt), v. 3, p. 119-169, 15 fig. (Dec. 28).

Charles, R. P.

(*55) 1949, *Essai d' étude phylogénique des gryphées liasiques:* Soc. Géol. France, Bull., ser. 5, v. 19 (1949), pt. 1-3, p. 31-41, 1 fig. (Nov. 25) (Dec.).

————, & Maubeuge, P.-L.

(*56) 1951, *Les huîtres plissées jurassiques de l'est du bassin parisien, (pt. 1):* Musée Histoire Nat. Marseille, Bull., v. 11 (1951), p. 101-119, 2 fig., pl. 1-3.

(*57) 1952a, *Les liogryphées du jurassique inférieur de l'est du bassin parisien:* Soc. Géol. France, Bull., ser. 6, v. 1 (1951), pt. 4-6, p. 333-350, 4 fig., 4 pl. (Jan. 16).

(*58) 1952b, *Les huîtres plissées jurassiques de l'est du bassin parisien (pt. 2):* Muséum Histoire Nat. Marseille, Bull., v. 12, p. 113-123, pl. 4-5.

(*59) 1953a, *Les liogryphées jurassiques de l'est du bassin parisien, II, Liogryphées du Bajocien:* Soc. Géol. France, Bull., ser. 6, v. 2 (1952), pt. 4-6, p. 191-195, 2 pl. (Feb. 13).

(*60) 1953b, *Révision des liogryphées du Musée d'Histoire Naturelle de Luxembourg:* Inst. Grand-Ducal de Luxembourg, sec. sci. nat., phys. & math., Archives, new ser., v. 20 (1951-53), p. 183-186, 1 fig.

Chave, K. E.

(*61) 1954, *Aspects of the biogeochemistry of magnesium, 1. Calcareous marine organisms:* Jour. Geology, v. 62, no. 3, p. 266-283, 16 fig. (May).

Cheltsova, N. A.

(*61a) 1969, *Znachenie mikrostruktury rakoviny melovykh ustrits dlya ikh sistematiki* [Significance of the microstructure of the shell of Cretaceous oysters for their systematics]: 87 p., 15 pl., Akad. Nauk SSSR (Moskva).

Chemnitz, J. H.

(*61b) 1780-95, *Neues systematisches Conchylien-Cabinet:* fortgesetzt (Bd. IV-XI) durch J. H. Chemnitz, etc., v. 4-11, Bauer & Raspe (Nürnberg).

Ćirić, B. M.

(*61c) 1951, *Nekoliko shkoljaka iz senonske faune rajićkog brda kod Guče (Dragčevo-Z. Srbija):* Prirod. Muz. Srpske Zemj'e, Glasnik, ser. A, v. 4, p. 61-66, pl. 1 [Quelques lamellibranches de la faune sénonienne de Jarićko Brdo, près de Guča (Dragačevo-Serbie occidentale); text in Serbian: Muséum Histoire Nat. Pays Serbe, Belgrade, Bull. ser. A].

Cizancourt, Marie de, & Cox, L. R.

(*61d) 1938, *Contribution à l'étude des faunes tertiaires de l'Afghanistan:* Soc. Géol. France, Mém., new ser., v. 17, Mém. 39, p. 1-44, pl. 1-5.

Clerc, M., & Favre, Jules

(*62) 1910-18, *Catalogue illustré de la collection Lamarck, pt. 1, Fossiles:* Muséum Histoire Nat. Genève, 10 p.+20 p. of Appendice, 117 pl. [Title page bears no authorships.]

Collier, Albert

(*63) 1959, *Some observations on the respiration of the American oyster Crassostrea virginica (Gmelin):* Univ. Texas, Inst. Marine Sci. Publ., v. 6, p. 92-108, 5 fig.

Conrad, T. A.

(*63a) 1832, *Fossil shells of the Tertiary formations of North America, illustrated by figures drawn on stone, from nature, v. 1, no. 2:* p. 21-28, pl. 7-14, Judah Dobson (Philadelphia) (Dec.).

(*63b) 1843, *Appendix C, List of fossils, etc.* [not seen], in J. N. Nicollet, 1843 (see *287a).

(*64) 1853, *Notes on shells, with descriptions of new species:* Philadelphia Acad. Nat. Sci., Proc., v. 6 (1852-53), no. 6, p. 199-200 [before Feb. 7, 1853].

(*64a) 1865a, *Catalogue of the Eocene and Oligocene Testacea of the United States:* Am. Jour. Conchology, v. 1, no. 1, p. 1-35 (Feb. 15 [Feb. 25]).

(*64b) 1865b, *Corrections and additions to Mr. Conrad's Catalogue of Eocene Mollusca, published in 1st number of this journal:* Same, v. 1, no. 2, two unnumbered pages following p. 190 (April 15).

(*64c) 1866, *Check list of the invertebrate fossils of North America. Eocene and Oligocene:* Smithsonian Misc. Coll. 200, iv+41 p. (May).

Coquand, Henri

(*65) 1862, *Géologie et paléontologie de la région sud de la province de Constantine:* Soc. émul. Provence, Mém., Marseille, 341 p.+ errata p., 59 fig. [8vo] and Atlas [entitled Géologie et paléontologie de la province de Constantine], 35 pl. [4°].

(*66) 1869, *Monographie du genre Ostrea, Terrain Crétacé:* text, 215 p., atlas, 74 pl., J.-B. Baillière & fils (Paris).

Cossmann, Maurice

(*66a) 1922, *Synopsis illustré des mollusques de l'Éocène et de l'Oligocène en Aquitaine:* Soc. Géol. France, Mém. Paléont., v. 24, fasc. 1-2, p. 113-220+4 p. of pl. explan., 7 pl.; Mém. no. 55, Suite et fin (June).

————, & Peyrot, Adrièn

(*67) 1914, *Conchologie néogénique de l'Aquitaine, suite:* Soc. Linnéenne de Bordeaux, Actes, v. 68, p. 5-210, 20 fig., and v. 67, pl. 11-22 (Aug. 1).

————, & Pissarro, G.

(*67a) 1904-13, *Iconographie complète des coquilles fossiles de l'Éocène des environs de Paris:* v. 1, 45 pl.+errata p. (1904-06); v. 2, 65 pl.+errata p. (1910-13), M. Pissarro (Paris).

Costa, E. M. da

(*68) 1776, *Elements of conchology; or, an introduction to the knowledge of shells:* viii+vi+318+errata p., 7 pl., Benjamin White (London).

Cotton, B. C.

(*68a) 1961, *South Australian Mollusca-Pelecypoda:* 363 p., 351 text fig., South Australian Branch of British Sci. Guild, Handbook of the Flora and Fauna of South Australia, Govt. Printer (Adelaide).

Couffon, Olivier

(*69) 1918, *Le Callovien du Chalet commune de Montreuil-Bellay (M.-&-L.):* Soc. Sci.

Angers, Bull., new ser., v. 47 (1917), p. 65-131, 8 fig., 4 pl.

Cox, J. C.

(*70) 1883, *On the edible oysters found in the Australian and neighbouring coasts:* Linnean Soc. New South Wales, Proc., v. 7, p. 122-133.

Cox, L. R.

(*71) 1924, *A Triassic fauna from the Jordan valley:* Annals & Mag. Nat. History, ser. 9, v. 14, p. 52-96, 1 fig., pl. 1-2.

(*71a) 1938, *Fossiles Éocènes du nord de l'Afghanistan:* in Marie de Cizancourt & L. R. Cox, 1938 (see *61d), p. 29-44, pl. 4-5.

Cragin, F. W.

(*71b) 1893, *A contribution to the invertebrate paleontology of the Texas Cretaceous:* Texas Geol. Survey, Ann. Rept. 4 (1892), p. i-iv and 139-294, pl. 24-46 (June).

Cumings, E. R.

(*72) 1903, *The morphogenesis of Platystrophia. A study of the evolution of a Paleozoic brachiopod:* Am. Jour. Sci., v. 165 or ser. 4, v. 15, no. 85, p. 1-48, 25 fig. (Jan.); no. 86, p. 121-136, fig. 26-27 (Feb.).

Cuvier, G. L. C. F. D.

(*73) 1817, *Le règne animal distribué d'après son organisation, pour servir de base à l'histoire naturelle des animaux et d'introduction à l'anatomie comparée, v. 2 contenant les reptiles, les poissons, les mollusques et les annélides:* xviii+532 p., Deterville (Paris).

————, & Brongniart, Alexandre

(*74) 1822, *La description géologique des couches des environs de Paris, parmi lesquelles se trouvent les gypses à ossements:* in B. Cuvier, Recherches sur les ossements fossiles, où l'on rétablit les caractères de plusieurs animaux dont les révolutions du globe ont detruit les espèces, new edit., v. 2, pt. 2, p. i-iv, 229-648, errata p., geol. map, pl. 1A-11, G. Dufour & E. d'Ocagne (Paris).

Dall, W. H.

(*75) 1880, *American work in the department of Recent Mollusca during the year 1879:* Am. Naturalist, v. 14, no. 6, p. 426-436.

(*76) 1898, *Contributions to the Tertiary fauna of Florida with especial reference to the silex beds of Tampa and the Pliocene beds of the Caloosahatchie River including in many cases a complete revision of and their American Tertiary species:* Wagner Free Inst. Sci. Philadelphia, Trans., v. 3, pt. 4, viii+p. 571-947, pl. 23-35 (Oct. 29).

(*77) 1914, *Notes on west American oysters:* Nautilus, v. 28, p. 1-3 (May).

Davis, H. C.

(*77a) 1950, *On interspecific hybridization in Os-*

trea: Science, v. 111, no. 2889, p. 522 (May 12).

Dechaseaux, Colette

(*78) 1934, *Principales espèces de Liogryphées liasiques, valeur stratigraphique et remarques sur quelques formes mutantes:* Soc. Géol. France, Bull., Notes et Mém., ser. 5, v. 4, no. 1-3, p. 201-212, 3 fig., pl. C-E (Sept. 18).

(*79) 1948, *Le genre Pachypteria de Koninck:* Same, Comptes Rendus Somm. des Séances, 1947, no. 15-16, p. 317-318 (Jan.).

Defrance, M. J. L.

(*80) 1821, *Gryphée. (Foss.):* p. 533-538, in Dictionnaire des Sciences Naturelles (edit. 2), v. 19, 540 p., L. G. Levrault (Paris) [Jan.].

(*81) 1821, *Huitres. (Foss.):* p. 20-33, in Dictionnaire des Sciences Naturelles, etc. [edit. 2], v. 22, 570 p., L. G. Levrault (Paris). [Dec., according to C. D. Sherborn, Index animalium, sec. 2, pt. 1, p. xliv.]

Delessert, Benjamin

(*82) 1841, *Recueil de coquilles décrit par Lamarck dans son Histoire naturelle des animaux sans vertèbres et non encore figurées:* 40 pl.+pl. explanations, Fortin, Masson & Cie (Paris).

Deshayes, G.-P.

(*83) 1824-37, *Description des coquilles fossiles des environs de Paris:* 814 p., atlas, lxv+101 pl., Chez l'auteur and others (Paris). [Dates of publication are given by R. Bullen Newton, Systematic list, etc.: v. 1, 1824 (p. 1-80), 1825 (p. 81-170), 1829 (p. 171-238), 1830 (p. 239-322), 1832 (p. 323-392); v. 2, 1824 (p. 1-80), 1825 (p. 81-146), 1832 (p. 147-290), 1833 (p. 291-426), 1835 (p. 427-498), 1837 (p. 781-814).]

(*84) 1860-66 (1856-65), *Description des animaux sans vertèbres découverts dans le bassin de Paris pour servir de supplement à la description des coquilles fossiles des environs de Paris comprenant une revue générale de toutes les espèces actuellement connues:* v. 1, 912 p.; v. 2, 968 p.; v. 3, 668 p., atlas v. 1, 89 pl.; v. 2, 107 pl. J. B. Baillière & Fils (Paris). [Publication dates of the livraisons are in v. 3, p. 668; Ostracea in text v. 2, p. 43-123, atlas v. 1, pl. 81-85; p. 1-120 dated Jan. 24, p. 121-192 dated May 2, 1861.]

Desmarest, M. E.

(*84a) 1850-61, *in J. C. Chenu, Encyclopédie d'histoire naturelle, . . .:* Tables générales, etc., Crustacés-Mollusques-Zoophytes, v. 4, p. ii+312, 40 pl., text fig., E. Girard & A. Boitte (Paris).

Diener, Carl

(*85) 1923, *Lamellibranchiata triadica:* Fossilium

Catalogus I, Animalia pt. 19, 257+2 index p. (Dec. 10).

Dodge, Henry

(*86) 1952, *A historical review of the mollusks of Linnaeus Part 1. The Classes Loricata and Pelecypoda:* Am. Museum Nat. History, Bull., v. 100, art. 1, 263 p. (Dec. 19).

Dollfus, G.-F.

(*87) 1903, *Étude géologique sur la Tunisie centrale, par M. L. Pervinquière* (review article): Jour. Conchyliologie, v. 51, pt. 3, p. 271-277.

(*88) 1915, *Recherches sur l'Ostrea gingensis et son groupe:* Soc. Géol. France, Bull., Comptes Rendus Sommaire des Séances, no. 10-12, p. 82-85 (June 7).

———, & Dautzenberg, Philippe

(*89) 1920, *Conchyliologie du Miocène moyen du bassin de la Loire, pt. 1: Pélécypodes (Suite et fin):* Soc. Géol. France, Mém. 27, Paleontologie v. 22, pt. 2-4, p. 379-500, pl. 34-51.

Douvillé, Henri

(*89a) 1879, *M. Douvillé présente à la Société, de la part de M. Bayle, l'atlas du IVe volume de l'Explication de la Carte géologique de la France:* Soc. Géol. France, Bull., ser. 3, v. 7 (1878-79), p. 91-92.

(*90) 1886, *Examen des fossiles rapportés du Choa par M. Aubry:* Soc. Géol. France, Bull., ser. 3, v. 14 (1885-86), no. 4, p. 223-241, pl. 12 (April).

(*91) 1904a, *Mollusques fossiles:* v. 3 (Études Géologiques), pt. 4 (Paléontologie, pt. 2), p. 191-380, pl. 25-50, in J. de Morgan, 1894-1905, Mission scientifique en Perse (1889-91), E. Leroux (Paris), 5 vol.

(*92) 1904b, *Les explorations de M. de Morgan en Perse:* Soc. Géol. France, Bull., ser. 4, v. 4, pt. 4, p. 539-553, 6 fig., pl. 13-14 (Dec. 12).

(*93) 1907, *Études sur les lamellibranches. Vulsellidés:* Annales Paléontologie, v. 2, pt. 3, p. 97-119, 11 fig., pl. 15-16 (Sept.).

(*94) 1910, *Observations sur les ostréidés:* Soc. Géol. France, Comptes Rendus, no. 13-14 (June 20), p. 118-119.

(*95) 1911, *Observations sur les ostréidés, origine et classification:* Same, Bull., ser. 4, v. 10 (1910), pt. 7, p. 634-645, pl. 10+11 (May 2).

(*96) 1916, *Les terrains secondaires dans le massif du Moghara à l'est de l'Isthme de Suez, d'après les explorations de M. Couyat-Barthoux. Paléontologie:* Acad. Sci. Inst. France, Mém., ser. 2, v. 54, no. 1, 184 p., 50 fig., 21 pl.

(*97) 1920, *Les Euostrea (groupe de l'O. edulis), les Gryphea (gr. de l'O. angulata), et les Crassostrea (gr. de l'O. virginiana); leurs origines:* Soc. Géol. France, Comptes Rendus Sommaires des Séances 1920, no. 7, p. 65-66 [Séance gén. ann. 12 april 1920].

*(*98)* 1936, *Le test des ostréidés du groupe de l'Ostrea cochlear (genre Pycnodonta, F. de W.) et test des rudistes:* Acad. Sci. Paris, Comptes Rendus hebdomaires des Séances v. 203, Sem. 2, no. 22, p. 1113-1117.

Dujardin, Félix
*(*99)* 1835, *Mémoire sur les couches du sol en Touraine et description des coquilles de la craie et des faluns:* Soc. Géol. France, Mém., v. 2, pt. 2, p. 211-311, pl. 15-20.

Duméril, A. M. C.
*(*100)* 1806, *Zoologie analytique, ou méthode naturelle de classification des animaux rendue plus facile à l'aide de tableaux synoptique:* xxxii+334 p. (Paris).

Dunker, Wilhelm
*(*101)* 1846a, *Diagnosen neuer Conchylien:* Zeitschr. Malakozoologie, Karl Theodor Menke, & Louis Pfeiffer (eds.), v. 3, p. 48 (March). [No illustrations.]

*(*102)* 1846b, *Ueber die in dem Lias bei Halberstadt vorkommenden Versteinerungen [pt. 1]:* Palaeontographica, v. 1, no. 1, p. 34-41, pl. 6 (Aug.).

Durve, V. S., & Bal, D. V.
*(*103)* 1961, *Some observations on shell-deposits of the oyster Crassostrea gryphoides (Schlotheim):* Indian Acad. Sci., Proc., v. 54, sec. B, no. 1, p. 45-55, 4 fig. (July).

Eaton, J. E., Grant, U. S., & Allen, H. B.
*(*104)* 1941, *Miocene of Caliente Range and. environs, California:* Am. Assoc. Petroleum Geologists, Bull., v. 25, p. 193-262, 14 fig., 8 pl. (Feb.).

Eberzin, A. G.
*(*105)* 1960, *Osnovy paleontologii spravochnik dlia paleontologov i geologov SSSR* [Fundamentals of paleontology as reference for paleontologists and geologists of USSR]: Mollyuski-Pantsirnye, Dvustvorchatye Lopatonogne Akad. Nauk SSR (Moscow), 300 p., 44 pl.

Eichwald, Eduard von
*(*106)* 1829, *Zoologia specialis quam expositis animalibus tum vivis, tum fossilibus potissimum Rossiae in Universum, et Poloniae in specie, in usum lectionum publicarum in Universitate Caesarea Vilnensi habendarum, v. 1:* vi+314 p., frontispiece+5 pl., Joseph Zawadzki (Wilno).

*(*106a)* 1871, *Geognostisch—palaeontologische Bemerkungen über die Halbinsel Mangischlak und die Aleutischen Inseln:* iii+200 p., 20 pl., Imper. Akad. Wiss. (St. Petersburg).

Ekman, Sven
*(*107)* 1934, *Indo-Westpazifik und Atlanto-*

Ostpazifik, eine tiergeographische Studie: Zoogeographica, v. 2, no. 3, p. 320-374, 11 fig. (Sept.).

Elsey, C. R.
*(*108)* 1935, *On the structure and function of the mantle and gill of Ostrea gigas (Thunberg) and Ostrea lurida (Carpenter):* Royal Soc. Canada, Trans., sec. 5, v. 29, p. 131-158, 1 fig., 5 pl.

Emmrich, A.
*(*109)* 1853, *Geognostische Beobachtungen aus den östlichen bayerischen und den angrenzenden österreichischen Alpen:* [K.-K.] Geol. Reichsanst., Jahrb., v. 4, no. 2, p. 326-394, fig. (unnumbered) (June).

Erdmann, Wilhelm
*(*110)* 1934, *Untersuchungen über die Lebensgeschichte der Auster. Nr. 5. Über die Entwicklung und die Anatomie der "ansatzreifen" Larve von Ostrea edulis mit Bemerkungen über die Lebensgeschichte der Auster:* Wissenschaft. Meeresunters.; Komm. zur Untersuchung der Deutschen Meere in Kiel und die Biol. Anst. auf Helgoland, new ser., Abt. Helgoland, v. 19, Abh. 6, 25 p., 5 fig., 8 pl. (Sept.).

Faujas-Saint-Fond, Barthélemi
*(*111)* [1802?], *Histoire naturelle de la montagne de Saint-Pierre de Maestricht:* 263 p., 54 pl., H. J. Jansen (Paris). [Published in the "An 7ème. de la Republique Française," according to title page.]

Finch, John
*(*112)* 1823, *Geological essay on the Tertiary formations in America:* Am. Jour. Sci., ser. 1, v. 7, p. 31-43.

Finlay, H. J.
*(*113)* 1928a, *Note by H. J. Finlay, D. Sc., p. 432,* in *J. Marwick, The Tertiary Mollusca of the Chatham Islands including a generic revision of the New Zealand Pectinidae:* New Zealand Inst., Trans. & Proc., v. 58, no. 4, p. 432-506 (March 19; publ. separately Feb. 28).

*(*114)* 1928b, *The Recent Mollusca of the Chatham Islands:* Same, v. 59, pt. 2, p. 232-286, pl. 38-43 (Aug. 31).

Fischer, Paul
*(*115)* 1864, *Note sur le genre Pernostrea:* Jour. Conchyliologie, v. 12 [=ser. 3, v. 4], no. 4, p. 362-368, pl. 15, (Oct. 1).

*(*116)* 1865, *Note sur une espèce nouvelle du genre Pernostrea:* Same, v. 13 [=ser. 3, v. 5], no. 1, p. 61-64.

*(*117)* 1880-87, *Manuel de conchyliologie et de paléontologie conchyliologique ou histoire naturelle des mollusques vivants et fossiles suivi d'un appendice sur les brachiopodes par D. P. Oehlert:* xxiv+1369 p., 1138 fig., frontispiece+23 pl., F. Savy (Paris)

(Sept. 21, 1880-June 15, 1887). [Fascicule 10, p. 897-1008, was published April 30, 1886.]

Fischer de Waldheim, Gotthelf

(*118) 1807, *Museum-Demidoff. Mis en ordre systématique et descrit par G. Fischer, v. 3, Végétaux et Animaux:* ix+330 p., pl. 1-6, Imprimerie de l'Université Impériale (Moscou).

(*119) 1829, *Sur les fossiles des corps organisés:* Soc. Imp. Nat. Moscou, Bull., v. 1, no. 2, p. 27-32, pl. 1.

(*120) 1835, *Lettre à M. le Baron de Férussac sur quelques genres de coquilles du Muséum Demidoff et en particulier sur quelques fossiles de la Crimée:* Same, Bull., v. 8, p. 101-119, pl. 1-5.

(*121) 1848, *Notice sur quelques fossiles du Gouvernement d'Orel:* Same [=Moskov. obshch. ispytat. prirody], v. 21, no. 4, p. 455-469, pl. 11.

Fleming, John

(*122) 1828, *A history of British animals, exhibiting the descriptive characters and systematical arrangement etc.:* xxiii+565 +corrigenda p., Bell & Bradfute (Edinburgh).

Forbes, Edward

(*122a) 1845, *On the fossil shells collected by Mr. Lyell from the Cretaceous formations of New Jersey:* Geol. Soc. London, Quart. Jour., v. 1, p. 61-64, text fig.

Fournel, Henri

(*122b) 1849, *Richesse minérale de l'Algérie accompagnée d'éclaircissements historiques et géographiques sur cette partie de l'Afrique septentrionale:* v. 1 texte, xviii+476 p., Imprimerie Nationale (Paris).

Frauscher, K. F.

(*123) 1886, *Das Unter-Eozän der Nordalpen und seine Fauna, Pt. 1, Lamellibranchiata:* K. Akad. Wiss. [Wien], Denkschr., Math. naturw. Kl. v. 51, pt. 2, p. 37-270, 3 tables on 8 p., 1 fig., 12 pl.

Frebold, Hans

(*124) 1940, *Untersuchungen über die Fauna und Stratigraphie des marinen Tertiärs von Ostturkestan:* Rept. Sci. Exped. northwestern Provinces of China under the leadership of Dr. Sven Hedin—The Sino-Swedish Exped. Pub. II, V. Invertebrate Paleontology, pt. 2, Stockholm; also Geol. Survey China, Palaeontologia Sinica, . . . Nanking, 35 p., 1 map, 7 pl. (July).

Frech, Fritz

(*125) 1905a, *Review of Otto M. Reis: Über Lithiotiden:* Neues Jahrb. Mineralogie, Geologie, Paläontologie (1904), v. 2, p. 326-328, 2 fig.

(*126) 1905b, *Zur Stellung von Lithiotis:* Centralbl. Mineralogie, Geologie, Paläontologie, p. 470.

(*127) 1906, *Bemerkungen zu G. Böhm's Artikel "Zur Stellung der Lithiotiden":* Same, p. 208-209.

Freneix, Suzanne

(*128) 1965, *Les bivalves du Jurassique moyen et supérieur du Sahara tunisien:* Annales de Paléontologie (Invertébrés), v. 51, pt. 1, p. 1-65, 10 fig., pl. 1-5 (before Aug. 3).

———, & **Busson, Georges**

(*128a) 1963, *Sur les faunes de Bivalves du Jurassique moyen et supérieur du Sahara tunisien:* Acad. Sci. (Paris), Comptes Rendus hebdom. des séances, v. 257, no. 9 (26 Août 1963), p. 1631-1633.

Gabb, W. M.

(*129) 1860, *Descriptions of new species of American Tertiary and Cretaceous fossils:* Acad. Nat. Sci. Philadelphia, Jour., ser. 2, v. 4, pt. 4, art. 14, p. 375-406, pl. 67-69 (Dec. 11).

(*130) 1861, *Description of new species of Cretaceous fossils from New Jersey, Alabama and Mississippi:* Same, Proc., v. 13, p. 318-330 (probably Dec.).

(*131) 1869, *Cretaceous and Tertiary fossils:* California Geol. Survey, Paleontology, v. 2, 299 p., illus.

Galtsoff, P. S.

(*132) 1930, *The fecundity of the oyster:* Science, v. 72, no. 1856, p. 97-98 (July 25).

(*133) 1955, *Recent advances in the studies of the structure and formation of the shell of Crassostrea virginica:* Natl. Shellfisheries Assoc., Proc., v. 45 (Aug., 1954), p. 116-135, 9 fig.

(*134) 1964, *The American oyster Crassostrea virginica Gmelin:* U.S. Bur. Commercial Fisheries, Fish. Bull., v. 64, iii+480 p., 400 fig.

———, & **Merrill, A. S.**

(*135) 1962, *Notes on shell morphology, growth, and distribution of Ostrea equestris Say:* Bull. Marine Sci. Gulf and Caribbean, v. 12, no. 2, p. 234-244, 5 fig. (June).

———, & **Smith, R. O.**

(*135a) 1932, *Stimulation of spawning and cross-fertilization between American and Japanese oysters:* Science, v. 76, no. 1973, p. 371-372 (Oct. 21).

Gardner, Julia

(*136) 1927, *New species of mollusks from the Eocene of Texas:* Washington Acad. Sci., Jour., v. 17, no. 14, p. 362-383, 44 fig. (Aug. 19).

Gekker [Hecker], R. F., Osipova, A. I. & Belskaya, T. N.

(*137) 1962, *Ferganskiy zaliv Paleogenovogo*

morya sredney Azii; ego istoriya, osadki, fauna, flora, usloviya ikh obitaniya i razvitie, v. 2 [Fergana Gulf of Paleogene sea in central Asia; its history, subsidence, fauna, flora, their ecology and development]: Akad. Nauk SSSR, Paleont. Inst., 332 p., 70 fig., 27 pl. (May 9).

George, T. N.
(*138) 1953, *Fossils and the evolutionary process:* British Assoc. Adv. Sci., v. 10, no. 38, p. 132-144 (Sept.).
(*139) 1962, *The concept of homoeomorphy:* Geologists Assoc. (London), Proc., v. 73, pt. 1, p. 9-64, 15 fig. (July).

Gesner, Johann
(*139a) 1758, *J. Gesneri . . . Tractatus physicus de petreficatis in duas partes distinctus, . . .:* 2 pts., 136 p. (Lugduni Batavorum).

Gevin, P.
(*140) 1947, *Sur la decouverte du genre Pachypteria de Koninck dans le Paléozoïque du bord S du Synclinal de Tindouf:* Soc. Géol. France, Comptes Rendus Somm. des Séances, no. 11-12, p. 243-245 (July).

Glibert, Maxime
(*141) 1936, *Faune malacologique des Sables de Wemmel, I. Pélécypodes:* Musée Royale Histoire Nat. Belgique, Mém. 78, 242 p., 7 pl. (Nov. 30).

———, & Van de Poel, Luc.
(*141a) 1965, *Les Bivalvia fossiles du Cénozoïque Étranger des collections de l'Institut Royal des Sciences Naturelles de Belgique; II Pteroconchida, Colloconchida et Isofilibranchida:* Inst. Royal Sci. Nat. Belgique, Mém., ser. 2, pt. 78, 105 p. (Nov. 15).

Gmelin, J. F.
(*142) [1791], *Caroli a Linné, etc., Systema naturae per regna tria naturae, etc.:* edit. 13, v. 1, pt. 6, p. 3021-3910, G. E. Beer (Lipsia [Leipzig]). [v. 1, pt. 6, not dated, but published before March 14, 1791.]

Goldfuss, G. A.
(*143) 1826-44, *Petrefacta Germaniae tam ea, quae in museo universitatis regiae Borussicae Friedericiae Wilhemiae Rhenanae servantur quam alia quaecunque in museis Hoeninghusiano Muensteriano aliisque extant, iconibus et descriptionibus illustrata. Abbildungen und Beschreibungen der Petrefacten Deutschlands und der angränzenden Länder unter Mitwirkung des Herrn Grafen Georg zu Münster:* pt. 1 (1826-33), x+252 p., pl. 1-71; pt. 2 (1834-40) [1833-41]; iii+312 p., pl. 72-165; pt. 3 (1841-44) [1841, 1844], iv+128 p., pl. 166-200, Arnz & Co. (Düsseldorf).

(*144) 1863, *Petrefacta Germaniae* etc. . . .: edit. 2, pt. 2 (text), List & Francke (Leipzig).

Gorizdro, Z. F.
(*145) 1915, *Materialy k" izucheniyu fauny tretichnykh" otlozheniy Turkestana* [Materials for the study of fauna of the Tertiary formations in Turkestan]: Soc. Imp. Nat. St. Pétersbourg, Travaux, v. 37, livr. 5, sec. Géologie Minéralogie, p. 1-56, pl. 1-2.

Gorodiski, A.
(*146) 1951, *Étude sur les Ostreidae du Nummulitique du Sénégal:* Soc. Géol. France, Bull., ser. 5, v. 20, no. 7-9, p. 353-374, pl. 18-19 (June).

Gould, A. A.
(*147) 1850, *Molluscs from the South Pacific:* Boston Soc. Nat. History, Proc., v. 3, pt. 22, p. 343-348 (Dec.).

Grabau, A. W.
(*148) 1936, *Early Permian fossils of China, pt. 2, Fauna of the Maping Limestone of Kwangsi and Kweichow:* Geol. Survey China, Palaeontologia Sinica, ser. B, v. 8, no. 4, 320 p., 31 pls., 9 p. of summary in Chinese.

Grave, Caswell
(*149) 1901, *The oyster reefs of North Carolina: a geological and economic study:* Johns Hopkins Univ. Circ., v. 20, no. 151, p. 50-53, 2 fig. (April).

Gray, J. E.
(*150) 1833, *Some observations on the economy of molluscous animals, and on the structure of their shells:* Royal Soc. London, Philos. Trans., v. 123, pt. 2, p. 771-819.
(*151) 1847, *A list of the genera of Recent Mollusca, their synonyma and types:* Zool. Soc. London, Proc., v. 15, p. 129-219.

Gregorio, Antonio de
(*152) 1884, *Studi su talune conchiglie mediterranee viventi e fossili con una revista del gen. Vulsella:* Soc. Malacol. Italiana, Boll., v. 10, p. 36-288, pl. 1-5.

Gümbel, C. W.
(*152a) 1861, *Geognostische Beschreibung des bayerischen Alpengebirges und seines Vorlandes:* Bavaria, K. Bayer. Oberbergamt, Geognost. Abt.; xx+950 p., text fig., 42 pl., Justus Perthes (Gotha).
(*152b) 1871, *Die sogenannten Nulliporen (Lithothamnium und Dactylopora) und ihre Betheiligung an der Zusammensetzung der Kalksteine, pt. 1: Die Nulliporen des Pflanzenreichs (Lithothamnium):* K. Bayer. Akad. Wiss., München, Abh. math.-phys. Kl., v. 11, pt. 1, p. 13-52, pl. 1-2.

Guilding, Lansdown
(*153) 1828, *Observations on the zoology of the*

Caribaean Islands: Zool. Jour., v. 3, no. 12, art. 61, p. 527-544 (Jan.-April).

Gunter, Gordon
(*154) 1948, *The genera of living oysters:* Anatomical Rec., v. 101, no. 4, p. 689 (Aug.).
(*155) 1950, *The generic status of living oysters and the scientific name of the common American species:* Am. Midland Naturalist, v. 43, no. 2, p. 438-449 (March) [June 1].
(*156) 1954, *The problem in oyster taxonomy:* Systematic Zoology, v. 3, no. 3, p. 134-137 (Sept.).

Haas, Fritz
(*157) 1938, *Bivalvia, Teil II, 2. Lieferung:* H. G. Bronn's Klassen und Ordnungen des Tierreichs, v. 3, Mollusca, pt. 3, p. 209-466, fig. 151-165: Akad. Verlag. (Leipzig).

Habe, Tadashige, & Kosuge, Sadao
(*158) 1966, *New genera and species of the tropical and subtropical Pacific molluscs:* Malacol. Soc. Japan, Venus (Japanese Jour. of Malacology), v. 24, no. 4, p. 312-341, pl. 29 (May).

Hall, James
(*159) 1856, *Description and notices of the fossils collected upon the route* [Whipple's reconnaissance near the 35th parallel]: U.S. Pacific Explor. Rept.: U.S. 33d Congress, 2nd sess., Senate Doc. 78, v. 3, pt. 4, p. 99-105, illus.

Hallam, Anthony
(*160) 1959a, *On the supposed evolution of Gryphaea in the Lias:* Geol. Mag., v. 96, no. 2, p. 99-108, 4 fig. (April 2).
(*161) 1959b, *The supposed evolution of Gryphaea:* Same, v. 96, no. 5, p. 419-420 (Nov. 3).
(*162) 1960, *On Gryphaea:* Same, v. 97, no. 6, p. 518-522, 2 fig. (Dec. 9).
(*163) 1962, *The evolution of Gryphaea:* Same, v. 99, no. 6, p. 571-574 (Dec. 12).

Haranghy, László, Balázs, András, & Burg, Miklós
(*163a) 1965, *Investigation on ageing and duration of life of mussels:* Acad. Sci. Hungarica (Budapest), Acta Biol., v. 16, no. 1, p. 57-67, 9 text fig.

Harris, G. D.
(*164) 1919, *Pelecypoda of the St. Maurice and Claiborne Stages:* Bull. Am. Paleontology, v. 6, no. 31, 269 p., 1 fig., 59 pl. (June 30).

Healey, Maud
(*165) 1908, *The fauna of the Napeng Beds or the Rhaetic beds of upper Burma:* Geol. Survey India, Mem., Palaeontologia Indica, new ser., v. 2, mem. 4, v+88 p., 1 fig., 9 pl.

Heilprin, Angelo
(*165a) 1884, *North American Tertiary Ostreidae:* U. S. Geol. Survey, Ann. Rept. 4 (1883), Appendix I, p. 309-316, pl. 62-72.

Heller, Theodor
(*166) 1926, *Die Fauna des obersilurischen Orthocerenkalks von Elbersreuth:* Geognost. Jahresh. (München, Bavaria, Oberbergamt, Geol. Landesunters.), v. 38 (1925), p. 197-278, fig., 4 pl.

Henderson, I. J.
(*167) 1935, *The lower Lias at Hock Cliff, Fretherne:* Bristol Naturalists' Soc., Ann. Rept. & Proc., ser. 4, v. 7, pt. 7 (1934), p. 549-564, fig. 3-6.

Hennig, Anders
(*168) 1897, *Revision af Lamellibranchiaterna i Nilssons "Petrificata Suecana Formationis Cretaceae":* Lunds Univ., Årsskrift, v. 33, 66 p., 3 pl.

Herdman, W. A., & Boyce, Rubert
(*169) 1899, *Oysters and disease:* Lancashire Sea-Fisheries, Mem. 1, ii+60 p., 8 pl.

Herrmannsen, A. N.
(*170) 1846-52, *Indicis generum malacozoorum primordia. Nomina subgenerum, generum, familiarum, tribuum, ordinum, classium; adjectis auctoribus, temporibus, locis systematicis atque literariis, etymis, synonymis. Praetermittuntur Cirripedia, Tunicata et Rhizopoda:* v. 1, pt. 1 (Sept. 1, 1846), i-xxvii+104 p.; pt. 2 (Dec. 1, 1846), p. 105-232; pt. 3 (March 1, 1847), p. 233-360; pt. 4 (April 18, 1847), p. 361-488; pt. 5 (May 25, 1847), p. 489-616; pt. 6 (July 17, 1847), p. 617-637; v. 2, pt. 1 (July 17, 1847), p. 1-104; pt. 2 (Sept. 8, 1847), p. 105-232; pt. 3 (Dec. 7, 1847), p. 233-352; pt. 4 (Feb. 18, 1848), p. 353-494; pt. 5 (Feb. 1849), p. 493-612; pt. 6 (March 1849), p. xxix-xlii+613-717; Supplementa et corrigenda (Dec. 1852), v+140 p., Theodor Fischer (Cassel).

Hertlein, L. G.
(*171) 1928, *Preliminary report on the paleontology of the Channel Islands, California:* Jour. Paleontology, v. 2, p. 142-157, pl. 22-25 (June).

———, & Allison, E. C.
(*172) 1966, *Additions to the molluscan fauna of Clipperton Island:* Veliger, v. 9, p. 138-140 (Oct. 1).

Hill, R. T., & Vaughan, T. W.
(*173) 1898, *The Lower Cretaceous gryphaeas of the Texas region:* U.S. Geol. Survey, Bull. 151, 138 p., 2 fig., 35 pl.

Hirase, Shintaro
(*174) 1930, *On the classification of Japanese*

oysters: Natl. Res. Council Japan, Japanese Jour. Zoology, v. 3, no. 1, p. 1-65, 95 text fig. (March 30).

Hoeninghaus, F. W.

*(*175)* 1829, *Verzeichniss der von Hoeninghaus des Museums der Universität zu Bonn . . . Petrificaten Sammlung:* (Crefeld). [Not seen.]

*(*176)* 1830, *Versuch einer geognostischen Eintheilung seiner Versteinerung-Sammlung nach Berathung* etc.: Jahrb. Mineralogie, Geognosie, Geologie, Petrefaktenkunde, v. 1, p. 226-245.

Hoese, H. D.

*(*177)* 1960, *Biotic changes in a bay associated with the end of a drouth:* Limnology & Oceanography, v. 5 (1960), no. 3, p. 326-336, 2 fig. (July).

Hollis, P. J. [Mrs. Pamela Peake]

*(*178)* 1963, *Some studies on New Zealand oysters:* Victoria Univ. Wellington, Zoology Publ. 31, 28 p., 5 fig., 3 pl. (March 15).

Howe, H. V.

*(*179)* 1937, *Large oysters from the Gulf Coast Tertiary:* Jour. Paleontology, v. 11, no. 4, p. 355-366, pl. 44 (June 7).

Hutchins, L. W.

*(*180)* 1947, *The bases for temperature zonation in geographical distribution:* Ecol. Mon., v. 17, no. 3, p. 325-335, 8 fig. (July).

Hutton, F. W.

*(*181)* 1873, *Catalogue of the Tertiary Mollusca and Echinodermata of New Zealand, in the collection of the Colonial Museum:* New Zealand Colonial Museum and Geol. Survey Dept., xvi+48 p. (probably June 18) (Wellington).

Huxley, Julian (ed.)

*(*182)* 1940, *The new systematics:* viii+583 p., illus., Clarendon Press (Oxford).

*(*183)* 1943, *Evolution the modern synthesis:* 645 p., illus., Harper & Brothers (New York & London).

Ihering, Hermann von

*(*183a)* 1902, *Historia de las ostras Argentinas:* Museo Nac. Buenos Aires, Anales, v. 7 or ser. 2, v. 4, p. 109-123, 9 text fig. (Jan. 9).

*(*184)* 1903, *Les mollusques des terrains crétaciques supérieurs de l'Argentine orientale:* Museo Nac. Buenos Aires, Anales, ser. 3, v. 2, p. 193-229, 2 pl. (Aug. 23).

*(*185)* 1907, *Les mollusques fossiles du Tertiaire et du Crétacée supérieur de l'Argentine:* Same, Anales, v. 14 (ser. 3, v. 7), xiii+611 p., 16 fig., 8 pl.

Imai, Takeo, Hatanaka, Masayoshi, Ryuhei, Sato, Sakai, Seiichi, & Yuki, Ryogo

*(*186)* 1950, *Artificial breeding of oysters in tanks:* Tohoku (Japan) Jour. Agr. Research, v. 1, p. 69-86.

Ingersoll, Ernest

*(*187)* 1881, *A report on the oyster-industry of the United States;* in United States of America, 10th Census, The history and present condition of the fishery industries, etc., sec. 10, mon. B, 250 p.

Iredale, Tom

*(*188)* 1936, *Australian molluscan notes, no. 2:* Australian Museum, Rec., v. 19, p. 267-340, pl. 20-24 (April 7).

*(*189)* 1939, *Mollusca Part I:* British Museum Nat. History, Great Barrier Reef Exped. 1928-29, Sci. rept., v. 5, no. 6, p. 209-425, 1 fig., pl. 1-7 (Feb. 25).

———, & Roughley, T. C.

*(*190)* 1933, *The scientific name of the commercial oyster of New South Wales:* Linnean Soc. New South Wales, Proc., v. 58, p. 278.

Jackson, R. T.

*(*191)* 1888, *The development of the oyster with remarks on allied genera:* Boston Soc. Nat. History, Proc., v. 23, pt. 4, p. 531-556, pl. 4-7 (April 4).

*(*192)* 1890, *Phylogeny of the Pelecypoda, The Aviculidae and their allies:* Same, Mem., v. 4, no. 8, p. 277-400, 53 fig., pl. 23-30 (July).

Jaworski, Erich

*(*193)* 1913, *Beiträge zur Stammesgeschichte der Austern:* Zeitschr. Induk. Abstamm.-u. Vererbungs., v. 9 (1913), p. 192-215, pl. 6-7.

*(*194)* 1926, *Bemerkungen über das Subgenus Flemingostrea Vredenburg 1916 und seine Beziehungen zu den Gryphaeen der obersten Kreide Patagoniens:* Centralbl. Mineralogie, v. 1926, Abt. B, no. 9, p. 314-318.

*(*195)* 1928, *Untersuchungen über den Abdruck der Mantelmuskulatur bei den Ostreiden und Chamiden und die sog. Cirrhenabdrücke:* Neues Jahrb. Mineralogie, Geologie, Paläontologie, Beil. v. 59, Abt. B, p. 327-356, 5 fig., pl. 20-24.

*(*195a)* 1935, *Review of Corroy, G., 1932, Le Callovien de la bordure orientale du Bassin de Paris:* Same, Referate, Jahrg. 1935, pt. 3, Hist. reg. Geologie, Paläontologie, p. 884-886.

*(*196)* 1951, *Review of H. B. Stenzel, Nomenclatural synopsis of supraspecific groups of the family Ostreidae:* Zentralbl. Geol-

ogie, Paläontologie, pt. 2, Hist. Geologie u. Paläontologie, v. 1951, no. 3, p. 648-650.

Jones, John
*(*197)* 1865, *On Gryphaea incurva and its varieties:* Cotteswold Naturalists' Field Club, Proc., v. 3, p. 81-95, pl. 1-5.

Jourdy, le Général E.
*(*198)* 1924, *Histoire naturelle des Exogyres:* Annales de Paléontologie, v. 13, 104 p., 8 fig., 11 pl.

Joysey, K. A.
*(*199)* 1959, *The evolution of the Liassic oysters Ostrea-Gryphaea:* Biol. Reviews, Cambridge Philos. Soc., v. 34, no. 3, p. 297-332, 17 fig. (Aug.).
*(*200)* 1960, *On Gryphaea:* Geol. Mag., v. 97, no. 6, p. 522-524, 1 fig. (Dec. 9).

Judd, J. W.
*(*201)* 1871, *On the anomalous mode of growth of certain fossil oysters:* Geol. Mag., v. 8, no. 8, p. 355-359, pl. 9 (Aug.).

Kanwisher, J. W.
*(*202)* 1955, *Freezing in intertidal animals:* Biol. Bull., v. 109, no. 1, p. 56-63, 3 fig. (Aug.).
*(*203)* 1959, *Histology and metabolism of frozen intertidal animals:* Same, v. 116, no. 2, p. 258-264, 3 fig. (April).

Kegel, Wilhelm
*(*204)* 1953, *Das Paläozoikum der Lindener Mark bei Giessen:* Hessische Landesamt Bodenforschung, Abhandl., no. 7, 3 p., 3 pl. (Oct. 1).

Kellogg, J. L.
*(*205)* 1892, *A contribution to our knowledge of the morphology of lamellibranchiate mollusks:* U.S. Fish Comm., Bull., v. 10 (1890), art. 15, p. 389-436, pl. 79-94.

Kieh, Yang
*(*206)* 1930, *Observations sur le genre Ostrea dans l'Eocène des regions mésogéennes:* Soc. Géol. France, Bull., ser. 4, v. 30, p. 77-100, pl. 6-7.

Kilian, W., & Reboul, P.
*(*207)* 1915, *Contribution à l'étude des faunes paléocretacées du Sud-Est de la France, pt. 1, La faune de l'Aptien inférieure des environs de Montélimar (Drôme) (Carrière de l'Homme de l'Armes):* France, Mém. pour servir à l'explication de la Carte géol. détaillée, 296 p., 15 pl.

Kiparisova, L. D.
*(*208)* 1936, *Verkhnetriasovye plastinchatozhabernye Kolymsko-Indigirskogo kraya* [Upper Triassic pelecypods from the Kolyma-Indigirka land]: Leningrad Vses. Arktich. Inst., Trudy [Arctic Inst. U.S.S.R. (Lenin-

grad), Trans.], v. 30, Geology, pt. 2, p. 71-136, 1 fig., 5 pl. [Russian with English summary on p. 115-128.]
*(*209)* 1938, *Plastinchatozhabernye Triasovykh otlozhenii SSSR* [Pelecypoda of the Triassic System of USSR]; Verkhnetriasovye plastinchatozhabernye Sibiri (Arkticheskoi i subarkticheskoi oblastei, Ussuriishogo Kraia i Zabaikalia) [Upper Triassic Pelecypoda of Siberia (Arctic and subarctic regions, Ussuri land and Transbaikalia)]: Tsentralnyi nauchno-issledov. geol.-razved. Inst. [=Central Geol. and Prospecting Inst.], Mon. po paleontologii SSSR [=Paleontology of USSR Mon.], v. 17, no. 1, 55 p., 8 pl.

Kitchin, F. L.
*(*210)* 1908, *The invertebrate fauna and palaeontological relations of the Uitenhage Series:* South African Museum, Annals, v. 7, pt. 2, p. 21-250, fig. 1, pl. 1-11 (Sept. 24).
*(*211)* 1912, *Palaeontological work; England and Wales:* Geol. Survey Great Britain and Museum Pract. Geology, Mem., Summ. of Prog. for 1911, p. 59-60.

Kittl, Ernst
*(*212)* 1907, *Die Triasfossilien vom Heureka Sund:* Vidensk.-Selsk. Kristiania, Rept. Second Norwegian Arctic Exped. "Fram" 1898-1902, v. 2, no. 7, 44 p., 3 pl. (June 29).

Klinghardt, Franz
*(*213)* 1922, *Vergleichende Anatomie der Rudisten, Chamen, Ostreen, pt. 2:* Archiv Biontologie, Gesell Naturf. Freunde Berlin, v. 5, no. 1, p. 21, pl. 3, fig. 6.
*(*213a)* 1929, *Entwicklungsgleichheiten (Convergenzen) zwischen Austern und Rudisten und die Ursachen ihrer Entstehung:* Neues Jahrb. Mineralogie, Geologie u. Paläontologie, Beil. Band 62, Abt. B (Geologie u. Paläontologie), p. 509-521, pl. 31-33.

Klipstein, August von
*(*214)* 1843, *Beiträge zur geologischen Kenntniss der östlichen Alpen:* x+312 p., 20 pl., Georg Friedrich Heyer (Giessen, Ger.). [This book has an extra title page: Mittheilungen aus dem Gebiete der Geologie und Palaeontologie, v. 1, Giessen, Georg Friedrich Heyer, 1845.]

Knight-Jones, E. W.
*(*215)* 1951, *Aspects of the setting behaviour of larvae of Ostrea edulis on Essex oyster beds:* Conseil Permanent Internat. pour l'Explor. de la Mer, Rap. et Procès-Verbaux des Réunions, v. 128, pt. 2, Contr. to sci. mtg., 1949, p. 30-34 (Feb.).

Koch, Antal [Anton]

(*216) 1896, _A Gryphaea Esterházyi (Pávay) előfordulásáról és elterjedéséről_ [_Über das Vorkommen und die Verbreitung der Gryphaea Esterhazyi Pávay_]: Magyar Kir. Földtani Intezet Hivatalos Közlönye Foldtani Közlöny [K. Ungarische Geol. Anstalt Geol Mitt.], v. 26, no. 11-12, p. 324-330 [p. 364-366] (Nov.-Dec.).

Koninck, L. G. de

(*217) 1851, _Description des animaux fossiles qui se trouvent dans le terrain carbonifère de Belgique:_ supplément (Liège). [Not seen.]

(*218) 1885, _Faune du Calcaire Carbonifère de la Belgique, pt. 5, Lamellibranches:_ Musée Royal Histoire Nat. Belgique, Annales, v. 11, 283 p.; atlas, 41 pl.

Korringa, Paul

(*219) 1941, _Experiments and observations on swarming, pelagic life and setting in the European flat oyster, Ostrea edulis L.:_ Archives Néerland. Zoologie, v. 5, livr. 1+2, p. 1-249, 24 fig. (Jan.).

(*220) 1951, _On the nature and function of "chalky" deposits in the shell of Ostrea edulis Linnaeus:_ California Acad. Sci., Proc., ser. 4, v. 27, no. 5, p. 133-158, 2 fig. (May 31).

(*221) 1952-53, _Recent advances in oyster biology:_ Quart. Rev. Biology, v. 27, no. 3 (Sept.), p. 266-308; v. 27, no. 4 (Dec.), p. 339-365 (Jan. 1953).

(*222) 1956, _Oyster culture in South Africa. Hydrographical, biological and ostreological observations in the Knysna Lagoon, with notes on conditions in other South African waters:_ Union South Africa, Dept. Commerce & Industries, Div. Fisheries, Inv. rept. 20, 86 p. (March).

Krach, Wilhelm

(*222a) 1951, _Małże z grupy Anisomyaria jury brunatnej okolic Krakowa (rodziny:Limidae, Ostreidae, Spondylidae, Aviculidae, Anomiidae)_ [Mollusks of the group Anisomyaria of the Brown Jura near Krakow]: Polski. Towarzystwo Geol., Rocznik [Soc. Géol. Pologne, Ann.], v. 20 (1950), p. 333-376, 3 text fig., pl. 11-13.

Krauss, Ferdinand

(*223) 1850, _Über einige Petrefacten aus der Unteren Kreide des Kaplandes:_ Acad. Caes. Leopold.-Carolin. Nat. Cur., Nova Acta, v. 22 (or, dec. 3, v. 3), pt. 2, p. 439-464, pl. 47-50.

Krenkel, Erich

(*224) 1915, _Die Kelloway-Fauna von Pópilani in Westrussland:_ Palaeontographica, v. 61, no. 5-6, p. 191-362, 26 fig., pl. 19-28 (Nov.).

Kříž, Jiří

(*225) 1966, _Rod Praeostrea Barrande, 1881, ze starších prvohor střední Evropy (Bivalvia)_ [The genus Praeostrea Barrande, 1881, from the older Paleozoic of central Europe (Bivalvia)]: Czechoslovakia. Národni Muz. Čas. Oddíl Přírodovědný Ročník 135 (1966), no. 1, p. 25-32, 4 fig., 2 pl.

Krumbeck, Lothar

(*226) 1913, _Obere Trias von Buru und Misól (Die Fogi-Schichten und Asphaltschiefer West Burus und der Athyridenkalk des Misól-Archipels):_ Beiträge Geologie Niederländisch-Indien von Georg Boehm, Abt. II, Abschnitt 1—Palaeontographica, suppl. v. 4, pt. 2, no. 1, 161 p., 11 pl.

Kulp, J. L., Turekian, K. K., & Boyd, D. W.

(*227) 1952, _Strontium content of limestone and fossils:_ Geol. Soc. America, Bull., v. 63, no. 7, p. 701-716 (June 30).

Kutassy, A.

(*228) 1931, _Lamellibranchiata triadica II:_ Fossilium Catalogus I: Animalia, pt. 51, p. i-iv, 261-477 (Nov. 16).

Lamarck, J. B. A. P. M. de

(*229) 1801, _Système des animaux sans vertèbres, ou table général des classes, des ordres et des genres de ces animaux:_ 432 p. (Jan.), Deterville (Paris).

(*230) 1806a, _Mémoire sur les fossiles des environs de Paris,_ etc.: Muséum Histoire Nat., Annales (Paris), v. 8, p. 156-166, pl. 35-37, 59-62 (between June and Sept.).

(*231) 1806b, _Mémoires sur les fossiles des environs de Paris,_ etc.: [collected reprints from Annales Muséum d'Histoire Nat. (Paris), v. 1-8], 284 p. and new pagination. [1806b, p. 261-271=1806a, p. 156-166].

(*231a) 1809, _Philosophie zoologique, ou exposition etc.,_ 2 vols. in one, facsimile reprint of edit. ✝, 1960: xxv+428 and 475 p., H. R. Engelmann and Wheldon & Wesley (Weinheim & Codicote).

(*232) 1819, _Histoire naturelle des animaux sans vertèbres, présentánt les caractères généraux et particuliers des ces animaux, leur distribution, leurs classes, leurs familles, leurs genres, et la citation des principales espèces qui s'y rapportent;_ etc.: v. 6, pt. 1, 343 p., chez l'auteur (Paris) (Feb.-June).

Lamy, Edouard

(*233) 1929-30, _Révision des Ostrea vivants du Muséum National d'Histoire Naturelle de Paris:_ Jour. Conchyliologie, v. 73 (ser. 4, v. 27), no. 1 (April 30, 1929), p. 1-46, 3 fig.; no. 2 (July 20, 1939), p. 71-108;

no. 3 (Oct. 30, 1929), p. 133-168; no. 4 (Feb. 28, 1930), p. 233-275, pl. 1.

Lea, Isaac
*(*234)* 1833, *Contributions to geology:* 227 p., 6 pl. (Dec. 3), Carey, Lea, & Blanchard (Philadelphia).

Lecointre, Georges, & Ranson, Gilbert
*(*235)* 1952, *Ostréidés:* p. 25-40, fig. 4-9, pl. 1-13, in Georges Lecointre, Récherches sur le Néogène et le Quaternaire marins de a côte Atlantique du Maroc; Maroc Service Géol., Notes et Mém. 99, v. 2, Paléontologie, 173 p., 12 fig., 28 pl.

Leenhardt, Henry
*(*236)* 1926, *Quelques études sur "Gryphea Angulata" (Huître du Portugal):* Inst. Océanogr. (Fondation Albert Ier, Prince de Monaco), Annales, new ser., v. 3, no. 1, p. 1-90, 36 fig. (Jan.).

Lees, G. M.
*(*237)* 1928, *The geology and tectonics of Oman and of parts of southeastern Arabia:* Geol. Soc. London, Quart. Jour., v. 84 (1928), pt. 4, no. 336, p. 585-670, 12 fig., pl. 41-51 (Dec. 31).

Lemoine, Paul
*(*238)* 1910, *Gryphaea angustata, fiche 200-200a; Gryphaea littuola, fiche 201-201a; Ostrea pennaria, fiche 202-202a:* Palaeontologia Universalis, ser. 3, pt. 2 (July 26).

Linné, Carl [Linnaeus, Carolus]
*(*239)* 1758, *Systema Naturae per tria regna naturae, etc.:* edit. 10, v. 1, 823 p. (Stockholm).

Lischke, C. E.
*(*240)* 1869-74, *Japanische Meeres-Conchylien ... Mit besonderer Rücksicht auf die geographische Verbreitung derselben:* Novitates Conchologicae, W. Dunker (ed.), Suppl. 4, 3 pts. in 1.

Lissajou, Marcel
*(*241)* 1923, *Étude sur la faune du Bathonien des environs de Mâcon:* Univ. Lyon, Travaux Lab. de Géologie Fac. Sci., mém.. 3, no. 3, p. 142 etc., pl. 25-33.

Logan, W. N.
*(*242)* 1898, *The invertebrates of the Benton, Niobrara and Fort Pierre Groups:* Kansas State Geol. Survey, v. 4, Paleontology, pt. 1, Upper Cretaceous, pt. 8, p. 431-488, pl. 86-120.

*(*243)* 1899a, *Some additions to the Cretaceous invertebrates of Kansas:* Kansas Univ. Quart., ser. A, v. 8, no. 2, p. 87-98, pl. 20-23 (April).

*(*244)* 1899b, *Contributions to the paleontology of the Upper Cretaceous Series:* Field Columbian Museum, pub. 36, geol. ser., v. 1, no. 6, p. 203-216, pl. 22-26 (April).

Loosanoff, V. L.
*(*245)* 1958, *Some aspects of behavior of oysters at different temperatures:* Biol. Bull., v. 114, no. 1, p. 57-70, 7 fig. (Feb.).

*(*246)* 1965, *The American or Eastern oyster:* U.S. Bur. Commercial Fisheries, Circ. 205, 36 p., 25 fig. (March).

Lund, E. J.
*(*247)* 1957a, *A quantitative study of clearance of a turbid medium and feeding by the oyster:* Univ. Texas Inst. Marine Sci. Publ., v. 4, no. 2, p. 296-312, 6 fig. (July).

*(*248)* 1957b, *Self-silting, survival of the oyster as a closed system, and reducing tendencies of the environment of the oyster:* Same, Publs., v. 4, no. 2, p. 313-319, 1 fig. (July).

Lycett, John
*(*249)* 1863, *Supplementary monograph on the Mollusca from the Stonesfield Slate, Great Oolite, Forest Marble, and Cornbrash:* Palaeontogr. Soc., v. 15 (1861), 129 p., pl. 31-45.

Maclennan, R. M., & Trueman, A. E.
*(*250)* 1942, *Variation in Gryphaea incurva (Sow.) from the Lower Lias of Loch Aline, Argyll:* Royal Soc. Edinburgh, Proc., Ser. B (Biology), v. 61 (1941-43), pt. 2, p. 211-232, 11 fig. (May 5).

Macnae, William, & Kalk, Margaret
*(*251)* 1958, *A natural history of Inhaca Island, Moçambique:* 163 p., 30 fig., 11 pl., Witwatersrand Univ. Press (Johannesburg).

Marceau, Francis
*(*252)* 1936, *Sur quelques propriétés spéciales des muscles adducteurs des mollusques acéphales on rapport avec leur disposition et leur structure:* Musée Royal Histoire Nat. Belgique, Mém., ser. 2, pt. 3, Mélanges Paul Pelseneer, p. 941-975, 19 fig. (April 30).

Marcou, Jules
*(*253)* 1962, *Notes on the Cretaceous and Carboniferous rocks of Texas:* Boston Soc. Nat. History, Proc., v. 8, p. 86-97.

Martin, Jules
*(*254)* 1860, *Paléontologie stratigraphique de l'Infra-Lias du Département de la Côte-d'Or etc.:* Soc. Géol. France, Mém., ser. 2, v. 7, pt. 1, p. 1-100, pl. 1-8.

*(*255)* 1865, *Zone à Avicula contorta ou étage Rhaetien:* Acad. Imp. Sci., Arts et Belles-Lettres de Dijon (France), Mém., ser. 2, v. 12 (1864), p. 246-250.

Marwick, John

(*255a) 1931, *The Tertiary Mollusca of the Gisborne District:* New Zealand Geol. Survey, Paleont. Bull. 13, 177 p., 18 pl. (Aug. 1).

Massy, A. L.

(*256) 1914, *Notes on the evidence of age afforded by the growth rings of oyster shells:* Dept. Agric. Tech. Instr. Ireland (Dublin), Fish. Br., Sci. Inv. (1913), no. 2, 12 p., 11 pl. (March).

Matheron, Philippe

(*257) 1843a, *Catalogue méthodique et descriptif des Corps organisés fossiles du département des Bouches-du-Rhône et lieux circonvoisins; etc.:* Soc. Stat. Marseille, Répertoire travaux, v. 6 (1842) [=ser. 2, v. 1], p. 81-341, pl. 1-41 (Jan.).

(*258) 1843b, *Reprint of Matheron 1843a* (May). [Repaginated as p. 1-269 but otherwise identical in text, except for slight addition to p. 269.]

Mattox, N. T.

(*259) 1949, *Studies on the biology of the edible oyster, Ostrea rhizophorae Guilding, in Puerto Rico:* Ecol. Mon., v. 19, no. 4, p. 340-356, 14 fig. (Oct.).

Mayer-Eymar, C. D. W.

(*259a) 1876, *Systematisches Verzeichniss der Versteinerungen des Parisian der Umgebung von Einsiedeln:* Beiträge zur Geol. Karte der Schweiz, Lief. 14, Abt. 2b, 100 p., 4 pl., 1 table.

(*260) 1889, *Diagnoses ostrearum novarum ex agris Aegyptiae nummuliticis:* Naturforsch. Gesell. Zürich, Vierteljahrsschr., v. 34, no. 3, p. 289-299.

McLean, R. A.

(*260a) 1941, *The oysters of the western Atlantic:* Notulae Naturae, Acad. Nat. Sci. Philadelphia, no. 67, 14 p., 4 pl. (Jan. 14).

McLearn, F. H.

(*261) 1937, *New species from the Triassic Schooler Creek Formation:* Ottawa Field-Naturalists' Club, Canadian Field-Naturalist, v. 51, no. 7, p. 95-98, pl. 1 (Oct. 7).

(*262) 1946, *Upper Triassic faunas in Halfway, Sikanni Chief, and Prophet River basins, northeastern British Columbia:* Canada Geol. Survey, Paper 46-25, 11 p. [mimeographed], fig. 1 [blue-line print], appendix of 1 p., 3 pl.

(*263) 1947, *Upper Triassic faunas of Pardonet Hill, Peace River foothills. British Columbia:* Same, Paper 47-14, 16 p. [mimeographed], fig. [blue-line print]; appendix, 2 p., 6 pl.

Medcof, J. C.

(*264) 1944, *Structure, deposition and quality of oyster shell (Ostrea virginica Gmelin):* Fisheries Research Board of Canada, Jour., v. 6, no. 3, p. 209-216, 3 fig. (May).

(*265) 1949, *Effects of sunlight exposure in rearing young oysters:* Same, Prog. Rept. Atlantic Coast Sta. 45, p. 6-11, 4 fig. (April).

Meek, F. B.

(*266) 1876, *A report on the invertebrate Cretaceous and Tertiary fossils of the Upper Missouri country:* U.S. Geol. Survey of the Territories, v. 9, lxiv+629 p., 85 fig., 45 pl.

Melleville, M.

(*266a) 1843, *Mémoire sur les sables tertiaires inférieurs du Bassin de Paris, etc.:* Ann. Sci. Géol. etc., M. A. Rivière (ed.), année 2, no. 2, 88 p., 10 pl.

Menke, C. T.

(*267) 1828, *Synopsis methodica molluscorum generum omnium et specierum earum, quae in Museo Menkeano adservantur, cum synonymia critica et novarum specierum diagnosibus:* xii+91 p., Henricus Gelpke (Pyrmont).

Menzel, R. W.

(*268) 1955, *Some phases of the biology of Ostrea equestris Say and a comparison with Crassostrea virginica Gmelin:* Univ. Texas Inst. Marine Sci. Publ., v. 4, no. 1, p. 69-153, 25 fig. (Sept.).

Mercier, Jean

(*269) 1929, *Apropos des variations de l'aire ligamentaire d'Ostrea wiltonensis Lyc. et du genre Pernostrea Munier-Chalmas:* Soc. Linnéenne Normandie, Bull., ser. 8, v. 1, année 1928, travaux orig., p. 3-7.

Mészáros, N., & Nicorici, E.

(*270) 1962, *Fauna din orizontul cu Gryphaea eszterhazyi de la Căpus (Reg. Cluj):* Acad. Répub. Popul. Roumaine, Comptes Rendus, v. 12, no. 9, p. 1043-1051 (Sept.).

Michelotti, Giovanni

(*270a) 1847, *Description des fossiles des terrains Miocènes de l'Italie septentrionale:* Natuurk. Hollands. Maatschapp. Wetensch. Haarlem, Verhhandel., ser. 2, v. 3, p. 81, pl. 3, fig. 6 [not fig. 3].

Middendorf, A. T. von

(*271) 1847-75, *Reise in den äussersten Norden und Osten Sibiriens während . . . 1843 und 1844 . . . auf Veranstaltung der Kaiserlichen Akademie der Wissenschaften zu St. Petersburg ausgeführt und . . . herausgegeben von A. T. v. Middendorf:* 4 v. and atlas (St. Petersburg). [v. 1, pt. 1, contains Fossil Mollusca by A. Graf Keyserling, 1848.]

Millar, R. H., & Hollis, P. J.
(*272) 1963, *Abbreviated pelagic life of Chilean and New Zealand oysters:* Nature, v. 197, no. 4866, p. 512-513 (Feb. 2).

Miller, H. W., Jr.
(*273) 1958, *Stratigraphic and paleontologic studies of the Niobrara Formation (Cretaceous) in Kansas:* 163 p., 11 text fig., 7 pl., unpubl. Ph.D. dissertation, Univ. Kansas (Lawrence, Kans.) (Aug.).

Mirkamalov, Kh. Kh.
(*273a) 1963, *Klassifikatsiya ekzogir* [Classification of the Exogyras]: Moskovsk Obshch., Byull., Ispytateley Prirody, new ser., v. 68, Otdel Geol., v. 38, no. 5, p. 152-153.
(*274) 1964, *K sistematicheskomu polozheniyu roda Amphidonta* [On systematic placement of genus *Amphidonta*]: Paleont. Zhurnal, no. 2, p. 149-152, pl. 14.
(*275) 1966, *Ekzogiry, ikh sistematika i znachenie dlya stratigrafii melovykh otlozheny yugo-zapadnykh otrogov gissarskogo khrebta* [Exogyras, their systematics and study of their stratigraphic occurrence in Cretaceous deposits of southwestern slopes of Gissarsk Range]: Inst. Geol. i Razved. nefty. i gazov. mestorozhd., Tashkent, 133 p., 4 fig., 21 pl.
(275a) 1969, *Ob obeme vida Rhynchostreon columbum (Lam.) i ego rodovoy prinadlezhnosti* (abstract) [On the species Rhynchostreon columbum (Lam.) and its generic placement (abstract)]: Moskov. Obshch. Ispyt. Prirody, Byull., n. ser., v. 74, otdel geol., v. 44, no. 6, noyabr-dekabr, vykhodit 6 raz v god, p. 150-151.

Miyake, M., & Noda, H.
(*275b) 1962(?), *Vitamin B group in the extracts of Mollusca—II. On vitamin B₆, inositol, pantothenic acid, biotin and niacin:* Japanese Soc. Sci., Fisheries Bull., v. 28, p. 597-601.

Mörch, O. A. L.
(*275c) 1850, *Catalogus conchyliorum quae reliquit C. P. Kierulf MD, DR. Nunc publica auctione X Decembris MDCCCL Hafniae dividenda:* 33⁻ p., 2 pl., Trieri(Hafniae).

Monterosato, Marchese di [Tommaso Allery di]
(*276) 1884, *Nomenclatura generica e specifica di alcune conchiglie mediterranee:* 152 p., Virzi (Palermo).
(*277) 1916, *Ostreae ed Anomiae del Mediterraneo:* Museo Civico Storia Nat. Giocomo Doria, ser. 3, v. 7 (47), p. 7-16, pl. 1-4.

Moore, H. F., & Danglade, Ernest
(*277a) 1915, *Condition and extent of the natural oyster beds and barren bottoms of Lavaca Bay, Texas:* U.S. Bureau Fisheries, Doc. 809, Appendix II to Rept. U.S. Commissioner of Fisheries for 1914, p. 1-45, pl. 1-5.

Moret, Léon
(*278) 1953, *Manuel de paléontologie animale:* edit. 3, xv+759 p., 274 fig., Masson & Cie (Paris).

Morris, J., & Lycett, John
(*279) 1853, *A monograph of the Mollusca from the Great Oolite, chiefly from Minchinhampton and the coast of Yorkshire, pt. 2. Bivalves:* Palaeontogr. Soc., Mon., v. 7, p. 1-80, pl. 1-8.

Moses, S. T.
(*280) 1927, *A preliminary report on the anatomy and life history of the common edible backwater oyster, Ostrea madrasensis:* Bombay Nat. History Soc., Jour., v. 32, p. 548-552.

Munier-Chalmas, E. C. P. A.
(*281) 1864, *Description d' un nouveau genre monomyaire du terrain jurassique:* Jour. Conchyliologie, v. 12 [=ser. 3, v. 4], no. 1, p. 71-75, pl. 3 (Jan. 1).

Neave, S. A.
(*281a) 1939-40, *Nomenclator zoologicus, A list of names of genera and subgenera in zoology from the tenth edition of Linnaeus 1758 to the end of 1935:* 4 vol. [v. 1, A-C (1939), 957 p.; v. 2, D-L (1939), 1025 p.; v. 3, M-P (1940), 1065 p.; v. 4, Q-Z (1940), 758 p.], Zool. Soc. London.

Nechaev [Netschajew], A. W.
(*282) 1894, *Die Fauna der Permablagerungen des östlichen Theils des europäischen Russlands:* Univ. Kazan, Obshchest. Estestvo., Trudy (Soc. Nat. Kazan, Mém.), v. 34, no. 6, p. 1-44.
(*282a) 1897, *Fauna eotsenovykh otlozheniy na Volgye mezhdu Saratovym i Tsaritsynym* [Eocene fauna in Volgian deposits near Saratov and Tsaritsyn]: Same, Trudy, v. 32, pt. 1, 247+iii p., 10 pl.

Nelson, T. C.
(*283) 1938, *The feeding mechanism of the oyster. I. On the pallium and the branchial chambers of Ostrea virginica, O. edulis and O. angulata, with comparisons with other species of the genus:* Jour. Morphology, v. 63, no. 1, p. 1-61, 21 text fig. (July).

Nestler, Helmut
(*284) 1965, *Entwicklung und Schalenstruktur von Pycnodonta vesicularis (LAM.) und Dimyodon nilssoni (V. HAG.) aus der Oberkreide:* Geologie, v. 14, no. 1, p. 64-76, pl. 1-3, 2 fig. (Jan.).

Newell, N. D.
(*285) 1937, *Late Paleozoic Pelecypoda: Pectina-cea:* Kansas State Geol. Survey, Publ., v. 10, 123 p., 42 fig.
(*286) 1960, *The origin of the oysters:* Internat. Geol. Congress, 21st session (Norden, 1960), Rept. pt. 22, p. 81-86.

Newton, R. B., & Smith, E. A.
(*287) 1912, *On the survival of a Miocene oyster in Recent seas:* Geol. Survey India, Rec., v. 42, pt. 1, p. 1-15, 8 pl. (April).

Nicollet, J. N.
(*278a) 1843, *Report intended to illustrate a map of the hydrographical basin of the upper Mississippi River:* U.S. 26th Congress, Second sess., Senate Doc. 237, 170 p., map [not seen]; or U.S. 28th Congress, Second sess., House Executive Doc. 52 (1845) [not seen].

Nikitin, S.
(*287b) 1894, *Review of G. Romanovsky: Materialien zur Geologie des Turkestans. III. Lieferung. Palaeontologischer Character der Sedimente im westlichen Tjan-Chan und in der Turan Niederung:* Neues Jahrb. Geologie, Mineralogie u. Paläontologie, Jahrg. 1894, v. 1, p. 171-172.

Nilsson, Sven
(*288) 1827, *Petrificata Suecana / Formationis Cretaceae,/descripta et iconibus illustrata/ Pars prior,/Vertebrata et Mollusca/sistens;* viii+39 p., 10 pl., Ex Officina Berlingiana (Londini Gothorum [Lund]). [Only part published.]
(*289) 1832, *Djur-petrifikater funna i Skånes Stenskolsbildning:* K. Svenska Vetensk. Akad. Stockholm, Handlingar 1831, p. 352-355, pl. 4.

Noetling, Fritz
(*290) 1880, *Die Entwicklung der Trias in Niederschlesien:* Deutsche Geol. Gesell., Zeitschr., v. 32, no. 2, p. 300-349, pl. 13-15 (April-June).

Oberling, J. J.
(*291) 1955a, *Shell structure of west American Pelecypoda:* Univ. California (Berkeley), Doctoral Dissertation, 407 p., 14 fig., 9 pl. (March 7).
(*292) 1955b, *Shell structure of west American Pelecypoda:* Washington Acad. Sci. Jour., v. 45, no. 4, p. 128-130, 2 fig. (April).
(*293) 1964, *Observations on some structural features of the pelecypod shell:* Naturf. Gesell. Bern (Switzerland), Mitteil., new ser., v. 20, p. 1-63, 3 fig., 5 pl. (Oct.).

Odum, H. T.
(*294) 1957, *Biogeochemical deposition of strontium:* Univ. Texas, Inst. Marine Sci. Publ., v. 4, no. 2, p. 38-114, 11 fig. (July).

Oken, L.
(*295) 1817, *Cuviers und Okens Zoologien neben einander gestellt:* Isis oder Encyclopaedische Zeitung, v. 8, no. 144-148, p. 1145-1185 [irregular pagination].

Olsson, A. A.
(*296) 1961, *Mollusks of the tropical eastern Pacific particularly from the southern half of the Panamic-Pacific faunal province (Panama to Peru):* 574 p., 86 pl., Paleont. Research Inst. (Ithaca, N.Y.). (March 10).

Oppenheim, Paul
(*296a) 1903, *Zur Kenntnis alttertiärer Faunen in Ägypten; 1. Lieferung: Der Bivalven erster Teil (Monomyaria, Heteromyaria, Homomyaria und Siphonida integripalliata) mit Taf. I-XVII:* Palaeontographica, v. 30, Abt. 3, p. 1-164, 9 text fig., 17 pl. (Dec.).

Orbigny, Alcide d'
(*297) 1835-47, *Voyage dans l'Amérique Méridionale (le Brésil, la République orientale de l'Uruguay, le République Argentine, la Patagonie, la République du Chili, la République de Bolivia, la République du Perou), exécuté pendant les années 1826-1833:* 7 v. (text)+2 v. (atlas) (Paris & Strasbourg). [Mollusques, 1847 in v. 5, pt. 3, 86 pl.]
(*298) 1842, *Coquilles et échinodermes fossiles de Colombie (Nouvelle Grenade), recueillis de 1821 à 1833, par M. Boussingault:* 64 p., 6 pl., P. Bertrand (Paris).
(*299) 1849-52, *Prodrome de paléontologie stratigraphique universelle des animaux mollusques et rayonnés, faisant suite au cours élémentaire de paléontologie, etc.:* 3 vol. (Paris).

Oria, M.
(*299a) 1933, *Observations sur des Ostreidae de l'Oxfordien de Normandie:* Soc. Linnéenne Normandie, Bull., ser. 8, v. 5 (Année 1932), Travaux Originaux, p. 19-76, pl. 1-4 (Oct. 24).

Orton, J. H.
(*300) 1928, *The dominant species of Ostrea:* Nature, v. 121, no. 3044, p. 320-321 (March 3).

——, & Amirthalingam, C.
(*301) 1927, *Notes on shell-depositions in oysters:* Marine Biol. Assoc. United Kingdom, Jour., new ser., v. 14 (1926-27), no. 4, p. 935-953, 3 fig. (May).

Paul, M. D.
(*302) 1942, *Studies on the growth and breeding of certain sedentary organisms in the Madras Harbour:* Indian Acad. Sci., Proc., v. 15, sec. B, p. 1-42, fig. 1-7, pl. 1.

Pávay, Alexis von
(*303) 1873, *Die geologischen Verhältnisse der Ungebung von Klausenburg:* K. Geol. Anst., Jahrb., v. 1, pt. 3, p. 351-441, 4 fig., pl. 6-12. [Same author and article as **304*.]

Pávay, Elektöl
(*304) 1871, *Kolozsvár környékének földtani viszonyai:* Magyar Kiralyi Földtani Intézet Evkönyve, v. 1, p. 327-462, pl. 6-12.

Payraudeau, B. C.
(*305) 1826, *Catalogue déscriptif et méthodique des annelides et de mollusques de l'Îsle de Corse, etc.:* 7+218 p., 8 pl. (Paris).

Pelseneer, Paul
(*306) 1896, *L'hermaphroditisme chez les mollusques:* Archives de Biologie, v. 14, p. 33-62, pl. 3-5.
(*307) 1906, *Mollusca:* in E. R. Lankester (ed.), A treatise on zoology, pt. 5, 355 p., 301 fig.; Adam & Charles Black (London). [Facsimile reprint Amsterdam, A. Asher & Co., 1965.]
(*308) 1911, *Les lamellibranches de l'expedition du Siboga, partie anatomique:* Siboga-Expeditie Mon. 53a, pt. 61, 125 p., 26 pl.

Pervinquière, Léon
(*309) 1910a, *Quelques observations sur la nomenclature des ostracés, à propos de la classification phylogénétique exposée par M. H. Douvillé:* Soc. Géol. France, Comptes Rendus Sommaires des Séances, no. 13-14 (June 20, 1910), p. 119-120.
(*310) 1910b, *Gryphaea columba; Gryphaea silicea; Gryphaea plicatula; Gryphaea distans; Gryphaea latissima; Gryphaea plicata; Gryphaea secunda; Ostrea carinata; Ostrea colubrina:* Palaeontologia Universalis, ser. 3, no. 2, fiche 190-198a (July 26).
(*311) 1911, *Quelques observations sur la nomenclature des ostracés, à propos de la classification phylogénétique exposée par. M. Henri Douvillé:* Soc. Géol. France, Bull., ser. 4, v. 10 (1910), pt. 7, p. 645-646 (May 2).
(*311a) 1912, *Études de paléontologie tunisienne, II, Gastropodes et lamellibranches des terrains crétacés:* Carte Géol. de la Tunisie, text vol., xiv+352 p., 16 text fig., atlas vol. 23 pl., Lamarre & Cie (Paris).

Pfannenstiel, Max
(*312) 1928, *Organisation und Entwicklung der Gryphäen:* Palaeobiologica, v. 1, no. 5-7, p. 381-418, 11 fig.

Philip, G. M.
(*313) 1962, *The evolution of Gryphaea:* Geol. Mag., v. 99, p. 327-344, 3 fig. (Sept. 7).

Philippi, Emil
(*314) 1898, *Beiträge zur Morphologie und Phy-*

logenie der Lamellibranchier: Deutsche Geol. Gesell., Zeitschr., v. 50, p. 597-622, 7 fig.

Philippi, R. A.
(*315) 1845-47, *Abbildungen und Beschreibungen neuer oder wenig gekannter Conchylien,* v. 2: Theodor Fischer (Cassel). [The volume is a collation of separately published issues, dating from 1845 to 1847; title page has date of 1847. The *Ostrea* issue is dated February 1846 and has only 2 pages (p. 81-82). Pages and plates of the various issues are not consecutively numbered.]

Pilsbry, H. A.
(*315a) 1890, *Ostrea gigas Thunberg:* Nautilus, v. 4, no. 8, p. 95 (Dec. 22).

Poisson, Henri
(*316) 1946, *Huîtres et ostréiculture à Madagascar:* Soc. Amis parc Bot. Zool. Tanarive, sec. océanogr. appl., cah. 3, 37 p., 7 pl.

Poli, J. X.
(*317) 1791-1827, *Testacea utriusque Siliciae eorumque historia et anatomie tabulis aeneis illustrata:* [In Latin, with Italian and French plate explanations.] v. 1 (1791), [6]+x+90+51+lxxiii p., frontispiece, pl. 1-18; v. 2 (1795), [4]+264+lxxvi p., pl. 19-39; v. 3, pt. 1 (1826) [issued posthumously by S. Delle Chiaje], [6]+xxiv+44+xlviii p., pl. 40-49; v. 3, pt. 2 (1827) [by S. Delle Chiaje], 56 p., pl. 50-57 (Parma, Italy).

Portlock, J. E.
(*318) 1843, *Report on the geology of the County of Londonderry and parts of Tyrone and Fermanagh:* xxi+errata+784 p., frontispiece (geol. map), 37+9 pl. (Jan.), Andrew Milliken (Dublin).

Preston, H. B.
(*319) 1916, *Report on a collection of Mollusca from the Cochin and Ennur backwaters:* Indian Museum Rec., v. 12, pt. 1, p. 27-39, 17 fig. (Feb. 29).

Prytherch, H. F.
(*319a) 1934, *The role of copper in the setting, metamorphosis, and distribution of the American oyster, Ostrea virginica:* Ecol. Mon., v. 4, no. 1, p. 47-107, 16 fig. (Jan.).

Quenstedt, F. A.
(*320) 1865-66, *Handbuch der Petrefaktenkunde:* (2nd edit.), viii+982 p., 183 fig.; atlas, 86 pl., H. Laupp (Tübingen). [p. 1-320, pl. 1-24 (1865); p. 321-640, pl. 25-51, 54-56 (1866); p. viii+641-982, pl. 52-53, 57-86 (1866).]

Rafinesque-Schmaltz, C. S.
(*321) 1815, *Analyse de la nature ou tableau de*

l'Univers et des corps organisés, etc.: 224 p. (Palermo).

Ranson, Gilbert

(*322) 1938, *Contribution à l'étude du développement de l'huître portugaise, Gryphaea angulata Lmk.:* Muséum Natl. Histoire Nat., Bull. (Paris), ser. 2, v. 10, p. 410-424, fig. 1-4.

(*323) 1939, *Le provinculum de la prodissoconque de quelques ostréidés:* Same, ser. 2, v. 11, p. 318-332, fig. 1-4, pl.

(*324) 1939-41, *Les huîtres et le calcaire I. Formation et structure des "chambres crayeuses." Introduction à la revision du genre Pycnodonta F. de W.:* Same, ser. 2, v. 11, p. 467-472; v. 12, p. 426-432, fig. 1-2; v. 13, p. 49-66, fig. 1-6, pl.

(*325) 1940, *La charnière de la dissoconque de l'huître:* Same, ser. 2, v. 12, p. 119-128, fig. 1-2.

(*326) 1941, *Les espèces actuelles et fossiles du genre Pycnodonta F. de W.: I. Pycnodonta hyotis (L.):* Same, ser. 2, v. 13, p. 82-92, fig. 1-6, pl.

(*327) 1942, *La prodissoconque de Pycnodonta cochlear (Poli):* Same, ser. 2, v. 14, p. 74-79, fig. 1-4.

(*328) 1943a, *La vie des huîtres:* 261 p., 19 pl. (Jan. 22), Gallimard (Paris).

(*329) 1943b, *Note sur la classification des ostréidés:* Soc. Géol. France, Bull., ser. 5, v. 12, p. 161-164.

(*330) 1948a, *Ecologie et répartition géographique des ostréides vivants:* Revue Scientifique, 86 année, p. 469-473.

(*331) 1948b, *Prodissoconques et classification des ostréidés vivants:* Musée Royal Histoire Nat. Belgique, Bull., v. 24, no. 42, 12 p., 7 fig.

(*332) 1949a, *Prodissoconques et classification des ostréidés vivants:* 13ᵉ Congrès Internat. Zoologie, Paris, Comptes Rendus, p. 454-455.

(*333) 1949b, *Écologie et répartition géographique des ostréidés vivants:* Same, Comptes Rendus, p. 455-456.

(*334) 1949c, *Prodissoconques et classification des ostréidés fossiles:* Same, Comptes Rendus, p. 565-566.

(*335) 1949d, *Note sur la répartition géographique des Ostréidés du genre Pycnodonta F. de W.:* Muséum Natl. Histoire Nat., Bull., ser. 2, v. 21, p. 447-452.

(*336) 1949e, *La chambre promyaire et la classification zoologique des ostréidés:* Jour. Conchyliologie (ser. 4, v. 42), v. 89, p. 195-200.

(*336a) 1949f, *Note sur trois espèces Lamarckiennes d' ostréidés:* Muséum Natl. Historie Nat., Bull., ser. 2, v. 21, no. 2, p. 248-254.

(*337) 1950, *La chambre promyaire et la classification zoologique des ostréidés:* Same, (ser. 4, p. 43), v. 90, p. 195-200 (Oct. 1).

(*338) 1951, *Observations morphologiques, biologiques, biogéographiques, géologiques et systématiques sur une espèce d'huître de Madagascar et d'Afrique du Sud: Gryphaea margaritacea (Lmk.):* Inst. Océanogr. Monaco, Bull., no. 983, 20 p., 8 fig. (Jan. 16).

(*339) 1952, *Les huîtres biologie-culture bibliographie:* Same, Bull. (Fondation Albert Ier, Prince de Monaco), no. 1001, 134 p. (Jan. 2).

(*340) 1960a, *Les prodissoconques (coquilles larvaires) des ostréidés vivants:* Same, v. 57, no. 1183, 41 p., 136 fig. (June 7).

(*341) 1960b, *Les ostréidés, les aviculidés et le problème de l'espèce:* Sciences (Paris), no. 8-9, p. 7-9, 5 fig. (unnumbered) (Oct.).

(*342) 1963, *Les huîtres et le calcaire:* Acad. Sci. Paris, Comptes Rendus, v. 257, no. 21, p. 3229-3230, 4 fig. (Nov. 18).

(*342a) 1967, *Les espèces d'huîtres vivant actuellement dans le monde, définies par leurs coquilles larvaires ou prodissoconques. Étude des collections de quelques-uns des grands musées d'histoire naturelle:* Pêches Maritimes, Revue travaux l'Inst., v. 31, no. 2, p. 127-199, 25 text fig.; no. 3, p. 205-274, text fig. 26-55 (June and Sept.).

Raulin, V., & Delbos, J.

(*343) 1855, *Extrait d'une monographie des Ostrea des terrains tertiaires de l'Aquitaine:* Soc. Géol. France, Bull., ser. 2, v. 12, p. 1144-1164.

Raymond, P. E.

(*344) 1925, *A new oyster from the Cretaceous of Cuba:* Boston Soc. Nat. History, Occas. Papers, v. 5, p. 183-185, pl. 7 (Dec. 18).

Reeside, J. B., Jr.

(*344a) 1929, *Exogyra olisiponensis Sharpe and Exogyra costata Say in the Cretaceous of the western Interior:* U.S. Geol. Survey, Prof. Paper 154-I, p. 267-278, pl. 65-69 (April 20).

Reis, O. M.

(*344b) 1906, *Bemerkungen zu G. Böhm's "Zur Stellung der Lithiotiden":* Centralbl. Mineralogie, Geologie u. Paläontologie, v. 1906, p. 209-217.

(*345) 1914, *Zur Morphologie der Austernschale:* Same, v. 1914, p. 169-170.

Rémond, Auguste

(*346) 1863, *Description of two new species of bivalve shells from the Tertiaries of Contra Costa County:* California Acad. Sci., Proc., v. 3, p. 13.

Renevier, E.
*(*347)* 1864, *Notices géologiques et paléontologiques sur les Alpes Vaudoises et les régions environnantes:* Soc. Vaudoise Sci. Nat., Mém., v. 8, no. 51, p. 39-97, pl. 1-3, 2 table insert.

Rengarten, V. P.
*(*348)* 1953, *O nekotorykh predstavitelyakh verkhiemelovoy fauny vostochnogo Priuralya* [On certain representatives of leading fauna of eastern Pre-Ural region]: Akad. Nauk SSSR, Voprosy petrografii i mineralogii, pt. 2, p. 474-484, 2 pl.

[Renier (S. A.)]
*(*349)* [1807], [Tavole per servire alle Classificazione e Connescenza degli Animali]: Rejected ICZN Opinion 427.

Reuss, A. E. von
*(*350)* 1840-44, *Geognostische Skizzen aus Böhmen:* 2 v. (Prague). [Not seen.]

Richardson, Linsdall
*(*351)* 1905, *The Rhaetic and contiguous deposits of Glamorganshire:* Geol. Soc. London, Quart. Jour., v. 61, p. 385-424, 5 fig., pl. 33 (Aug.).

Risso, J. A.
*(*352)* 1826, *Histoire naturelle des principales productions de l'Europe méridionale et particulièrement des celles des environs de Nice et des Alpes maritimes:* v. 4 (of 5), vii+439 p., 12 pl., F.-G. Levrault (Paris).

Roche, Jean, Ranson, Gilbert, & Eysseric-Lafon, Marcelle
*(*353)* 1951, *Sur la composition des scléroprotéines des coquilles des mollusques (conchiolines):* Soc. Biologie, Comptes Rendus, v. 145, no. 19-20, p. 1474-1477.

Röding, P. F. [or Bolten, J. F.]
*(*354)* 1798, *Museum Boltenianum sive catalogus cimeliorum; Pars secunda continens Conchylia:* viii+199 p., J. C. Trapii (Hamburg) (Dec. 31.). [The introduction by Lichtenstein is dated Sept. 10, 1798.]

Roemer, Ferdinand
*(*355)* 1849, *Texas,* etc.: xiv+464 p., map (Bonn).
*(*356)* 1851, *Ueber einige neue Versteinerumgen aus dem Muschelkalke von Willebadessen:* Palaeontographica, v. 1, pt. 6, p. 311-315, 340, pl. 36 (July).
*(*357)* 1852, *Die Kreidebildungen von Texas und ihre organischen Einschlüsse:* vi+100 p., 11 pl., Adolph Marcus (Bonn).

Rollier, Louis
*(*358)* 1911, *Les faciès du Dogger ou Oolithique dans le Jura et les régions voisines:* Fondation Schnyder von Wartensee à Zurich, Mém. 18, v+352 p., 56 fig., Georg & Cie (Geneva and Basel).
*(*359)* 1917, *Fossiles nouveaux ou peu connus des terrains secondaires (Mésozoïques) du Jura et des contrées environnantes, tome 1, pt. 6:* Soc. Paléont. Suisse, Mém., v. 42, p. 501-634+errata page, pl. 33-40.

Roman, Frédéric
*(*360)* 1940, *Listes raisonnées des faunes du Pliocène et du Miocène de Syrie et du Liban:* Haut-Comm. Républ. Française Syrie Liban, Service des travaux publics, Sec. d'études géol., Notes et Mém., v. 3, p. 353-410, pl. 1-5.

Romanovskiy, G. D.
*(*360a)* 1878-90, *Materialy dlya geologiy Turkestanskago kraya* [Materials for the geology of the Turkestanian region]: Acad. Impér. Sci., St. Petersbourg, v. 1, viii+167 p., 30 pl. (1878); v. 2, xii+161 p., 27 pl. (1884); v. 3, x+165 p., 23 pl. (1890).
*(*360b)* 1879, *Dva novykh vida, iz semeystva ustrichnykh rakovin, naydennykh v Ferganskoy Oblasti* [Two new species of the oyster family from the Fergansk Province]: Imper. Mineral. Obshchest. St. Petersburg, Zapiski [Russian Imper. Mineral. Soc., St. Petersburg, Trans.], ser. 2, v. 14, p. 150-154, 2 text fig.
*(*360c)* 1882, *Ferganskiy yarus melovoy pochvy i paleontologicheskiy ego kharakter* [Fergansk Stage of Cretaceous and its paleontological character]: Same, Zapiski, ser. 2, v. 17, p. 35-60, pl. 1-8.
*(*360d)* 1883, [Untitled report]: Same, ser. 2, v. 18, p. 25].

Roughley, T. C.
*(*361)* 1931, *Giant oysters:* Nature, v. 127, no. 3196, p. 165, fig. 1.
*(*362)* 1933, *The life history of the Australian oyster (Ostrea commercialis):* Linnean Soc. New South Wales, Proc., v. 58, pt. 3-4 (no. 247-248), p. 279-333, 2 fig., pl. 10-27 (Sept. 15).

Rudwick, M. J. S.
*(*363)* 1964, *The function of zigzag deflexions in the commissure of fossil brachiopods:* Palaeontology, v. 7, pt. 1, p. 135-171, 14 fig., pl. 21-29 (April 14).
*(*364)* 1965, *Sensory spines in the Jurassic brachiopod Acanthothiris:* Same, v. 8, pt. 4, p. 604-617, pl. 84-87 (Dec. 15).

Russell, L. S., & Landes, R. W.
*(*365)* 1940, *Geology of the southern Alberta Plains:* Canada Geol. Survey, Mem. 221, Pub. 2453, iv+223 p., 21 fig., 11 pl.

Rutsch, R. F.
*(*366)* 1955, *Die fazielle Bedeutung der Crasso-*

streen (*Ostreidae, Mollusca*) *im Helvétien der Umgebung von Bern:* Eclogae Geol. Helv., v. 48, p. 453-464.

Sacco, Federico
(*367) 1897a, *I molluschi dei terreni terziarii del Piemonte e della Liguria:* Musei Zoologia Anatomia comp. R. Univ. Torino, Boll., v. 12, no. 298, p. 99-102 (June 11).
(*368) 1897b, *Pelecypoda (Ostreidae, Anomiidae e Dimyidae):* of L. Bellardi & Federico Sacco, 1872-1904, I molluschi dei terreni Terziarii de Piemonte e della Liguria, 30 pts. separately paged, pt. 23, 66 p., 11 pl. (June, 1897), Carlo Clausen (Torino).

Saint-Seine, Roseline de
(*369) 1952, *Mimétisme ou "pseudomorphose" chez les lamellibranches fixé sur échinides:* Soc. Géol. France, Bull., ser. 6, v. 1 (1951), no. 8, p. 653-656, pl. 24-25 (June).

Saleuddin, A. S. M.
(*369a) 1965, *The mode of life and functional anatomy of Astarte spp. (Eulamellibranchia):* Malacol. Soc. London, Proc., v. 36 (1965), pt. 4, p. 229-257, 6 text fig. (April).

Salisbury, A. E., & Edwards, M. A.
(*369b) 1959, Mollusca: Zoological Record, v. 93, sec. 9 (1956), 149 p.

Saville-Kent, William
(*370) 1893, *The Great Barrier Reef of Australia; its products and potentialities:* xvii+387 p., 48+16 pl., text fig., W. H. Allen & Co. (London).

Say, Thomas
(*371) 1820, *Observations on some species of zoophytes, shells, etc. principally fossil:* Am. Jour. Sci., ser. 1, v. 2, p. 34-45 (Nov.).
(*372) 1834, *American conchology, or descriptions of the shells of North America illustrated by coloured figures from original drawings executed from nature:* v. 1, no. 6, 42 p. (unnumbered), pl. 51-60 (April), School Press (New Harmony, Ind.).

Scalia, Salvatore
(*373) 1912, *La fauna del Trias superiore del grupo di M.te Judica, pt. 2:* Accad. Gioenia Sci. Nat. Catania Atti, Anno 89, ser. 5, v. 5, mem. 8, 58 p., 3 pl.

Schäfle, Ludwig
(*374) 1929, *Ueber Lias- und Doggeraüstern:* Geologische u. Palaeontologische Abhandl., new ser., v. 17 (=v. 21 of entire serial), no. 2, 88 p., 12 fig., 6 pl.

Schauroth, Carl von
(*374a) 1865, *Verzeichniss der Versteinerungen im Herzogl. Naturaliencabinet zu Coburg*

(*No. 1-4328*) *mit Angabe der Synonymen und Beschreibung vieler neuen Arten, sowie der letzteren Abbildung auf 30 Tafeln:* xv+327 p., 30 pl., Dietz'sche Hofbuchdruckerei (Coburg).

Schlotheim, E. F. von
(*375) 1813, *Beiträge zur Naturgeschichte der Versteinerungen in geognostischer Hinsicht:* Taschenbuch Gesammte Mineralogie, etc., C. C. Leonhard (ed.), v. 7, Abh., p. 3-134, pl. 1-4.
(*376) 1820, *Die Petrefactenkunde auf ihrem jetzigen Standpunkte durch die Beschreibung seiner Sammlung versteinerter und fossiler Überreste des Thier- und Pflanzenreichs der Vorwelt erläutert:* lxii+437 p. [text volume only], Becker (Gotha).
(*376a) 1823, *Nachträge zur Petrefactenkunde, Zweyte Abtheilung:* i+114 p., text; pl. 22-37, atlas, Becker (Gotha).

Schmidt, Herta
(*377) 1937, *Zur Morphologie der Rhynchonelliden:* Senckenbergiana, v. 9, p. 22-60.

Schmidt, Martin
(*378) 1928, *Die Lebewelt unserer Trias:* 461 p., 1220 fig., Hohenlohe'sche Buchhandlung, Ferdinand Rau (Öhringen).

Schmidt, W. J.
(*379) 1931, *Über die Prismenschicht der Schale von Ostrea edulis L.:* Zeitschr. Morphologie u. Ökologie Tiere, v. 21, p. 789-805, 15 fig.

Schweigger, A. F.
(*379a) 1820, *Handbuch der Naturgeschichte der skelettlosen ungegliederten Thiere:* xvi+776 p., Dyk (Leipzig).

Seguenza, Giuseppe
(*379b) 1882, *Studi geologici e paleontologici sul cretaceo medio dell' Italia meridionale:* [R.] Accad. Lincei, Atti, ser. 3, Cl. Sci. Fisiche etc., Mem., v. 12, p. 65-214, 21 pl.

Seilacher, Adolph
(*380) 1960, *Epizoans as a key to ammonoid ecology:* Jour. Paleontology, v. 34, p. 188-193, 3 fig. (Feb. 12).

Serres, Marcel de
(*381) 1843, *Observations sur les grandes huîtres fossiles des terrains tertiaires des bords de la Méditerranée:* Annales Sci. Nat., ser. 2, v. 20, Zoologie, p. 142-168, pl. 2-3 (Sept.).

Sharp, Benjamin
(*382) 1888, *Remarks on the phylogeny of the Lamellibranchiata:* Acad. Nat. Sci. Philadelphia, Proc., v. 40, pt. 4, p. 121-124 (May 8).

Sharpe, Daniel
(*382a) 1849, *On the Secondary District of Portu-*

gal which lies on the north of the Tagus: Geol. Soc. London, Quart. Jour., v. 6 (1850), p. 135-195, 7 text fig., pl. 14-26 (Nov. 21).

Shumard, B. F.
(*383) 1860, *Descriptions of new Cretaceous fossils from Texas:* Acad. Sci. St. Louis, Trans., v. 1, p. 590-610.
(*383a) 1861, *Descriptions of new Cretaceous fossils from Texas:* Boston Soc. Nat. History, Proc., v. 8, p. 188-205 (Sept. 4).

Simpson, G. G.
(*384) 1950, *The meaning of evolution; A study of the history of life and of its significance for man:* xv+364 p., 38 fig., 4th printing, Yale Univ. Press (New Haven).
(*385) 1953, *The major features of evolution:* Columbia Biol. Ser., no. 17, xx+434 p., 52 fig., Columbia Univ. Press (New York).

Skarlato, O. A.
(*386) 1960, *Bivalve mollusks of the far eastern seas of the U.S.S.R. (Otryad Dysodonta):* Akad. Nauk SSSR, Opredel. po faune SSSR, Izdavayemye Zool. Inst. SSSR, no. 71, 150+2 p., 61 fig., 17 pl.

Smith, William
(*387) 1816-19, *Strata identified by organized fossils:* p. 1-16, 9 pl. (1816); p. 7-24, 5 pl. (1817); p. 25-32, 5 pl. (1819).

Sohl, N. F., & Kauffman, E. G.
(*388) 1964, *Giant Upper Cretaceous oysters from the Gulf Coast and Caribbean:* U.S. Geol. Survey, Prof. Paper 483-H, iv+31 p., 3 fig., 5 pl.

Sokolov, D. V.
(*388a) 1910, *K voprosu o Ferganskom yaruse* [The question of the Fergana Stage]: Moskov. Obshchest. Ispyt. Prirody, Byull., v. 23 (1909), p. 44-93, text fig.

Solander, D. C.
(*388b) See **Brander, Gustav.**

Someren, V. D. van, & Whitehead, P. J.
(*389) 1961, *An investigation of the biology and culture of an East African oyster Crassostrea cucullata:* Fishery Publ. 14, 36 p., 5 fig., 3 maps, 5 pl., Colonial Office (London).

Sowerby, G. B., Jr.
(*389a) 1839, *A conchological manual:* (edit. 1), v+130+errata p., 2 tables, frontispiece and unnumbered pls., G. B. Sowerby (London).
(*390) 1842, *A conchological manual:* (edit. 2), vii+313 p., 98 fig., 2 table, 26 pl. (unnumbered), frontispiece, H. G. Bohn, (London).
(*391) 1852, *A conchological manual:* (edit. 4),

vii+337 p., 98 fig., 2 table, 28 pl., frontispiece, H. G. Bohn (London).
(*392) 1870-71, *Monograph of the genus Ostraea:* in L. A. Reeve, 1843-78, Conchologia Iconica; or illustrations of the shells of molluscous animals (20 v.), v. 18, 33 pl.+index (2 p.) (Oct. 1870-Nov. 1871), L. Reeve & Co. (London).

Sowerby, James, & Sowerby, J. de C.
(*393) 1812-46, *The mineral conchology of Great Britain; or, coloured figures and descriptions of those remains of testaceous animals or shells,* etc.: 7 v., 113 pts., 1295 p., 648 pl. (London).

Staff, Hans von, & Reck, Hans
(*394) 1911, *Die Lebensweise der Zweischaler des Solnhofener lithographischen Schiefers:* Gesell. Naturf. Freunde Berlin, Sitzungsber., v. 1911, p. 157-175, pl. 6-11.

Stanton, T. W.
(*395) 1947, *Studies of some Comanche pelecypods and gastropods:* U.S. Geol. Survey, Prof. Paper 211, 256 p., 1 fig., 67 pl.

Steininger, Jean
(*396) 1834, *Observations sur les fossiles du calcaire intermédiaire de l'Eifel:* Soc. Géol. France, Mém., ser. 1, v. 1, pt. 2, p. 331-371.

Steininger, Johann
(*397) 1831, *Bemerkungen über die Versteinerungen, welche in dem Übergangs-Kalkgebirge der Eifel gefunden werden, etc.:* 44 p., Gymnasium Programm (Trier). [Not seen.] [Translated into French and published again as Steininger, 1834.]

Stenzel, H. B.
(*397a) 1945a, *Stratigraphic significance of the Patagonian Odontogryphaeas:* Geol. Soc. America, Bull., v. 56, no. 12, pt. 2, p. 1202 (Dec.).
(*397b) 1945b, *Oysters of the Odontogryphaea—Flemingostrea stock:* Same, v. 56, no. 12, pt. 2, p. 1202 (Dec.).
(*398) 1947, *Nomenclatural synopsis of supraspecific groups of the family Ostreidae (Pelecypoda, Mollusca):* Jour. Paleontology, v. 21, p. 165-185 (March) [April 21].
(*399) 1949, *Successional speciation in paleontology: The case of the oysters of the sellaeformis stock:* Evolution, v. 3, p. 34-50, 8 fig. (March). [Reprinted as Univ. Texas, Bur. Econ. Geology Rept. Inv. 3.]
(*400) 1959, *Cretaceous oysters of southwestern North America:* Cong. Geol. Internac., XXa sesión, Ciudad de México, 1956, El sistema Cretacico, v. 1, p. 15-37, 19 fig.
(*401) 1962, *Aragonite in the resilium of oysters:*

Science, v. 136, no. 3522, p. 1121-1122 (June 29).

(*402) 1963a, *A generic character, can it be lacking in individuals of the species in a given genus?:* Systematic Zoology, v. 12, no. 3, p. 118-121, 2 fig. (Sept. 16).

(*403) 1963b, *Aragonite and calcite as constituents of adult oyster shells:* Science, v. 142, no. 3589, p. 232-233, 1 fig. (Oct. 11).

(*404) 1964, *Oysters: Composition of the larval shell:* Same, v. 145, no. 3628, p. 155-156, 2 fig. (May 16).

(*405) 1971, *Nomenclatural clarifications of some generic and subgeneric names in family Ostreidae (Bivalvia):* this volume, p. N1200.

———, Krause, E. K., & Twining, J. T.

(*406) 1957, *Pelecypoda from the type locality of the Stone City Beds (Middle Eocene) of Texas:* Univ. Texas, Publ. 5704, 237 p., 31 fig., 22 pl. (Feb. 15).

Stephenson, L. W.

(*407) 1914, *Cretaceous deposits in the eastern Gulf region and species of Exogyra from the eastern Gulf region and the Carolinas:* U.S. Geol. Survey, Prof. Paper 81, 77 p., 2 fig., 21 pl., 8 table.

(*408) 1929, *Two new mollusks of the genera Ostrea and Exogyra from the Austin Chalk, Texas:* U.S. Natl. Museum, Proc., v. 76, no. 2815, art. 18, 6 p., 3 pl. (Dec. 23).

(*409) 1941, *The larger invertebrate fossils of the Navarro Group of Texas (exclusive of corals and crustaceans and exclusive of the fauna of the Escondido Formation):* Univ. Texas, Publ. 4101, 641 p., 13 fig., 95 pl., 6 table (Oct.).

(*409a) 1945, *A new Upper Cretaceous oyster from deep wells in Mississippi:* Jour. Paleontology, v. 19, no. 1, p. 72-74, 7 text fig. (Jan. 25).

Strausz, Laszlo

(*410) 1928, *Geologische Facieskunde:* K. Ungarische Geol. Anst. Jahrb., v. 28, p. 72-272 (Sept. 10). [Serial also listed as Magyar Királyi Földtani Intézet Házinyomdája.]

Suter, Henry

(*411) 1917, *Descriptions of new Tertiary Mollusca occurring in New Zealand, accompanied by a few notes on necessary changes in nomenclature, pt. 1:* New Zealand Geol. Survey, Palaeont. Bull. 5, vii+93 p., errata slip, frontispiece+13 pl.

Swainson, William

(*412) 1835, *The elements of modern conchology: briefly and plainly stated, for the use of students and travellers:* viii+62 p. (London).

(*413) 1840, *A treatise on malacology, or shells and shellfish:* The Cabinet Cyclopaedia conducted by Dionysius Lardner etc., Natural History, viii+419 p., 129 fig., frontispiece, Longman, Orme, Brown, Green, & Longmans and John Taylor (London).

Swinnerton, H. H.

(*414) 1932, *Unit characters in fossils:* Cambridge Philos. Soc., Biol. Reviews, v. 7, p. 321-335, fig. 4 (Oct.).

(*415) 1939, *Palaeontology and the mechanics of evolution:* Geol. Soc. London, Quart. Jour., v. 95 (1939), pt. 2, p. xxxiii-lxx, 10 fig. (May 26).

(*416) 1940, *The study of variation in fossils:* Same, v. 96 (1940), pt. 3, p. lxxvii-cxviii, 13 fig. (Nov. 30).

(*417) 1959, *Concerning Mr. A. Hallam's article on Gryphaea:* Geol. Mag., v. 96, p. 307-310 (Aug. 7).

(*418) 1964, *The early development of Gryphaea:* Same, v. 101, p. 409-420, 3 fig. (Oct. 31).

Sylvester-Bradley, P. C.

(*419) 1952, *Proposed use of the plenary powers to validate the trivial name "knorri" Voltz, 1828 etc.:* Bull. Zool. Nomencl., v. 6, p. 7, 201-202.

(*419a) 1958, *The description of fossil populations:* Jour. Paleontology, v. 32, no. 1, p. 214-235, 16 text fig. (Jan. 27).

Tate, Ralph, & Blake, J. F.

(*419b) 1876, *The Yorkshire Lias:* viii+475+xii p., illus., 23 pl., 2 maps, J. Van Voorst (London).

Termier, Henri, & Geneviève

(*420) 1949, *Role des Aviculopectinidae dans la morphogenèse des dysodontes mésozoiques:* Muséum Natl. Histoire Nat., Bull., ser. 2, v. 21, p. 292-299, 12 fig.

Terquem, M. O.

(*420a) 1855, *Paléontologie de l'étage inférieur de la formation liasique de la Province de Luxembourg, Grand-Duché (Hollande), et de Hettange, du Département de la Moselle:* Soc. Géol. France, Mém., ser. 2, v. 5, pt. 2, Mém. 3, p. 219-343, pl. 12-26.

Thiele, Johannes

(*421) 1934, *Handbuch der systematischen Weichtierkunde:* pt. 3, p. 779-1022, fig. 784-893, Gustav Fischer (Jena).

Thompson, Sir D'A. W.

(*422) 1917, *On growth and form:* edit. 1, xv+793 p., illus. (Cambridge, Eng.).

Thompson, T. G., & Chow, T. J.

(*423) 1955, *The strontium-calcium atom ratio in carbonate-secreting marine organisms:* Papers in Marine Biology and Oceanoggraphy, Deep-Sea Research, suppl. to v. 3, p. 20-39.

Thomson, J. M.
(*424) 1954, *The genera of oysters and the Australian species:* Australian Jour. Mar. & Freshwater Research, v. 5, p. 132-168, pl. 1-11 (March).

Thunberg, C. P.
(*425) 1793, *Techning och beskrifning pa en stor Ostronsort ifran Japan:* K. Svenska Vetensk. Akad., Handlingar, v. 14, p. 140-142.

Tozer, E. T.
(*426) 1961, *Triassic stratigraphy and faunas, Queen Elizabeth Islands, Arctic Archipelago:* Geol. Survey Canada, Mem. 316, 116+7 p., 10 fig., 30 pl.

———, **& Thorsteinsson, R.**
(*427) 1964, *Western Queen Elizabeth Islands, Arctic Archipelago:* Geol. Survey Canada, Mem. 332, xviii+242 p., 20 fig., 55 pl., 1 map.

Traub, Franz
(*427a) 1938, *Geologische und paläontologische Bearbeitung der Kreide und des Tertiärs im östlichen Rupertiwinkel, nördlich von Salzburg:* Palaeontographica, v. 88, Abt. A, pt. 1-3, p. 1-114, 2 text fig., pl. 1-8, geol. map, 3 profiles (June).

Troelsen, J. C.
(*428) 1950, *Contributions to the geology of Northwest Greenland, Ellesmere Island and Axel Heiberg Island:* Meddel. Grønland, v. 149, no. 7, 86 p., 17 fig., map (March 22).

Trueman, A. E.
(*429) 1922, *The use of Gryphaea in the correlation of the Lower Lias:* Geol. Mag., no. 696, v. 59, p. 256-268, 7 fig. (June).

(*430) 1940, *The meaning of orthogenesis:* Geol. Soc. Glasgow, Trans., v. 20, pt. 1 (1937-40), no. 6, p. 77-95, 2 fig. (March 7).

Trueman, E. R.
(*431) 1951, *The structure, development, and operation of the hinge ligament of Ostrea edulis:* Quart. Jour. Micros. Sci., ser. 3, v. 92, pt. 2, p. 129-140, 8 fig. (June).

Tryon, G. W., Jr.
(*432) 1882-84, *Structural and systematic conchology: An introduction to the study of the Mollusca:* v. 1 (1882), viii+312 p., 1 map; v. 2 (1883), 430 p.; v. 3 (1883), 453 p., 140 pl., author (Philadelphia).

Turekian, K. K., & Armstrong, R. L.
(*433) 1961, *Chemical and mineralogical composition of molluscan shells from the Fox Hills Formation, South Dakota:* Geol. Soc.

America, Bull., v. 72, p. 1817-1828 (Dec. 26).

Tzankov, V.
(*434) 1932, *Mollusques fossiles de la Craie Supérieure dans la Bulgarie du Nord:* Bulgarische Geol. Gesell. Zeitschr. (Sofia), v. 4, p. 46-78, pl. 1-7.

Verneuil, Édouard de
(*435) 1845, *Géologie de la Russie d'Europe et des montagnes de l'Oural:* in R. I. Murchison, Edouard de Verneuil, & Alexander von Keyserling, 1845, The Geology of Russia in Europe and the Ural Mountains, 2 v., Paléontologie, v. 2, pt. 3, xxxii+512 p., 43+A-G pl., John Murray (London). [v. 2, pt. 3, is in French.]

Vokes, H. E.
(*435a) 1967, *Genera of the Bivalvia: A systematic and bibliographic catalogue:* Bull. Am. Paleontology, v. 51, no. 232, p. 105-394.

Volkova, N. S.
(*435b) 1955, *Polevoi atlas kharakternykh kompletsov fauny tretichnykh otlozhenii tsentralnogo Predkavkazya* [Field atlas of characteristic Tertiary fauna in Central Predavkazya]: Vses. Nauch.-Issled. Geol. Inst. [VSEGEI], Moscow, 162 p., illus.

Voltz, P. L.
(*436) 1828, *Uebersicht der Petrefakten der beiden Rhein-Departemente:* in J. F. Aufschlager, Das Elsass oder die Departemente des Ober- und Niederrheins, 64 p. (Aug.).

Vredenburg, E. W.
(*437) 1916, *Flemingostrea, an eastern group of Upper Cretaceous and Eocene Ostreidae: with descriptions of two new species:* Geol. Survey India, Records, v. 47, pt. 3, p. 196-203, pl. 17-20 (Aug.). [Pt. 3 was received in U.S. Geol. Survey Library April 23, 1917.]

Vyalov [Vialov], O. S.
(*438) 1936, *Sur la classification des huîtres:* Acad. Sci. URSS, Comptes rendus (Doklady), new ser., v. 4 (13), no. 1 (105), p. 17-20 (after Aug. 1).

(*439) 1937a, *Sur la classification des ostréidés et leur valeur stratigraphique:* Internatl. Congress Zoology, 12th sess. Lisbon, 1935, Comptes rendus, sec. 8, v. 3, p. 1627-1639.

(*439a) 1937b, *Rukovodyashchie ustritsy paleogena Fergany* [Index oysters of the Paleogene of Fergana]: Geol. Razved. Sluzhby tresta "Sredazneft," Trudy, Vyp. pervyy, Tashkent, Izdatel. Kom. Nauk Uzbek. SSR, 48 p., 33 pl., + index p.

(*440) 1945, *New oysters from the Paleogene of*

the Trans-Caspian region: Acad. Sci. URSS, Comptes Rendus (Doklady), new ser., v. 48 (158), no. 3, p. 200-203, 6 fig.

(*441) 1946, *Triasovi ustritsi SRSR* [Triassic oysters from SRSR]: Lvovskogo Derzhavnogo Univ. Ivana Franka naukovi zapiski, v. 2, ser. geol., no. 3, p. 22-54, 1 fig., 3 pl.

(*442) 1948a, *Printsiply klassifikatsii semeystva Ostreidae* [Principles of classification of the family Ostreidae]: Lvov. geol. obshest. trudy pri gosudarst. Univ. Im. Ivan Franko, Paleont. ser., no. 1 (1948), p. 30-40.

(*443) 1948b, *Paleogenovye ustritsy Tadzhikskoi depressii* [Paleogene oysters of the Tadzhik basin]: Vses. Neft. Nauch.-Issled. Geol.-Razv. Inst. (VNIGRI), Trudy, new ser., no. 38, 94 p., 38 pl. (Nov. 5).

(*443a) 1948c, *Paleogenovykh ustritsakh iz Kashgara* [On Paleogene oysters from Kashgara]: Akad. Nauk SSSR, Doklady, v. 62, no. 3, p. 381-384, 1 text fig.

(*443b) 1965, *Nekotoryye paleogenovyye ustritsy* [Some paleogene oysters]: Lvov. Geol. Obshchest. Paleont. Sbornik [Lvov, Ukrainskaya S.S.R.], v. 2, pt. 1, p. 5-13, 4 pl. (July 15).

———, & Solun, V. I.

(*443c) 1957, *Zarozdenie Turkestanskikh Fatina v Alayskom yaruse* [Considerations of Turkestanian Fatina in Alayska Stage]: Voprosy Paleobiogeog. i Biostratig. (Vses. Paleont. Obshch., Sess. I, Trudy), p. 191-197, 1 text fig.

Waagen, Lukas

(*444) 1907, *Die Lamellibranchiaten der Pachycardientuffe der Seiser Alm nebst vergleichend paläontologischen und phylogenetischen Studien:* K.-K. Geol. Reichsanst. Abhandl., v. 18, no. 2, i+180 p., 19 fig., pl. 25-34 (April).

Wada, S. K.

(*445) 1953, *Larviparous oysters from the tropical west Pacific:* Rec. Oceanogr. Works Japan, new ser., v. 1, p. 66-72 (Dec.).

Walne, P. R.

(*446) 1963, *Breeding of the Chilean oyster (Ostrea chilensis Philippi) in the laboratory:* Nature, v. 197, no. 4868, p. 676 (Feb. 16).

(*447) 1964, *Observations on the fertility of the oyster (Ostrea edulis):* Marine Biol. Assoc. United Kingdom, Jour., new ser., v. 44, p. 293-310, 9 fig. (June).

Wells, H. W.

(*448) 1961, *The fauna of oyster beds, with special reference to the salinity factor:* Ecol. Mon., v. 31, p. 239-266, 7 fig.

Westermann, G. E. G.

(*449) 1962, *Succession and variation of Monotis and the associated fauna in the Norian Pine River bridge section, British Columbia (Triassic, Pelecypoda):* Jour. Paleontology, v. 36, p. 745-792, 19 fig., pl. 112-118 (July 16).

Westoll, T. S.

(*450) 1950, *Some aspects of growth studies in fossils:* Royal Soc. London, Proc., ser. B, no. 889, v. 137, p. 490-509, fig. 22-26 (Nov. 28).

Whidborne, G. F.

(*450a) 1883, *Notes on some fossils, chiefly Mollusca, from the Inferior Oolite:* Geol. Soc. London, Quart. Jour., v. 39, p. 487-540, pl. 15-19.

White, C. A.

(*451) 1884, *A review of the fossil Ostreidae of North America; and a comparison of the fossil with the living forms:* U.S. Geol. Survey, Ann. Rept. 4 (1883), p. 273-308 pl. 34-61.

(*451a) 1887, *Contribuições á paleontologia do Brazil* . . . [Contributions to the paleontology of Brazil; comprising descriptions of Cretaceous invertebrate fossils, mainly from the provinces of Sergipe, Parnambuco, Para and Bahia]: Museo Nac. Rio de Janeiro, Arch., v. 7, 273 p.+v, 28 pl., errata slip.

Winchell, Alexander

(*452) 1865, *Descriptions of new species of fossils, from the Marshall Group of Michigan, and its supposed equivalent, in other states; with notes on some fossils of the same age previously described:* Acad. Nat. Sci. Philadelphia, Proc., v. 17 (1865), p. 109-133 (June) [before Oct. 16].

Winkler, T. C.

(*452a) 1863-67, *Catalogue systématique de la collection paléontologique:* viii+697 p., Livr. 1, 1863; Livr. 2, 1864; Livr. 3, 1865; Livr. 4, 1865; Livr. 5, 1866; Livr. 6, 1867; Musée Teyler, Teyler's Stichting (Haarlem, Neth.).

Winslow, Francis

(*453) 1882, *Report on the oyster beds of the James River, Va., and of Tangier and Pocomoke Sounds, Maryland and Virginia:* U.S. Coast & Geodetic Survey, Rept. for 1881, Appendix 11, 87 p., 22 pl., 3 map.

Wöhrmann, S. Freiherr von

(*454) 1889, *Die Fauna der sogenannten Carditaund Raibler-Schichten in den Nordtiroler und bayerischen Alpen:* K.-K. Geol. Reichsanst. Jahrb., v. 39 (1889), no. 1+2, p. 181-258, 5 fig., pl. 5-10 (July 1).

Wood, S. V.
(*455) 1861, *A monograph of the Eocene Mollusca, or, descriptions of shells from the older Tertiaries of England, pt. 1, Bivalves:* Palaeontogr. Soc. 1859, v. 13, p. 1-74, pl. 1-13 (Dec.).

Woods, Henry
(*456) 1913, *A monograph of the Cretaceous Lamellibranchia of England, v. 2, pt. 9:* Palaeontogr. Soc. 1912, v. 66, p. 341-473, pl. 55-62 (Feb.).

Yonge, C. M.
(*458) 1926, *Structure and physiology of the organs of feeding and digestion in Ostrea edulis:* Marine Biol. Assoc. United Kingdom, Jour., new ser., v. 14 (1926-27), no. 2, p. 295-386, 42 fig.
(*459) 1953, *The monomyarian condition in the Lamellibranchia:* Royal Soc. Edinburgh, v. 62, pt. 2, p. 443-478, 13 fig. (April 7).
(*460) 1960, *Oysters:* xiv+209 p., 72 fig. 17 pl., Collins (London).

Zaprudskaya, M. A.
(*460a) 1953, *Plastinchatozhaberiye mollyuski nizhnego turona Alayskogo khrebta* [Lamellibranch mollusks of the lower Turonian Alaysk Range]: Vses. Neft. Nauch.-Issled. Geol.-Razv. Inst. (VNIGRI), Trudy, new ser., v. 73, p. 21-61, 15 pl.

Záruba, Bořivoj
(*461) 1965, *Beitrag zur Kenntnis der Art Exogyra sigmoidea Reuss, 1844 (Ostreidae) aus der Brandungsfazien der Böhmischen Kreideformation:* Narod. Musea Praze Sborník (Musei Natl. Pragae Acta), v. 21B (1965), no. 1, p. 11-40, 12 fig., pl. 1-8 (April 20).

(*461a) 1966, *O některých ústřicích z Texaské svrchní křídy* [On some oysters from the Texas Upper Cretaceous]: Same, Casopis, Oddíl Přírod., v. 135 (1966), no. 3, p. 191-192, pl. 1 (13).

Zenkevitch, L.
(*462) 1963, *Biology of the seas of the U.S.S.R.* (transl. by S. Botcharskaya): 955 p. 427 fig. Interscience Publishers (New York).

Zeuner, Friedrich
(*463) 1933a, *Lage der Gryphaea arcuata in Sediment:* Centralbl. Mineralogie, Geologie, Paläontologie, Abt. B. (Geol. u. Paläont.), Jahrg. 1933, p. 568-574, no fig.
(*464) 1933b, *Die Lebensweise der Gryphäen:* Palaeobiologica, v. 5, no. 3, p. 307-320, pl. 18.

Ziegler, Bernhard
(*464a) 1969, *Über Exogyra virgula (Lamellibranchiata, Oberjura):* Eclogae Geol. Helvetiae, v. 62, no. 2, p. 685-696, 12 fig., 6 pl. (Dec.).

Zimmermann, E. H.
(*465) 1886, *Ein neuer Monomyarier aus dem ostthüringischen Zechstein (Prospondylus Liebeanus):* K. Preuss. Geol. Landesanst. Bergakad., Jahrb. 1885, p. 105-119, pl. 2.

Zittel, K. A.
(*466) 1864, *Fossile Mollusken und Echinodermen aus Neuseeland:* in Ferdinand von Hochstetter et al., 1864-1866, Reise der Österreichischen Fregatte Novara um die Erde in den Jahren 1857, 1858, 1859 unter den Befehlen des Commodore B. von Wüllersdorf-Urbair, Geol. Theil, 2 v., K. Akad. Wiss. Wien, v. 1, pt. 2, p. 17-68, pl. 6-15 [Geol. Theil is in v. 1, pt. 2, and v. 2, pt. 2].

FAMILIES DOUBTFULLY RELATED TO OYSTERS

Family CHONDRODONTIDAE Freneix, 1959

[Materials for this family prepared by † L. R. Cox and H. B. Stenzel]

Oyster-like, suborbicular, subtrigonal or linguiform, commonly elongated dorsoventrally, compressed, thick-shelled, inequivalve, sessile, attached by more strongly convex LV, which Freneix & Lefèvre (1967) considered to be RV; both valves radially plicated or smooth; dimyarian, with distinct pallial line well separated from valve margins; edentulous; umbonal angle of each valve very acute and occupied by triangular plate resembling ligament area of *Ostrea;* in LV this plate is continued ventrally by a projecting linguiform process, recess below receiving dorsally pointed hooklike process, which projects from wall of RV, these processes interpreted as chondrophores between which internal ligament extended (i.e., it was attached to undersurface of process of LV); posterior adductor scar located on shell wall; shell subnacreous. [Usually occurs in rudist-bearing limestones.] *L.Cret.(Alb.)-U.Cret.(Campan.).*

Douvillé (1902) observed symmetrically arranged markings (Fig. J149,*1c*) on upper part of plate occupying umbonal angle and considered these to be bifid anterior adductor insertions. STANTON (1947) rejected this interpretation, pointing out that a muscle located in this position would be ineffective, and stated that he had observed no comparable markings in specimens of *Chondrodonta* from Texas.

FRENEIX & LEFÈVRE (1967) placed this enigmatic family in the Pectinacea, as did STANTON (1947), who carefully discussed the evidence and all published opinions to the contrary. STENZEL believes that the Chondrodontidae do not fit in the Ostreacea.

Chondrodonta STANTON, 1901, p. 301 [*Ostrea munsoni* HILL, 1893, p. 105; OD]. Characters of family. *L.Cret.(Alb.)-U.Cret.(Santon.-Campan.)*, N.Am.(Texas-Mexico)-Eu.(France-Port.-Italy-Dalmatia)-SW.Asia (Turkey-Syria-Israel-Sinai-Iran)-E.Afr.(Somalia).

C. (Chondrodonta). Ligament simple, not subdivided. *L.Cret.(Alb.)-U.Cret.(Turon.)*, N.Am.(Texas-Mexico)-Eu. (France-Port.-Italy-Yugo.)-SW.Asia (Iran-Lebanon-Israel-Sinai)-E.Afr.(Somalia).——FIG. J149,*1a,b*. *C. (C.) joannae* CHOFFAT), U.Cret.(Turon.), Port.; *1a*, LV ext., ×0.7 (Choffat, 1902); *1b*, dorsoventral sec. perpendicular to commissure through dorsal part of both valves (LV on left, umbonal region at top), ×1.5(Douvillé, 1902).

C. (Freneixita) STENZEL, herein [*nom. subst. pro Chondrella* FRENEIX & LEFÈVRE, 1967, p. 764 (*non* PEASE, 1871, p. 465)] [*Ostrea desori* COQUAND, 1869, p. 117; OD]. Internal ligament subdivided into more superficial part inserted in RV on upper face of chondrophore in umbilical region. *U.Cret.(Cenoman.-Turon.)*, Eu.(France-Italy-Yugo.).——FIG. J149,*1c*. **C. (F.) desori* (COQUAND), U.Cret.(Cenoman.), France (Angulême); dorsal part of LV int. showing (near top of figure in middle) markings interpreted by DOUVILLÉ as bifid adductor scar and (middle of figure) linguiform process regarded as chondrophore, ×2 (Douvillé, 1902).

C. (Cleidochondrella) FRENEIX & LEFÈVRE, 1967, p. 765 [**C. (Cleidochondrella) elmaliensis*; OD]. Lamina of internal ligament, developed only in RV, is inserted into int. of double chondrophore, lower part of which grows hollow as a cupula, whereas upper part is opercular. *U.Cret.(Santon-Campan.)*, Turkey(Taurus Mts.).

REFERENCES

Douvillé, Henri

(1) 1902, *Sur le genre Chondrodonta Stanton;* Soc. Géol. France, Bull., ser. 4, v. 2, p. 314-318, pl. 11.

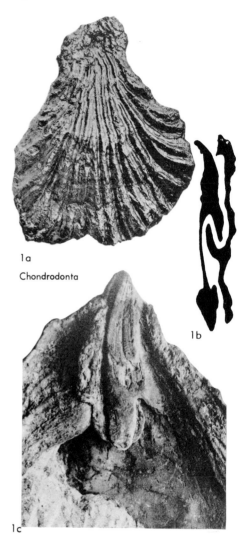

1a
Chondrodonta

1b

1c

FIG. J149. Chondrodontidae· (p. *N1198*).

Freneix, Suzanne

(2) 1959, *Lamellibranches du Crétacé Supérieur de France* [*Protobranches, Prionodontes, Dysodontes (pars)*]: 84ème Cong. des Soc. Savantes de Paris et des Départements, Comptes Rendus, Dijon 1959, Sec. des Sci., Sous-Sec. de Géologie, Colloque sur le Crétacé supérieur français, p. 175-248, Gauthiers-Villars (Paris).

———, & Lèfevre, Roger

(3) 1967, *Deux espèces nouvelles de Chondrodonta et Neithea (Bivalves) du Sénonien du Taurus Lycien (Turquie):* Soc. Géol.

France, Bull., ser. 7, v. 9, p. 762-776, pl. 26-29a (Oct.).

Pease, W. H.
(4) 1871, *Catalogue of the land-shells inhabiting Polynesia, with remarks on their synonymy, distribution, and variation, and descriptions of new genera and species:* Zool. Soc. London, Proc. 1871, no. 29, p. 449-477.

Stanton, T. W.
(5) 1947, *Studies of some Comanche pelecypods and gastropods:* U.S. Geol. Survey, Prof. Paper 211, 256 p., 1 text fig., 67 pl., 2 charts.

? Family LITHIOTIDAE Reis, 1903

[*nom. Latine redditum et transl.* Cox, herein (*pro* "Unterfamilie Lithiotiden" REIS, 1903)] [Materials for this family prepared by † L. R. Cox]

Large, thick-shelled, oblong, much elongated dorsoventrally, compressed, slightly to moderately inequivalve, with general resemblance to *Crassostrea* [Ostreidae] but attached possibly by RV, which is more convex than LV; umbones very acute, curved in some specimens, either to front or rear; hinge edentulous; ligamental area large, greatly elongated dorsoventrally, differing from that of Ostreidae in absence or narrowness of median groove for fibrous ligament; monomyarian, commonly with thin internal buttress in each valve passing from lower margin of ligamental area to posterior margin of adductor scar; ostracum formed of lamellar calcite together with prismatic calcite developed as intercalated layers or as masses of radially disposed crystals surrounding tubular vesicles. *L.Jur.(L.Lias.).*

The shells included in the genera *Lithiotis* and *Cochlearites* occur in very hard limestone, from which no perfect specimens have yet been extracted. The account here given is based on observations and reconstructions of REIS, who considered that these genera were related to the "toothless spon-

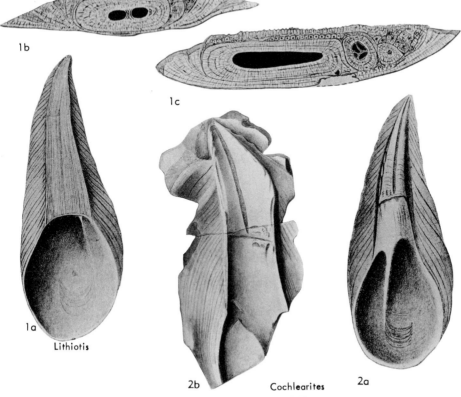

1b

1c

1a

Lithiotis

2b   Cochlearites   2a

FIG. J150. Lithiotidae (p. *N*1200).

dylids," that is, to the group of genera included herein in the new family Terquemiidae (p. *N*380). Böhm (1892) strongly contested this view, maintaining that Reis's observations were unreliable and that both genera were founded on a single species not separable from the Ostreidae. The material examined by the compiler of the present account has proved inadequate to allow him to express any strong opinion on the matter. Prismatic calcite and internal vesicles occur in some shells belonging to the Ostreidae, but the absence or narrowness of a median groove on the ligamental area seems to distinguish the present forms from that family.

Lithiotis Gümbel, 1874, p. 48 [*L. problematica;* M]. Moderately inequivalve; ligamental area striated in dorsoventral direction but without median groove for fibrous ligament; internal buttresses weak or absent; interior of wall of umbonal cavity formed of calcite prisms all oriented perpendicularly to its surface; minor tubular cavities, extending dorsoventrally and similarly surrounded, present within other parts of shell wall. *L.Jur.(L. Lias.)*, Eu.(N.Italy-Croatia)-SW.Asia(?Iran).——Fig. J150,1. *L. problematica*, Italy (Verona prov.); 1a, RV int. (reconstr.) showing striated ligament area lacking median groove, *ca.* ×0.2;

1b,c, transv. secs. through upper part of valve in 2 different specimens, showing internal cavities (black); 1b showing 2 main cavities which have united (in more ventral section, 1c) to form umbonal cavity, while minor tubelike cavities in ostracum are well seen in 1c; striated ligamental area is seen in section along top of each figure; ×1.3 (all Reis, 1903).

Cochlearites Reis, 1903, p. 2 [*Trichites loppianus* Tausch, 1890, p. 18; M] [=*Chochlearites* Reis, 1923 *(nom. null.)*]. Only slightly inequivalve; ligamental area with narrow median groove varying in length and not extending to its lower margin; internal buttresses well developed; ostracum without internal vesicles. *L.Jur.(L.Lias.)*, Eu. (N. Italy).——Fig. J150,2. *C. loppianus* (Tausch), Verona prov.; 2a, LV (reconstr.) (upper valve according to Reis) int., *ca.* ×0.2; 2b, dorsal part of LV int., ×0.3 (both Reis, 1903).

REFERENCES

Böhm, Georg

(1) 1892, *Lithiotis problematica, Gümbel:* Naturf. Gesell. zu Freiburg im Breisgau, Berichte, v. 6, no. 3, p. 55-80, pl. 2-4.

Reis, O. M.

(2) 1903, *Ueber Lithiotiden:* K. K. Geol. Reichsanst. Wien, Abhandl., v. 17, no. 6, p. 1-44, 4 text fig., 7 pl. (Oct. 31).

NOMENCLATURAL CLARIFICATIONS OF SOME GENERIC AND SUBGENERIC NAMES IN FAMILY OSTREIDAE (BIVALVIA)

INTRODUCTION

Stray generic or subgeneric names not yet discovered or not yet investigated thoroughly as to their availability and validity in zoological nomenclature are potentially disruptive. According to the rules of priority of the *International Code of Zoological Nomenclature* (ref. 13) some of these names may have to be accepted, and thus they may displace better known names in use today.

Such threats of displacement should be avoided. It is therefore necessary to search for names of this sort and to dispose of them, if possible. Unfortunately, much time and effort must be spent on these investigations, even though necessary.

The present discussion is part of a continuing effort to distinguish and clean up genus-group names in the family Ostreidae (Stenzel, 1947; 1959). Its purpose is to settle the nomenclatural status of *Rastellum* Faujas-Saint-Fond, 1799 [?1802], of *Cristacites* (corrected form of *"crist."*) von Schlotheim, 1820, and of *Cristacites* von Schlotheim, 1823.

ACKNOWLEDGMENTS

The late Mr. Noel K. Brown, Jr., of Houston, Texas, pointed out to me the Silvestri article and lent me his copy of it. Dr. W. I. Follett, Curator of Fishes of the California Academy of Sciences, and Dr. A. Myra Keen of Stanford University discussed with me some of the questions pertaining to *Rastellum*. The library at the Academy of Natural Sciences of Philadelphia was used extensively.

Dr. L. B. Holthuis of the Rijksmuseum van Naturlijke Historie at Leiden, Netherlands, kindly furnished quotations from old literature in his private library. Dr. C. O. van Regteren Altena, curator at Teyler's Museum of Haarlem, Netherlands, has given invaluable help by pointing out to me old, overlooked literature and by kindly lending his copy of Pasteur's volume 1. He arranged the loan of Faujas' type specimens from that museum. I am much obliged to Teyler's Museum in Haarlem for the generous loan of two type specimens and other specimens in their collection.

My former colleagues, Mr. R. Wright Barker, Dr. B. F. Perkins, and Dr. R. J. Stanton, read and criticized the manuscript. All these have helped to improve this discussion and are thanked accordingly.

Liberal support of my work on oysters and many facilities have been furnished by the Shell Development Company, a Division of Shell Oil Company, and are gratefully acknowledged here.

PART 1—RASTELLUM

One of the byproducts of bloody wars set off by the French Revolution was a monograph on the geology of the St. Pietersberg area in southern Netherlands. It contains an elongate hill about 1.5 km. wide from east to west, more than 4 km. long, and about 110 m. above sea level. It rises steeply from the left bank of the Maas River, 1 km. south of the center of the ancient city of Maastricht in Zuid Limburg. The flat-lying calcareous strata forming the hill are the type locality of the Maastrichtian Stage (Late Cretaceous) and have furnished many interesting animal remains, for instance, the type species of the swimming reptile *Mosasaurus*. The rock, called "tuffeau de Maestricht" in French geological literature, has been quarried for hundreds of years and supplies today a large cement plant. The official Netherlands spelling of the city is Maastricht. A tuffeau, according to French usage, is a crumbly limestone. In modern terms, the rock is a crumbly, porous lime grainstone.

In 1795 the French army took the fortress Maastricht. Faujas-Saint-Fond, professor at the Muséum National d'Histoire Naturelle in Paris, accompanied the army and investigated the St. Pietersberg area. The fossils collected were deposited by him at the Muséum (ref. 7, p. 111), and some of them were described by Lamarck (1801, p. 400) as *Planospirites ostracina* (see Fig. J96). Various fossils illustrated by Faujas ended up at the Teyler's Museum in Haarlem, Netherlands (Winkler, 1863-67, p. 251, 253; van Regteren Altena, 1957, p. 96, 110, and 1963).

ORIGINAL USAGE OF NAME

The present inquiry concerns the generic name *Rastellum* as used in a monograph with title page given as follows: *Histoire Naturelle | de | la Montagne de Saint-Pierre | de Maestricht, | par B. Faujas-Saint-Fond, | Administrateur et Professeur de Geologie au Muséum National | d'Histoire Naturelle de Paris. | A Paris, | Chez H. J. Jansen, Imprimeur-Libraire, Rue des Saints-Pères, no. 1195. | An 7ème. de la Republique Françoise.*

Few libraries in the United States possess this magnificent book, which is not rare in the antique book trade. The book gives the author's name as Faujas-Saint-Fond; elsewhere it is given as Faujas de Saint Fond. The author was a French nobleman, born 1741, and at one time the "King's Commissioner for the Mines," but titles of nobility had been abolished by the French Republic. Within the text of the book the author refers to himself as Faujas (ref. 7, p. 111, 127, 129-136).

BINOMINAL OR NOT?

The first question to settle is whether the above-cited work is binominal or not. Faujas was a professor of geology at the Muséum National d'Histoire Naturelle in Paris and a contemporary and colleague of Cuvier and of Lamarck, who was a good friend of his as is indicated by the footnote on p. 136, where he stated: "This work [by Lamarck], which is about to appear in the immediate future and the leaves of which Lamarck has kindly furnished me in order to make exact citations available to me, has for its title: *Systême des animaux sans vertèbres, ou Tableau général des classes, des ordres et des genres de ces animaux,* etc. (in —8°). Par Lamarck, Paris, chez Deterville, rue du Battoir." [translated from the French]. In his work Faujas showed thorough familiarity with the works of Linné, Bruguière, Cuvier, Lamarck, Latreille, and other prominent binominal authors (7, p. 23, 24, 180, 196, 231); he quoted these authors many times with meticulous care and obvious respect. Faujas showed great admiration for Linné ("Linné arrived, and this extraordinary man, born with a bold and methodical intellect . . ." (7, p. 23) and outlined Linné's binominal nomenclatural method with evident approval.

Faujas' work contains a very capable description of the Pietersberg south of Maastricht in Zuid Limburg, Netherlands, and its outstanding fossil remains (Fig. J151-

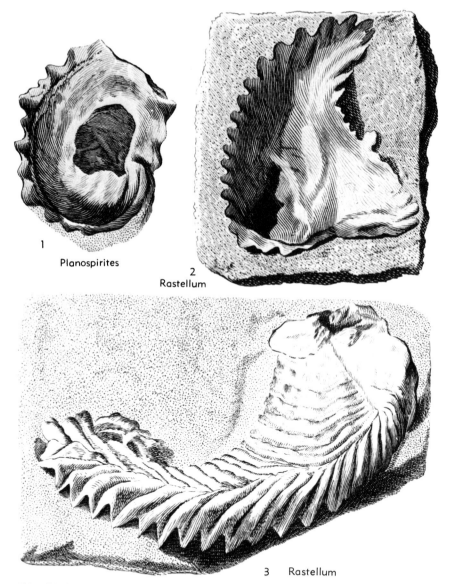

1

Planospirites

2
Rastellum

3   **Rastellum**

Fig. J151. Ribbed oysters from tuffeau de Maestricht, U.Cret. (Maastricht.), St. Pietersberg south of Maastricht, Zuid Limburg, southern Netherlands as figured by Faujas (1799 [?1802]).

1. *Planospirites ostracina* Lamarck, 1801, LV int., ×1 (*"une espèce de rastellum"* of Faujas, pl. 28, fig. 5).
2. *Rastellum macropterum* (J. de C. Sowerby, 1824) (*sensu* Winkler, 1863-67); LV int., ?×1 (*"Rastellum de forme presque triangulaire"*

of Faujas, pl. 28, fig. 7, here regarded as illustration of type species of *Rastellum* Faujas, 1799 [?1802]).
3. Probably *Rastellum macropterum* (J. de C. Sowerby, 1824), LV ext. (*"gryphite de forme allongée"* of Faujas, pl. 24, fig. 1).

Fig. J152. Ribbed oysters from Upper Cretaceous (Maastricht.) of Maastricht as figured by Faujas (1799 [?1802]).—*1.* Probably *Rastellum macropterum* (J. de C. Sowerby, 1824); LV ext., ?×0.8 (*"une gryphite plus grande encore que celle de la figure 1"* of Faujas, pl. 24, fig. 2).

J-153). The work is informative, even by modern standards, and meticulous, as can be seen by the explicit and complete way most of the references to pertinent literature were handled (7, p. 22, footnotes). All these facts prove that Faujas was a competent, up-to-date scientist. To assume from any minor evidence that he was not binominal would be a gross misunderstanding of this geologist and his work.

Nonbinominal zoological names are found in this work only in three places. 1) A lengthy quotation from Paul de Lamanon, set off from the running text by repeated quotation marks, which contains a reference to a supposed fossil turtle (7, p. 90). 2) A reference to *"Testudo marina vulgaris,* Ray." (7, p. 92, footnote), which is firmly placed in the synonymy of *Testudo mydas* Linné. 3) A reference (7, p. 173) to

Echinometra digitata secunda rotata vel cidaris Mauri figured by Rumphius, which is firmly placed in the synonymy of *Turban maure,* a vernacular name for pl. 136, fig. 6-8 of Bruguière. None of these show any sign of approval as specific names by Faujas; they and other references to nonbinominal authors are simply synonymy references or quotes from ancient literature and cannot be used to disqualify the work from being binominal.

ORIGINAL DEFINITION

A definition of the genus *Rastellum* as given by Faujas (7, p. 167) is here freely translated from the French: *"Figure 5,* is a species of *rastellum,* which [genus] appears to occupy the middle between the oyster and the gryphite. I think it will be convenient to separate the gryphites and the shells

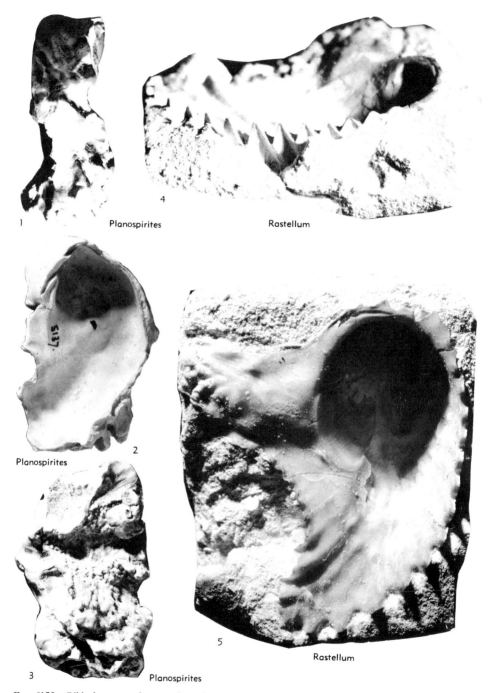

FIG. J153. Ribbed oysters from tuffeau de Maestricht, U.Cret.(Maastricht), St. Pietersberg south of Maastricht, Zuid Limburg, southern Netherlands, figured by FAUJAS (1799[?1802]) (Stenzel, n; courtesy of C. O. VAN REGTEREN ALTENA, Teyler's Museum, Haarlem, Netherlands).

known under the name of *rastellum*, from the oysters" (see Fig. J96,*1;* J153,*1-3*). Practically the same definition is given in FAUJAS' index (7, p. 262): "RASTELLUM, which seems to occupy the middle between the oyster and the gryphite, pl. 28, fig. 5, p. 167. Another one of nearly triangular form, *ibid.* [evidently refers to fig. 7]" (Fig. J138, Fig. 151,*2;* J152; J153,*4-5*).

These statements qualify as definitions of *Rastellum* under Article 12 of the *Code* and also foreshadow the separation of the gryphites from the genus *Ostrea,* which LAMARCK (1801, p. 398-399) later performed by formally proposing the genus *Gryphaea.* Also, the listing of *Rastellum* as a generic name in the index (7, p. 262) alone fulfills all obligatory conditions under *Code* Articles 11(c)(ii) and 16(a)(ii) to make it available as a published generic name. Therefore, it must be concluded that the generic name *Rastellum* FAUJAS was proposed in a nomenclaturally correct fashion and must be accepted as available.

ETYMOLOGY

The Latin noun *rastrum,* of neuter gender, is applied to a toothed hoe or rake. *Rastellum* is its diminutive. The name is appropriate and descriptive.

CAPITAL LETTER

The word *Rastellum* is found (7) in three places on p. 167 and in one place in the index on p. 262. It is spelled on p. 167 with a lower-case *r* in two places, both in the middle of a sentence, and with a capital *R* in one place, at the beginning of a sentence; it has a capital *R* in the index, as have all other words. Whether FAUJAS intended to spell the generic name as *rastellum* or whether the lower-case *r* is a misprint, repeated twice, is difficult to tell. However, the following considerations are pertinent.

The new *International Code of Zoological Nomenclature* does not state anywhere that a generic name must start with a capital letter to be available when it is first published. The *Code* merely requires that *rastellum* once published has to be corrected to *Rastellum* (see Articles 17(6) and 28).

FAUJAS spelled nearly all generic names in his book with a capital letter, only ten times using lower-case initial letters (all references, 7): *cerithium denticulatum* Bruguière (p. 30, footnote); *cerithium hexagonum* Bruguière (p. 31, footnote); *anomia pectiniforme* Gmelin (p. 104); *loligo calmar* Lamarck (p. 112); *rastellum* (twice on p. 167); *orthocera* Lamarck (p. 199); *gorgonia ceratophyta* Linné (p. 202); *flabellum veneris* Linné (p. 202); *flustra foliacea* Ellis and Solander (p. 203).

However, we find *Anomia pectiniformis* (7, p. 164) and *Flustra* (7, p. 201), and the two species of *cerithium* are in a list of seven names, of which the other five begin with a capital letter. One may safely conclude that the lower-case letters at the beginning of generic names are either misprints or *lapsus calami* by the author. The same conclusion applies to *"Trocus aglutinans"* (7, p. 31) for *Trochus agglutinans* LAMARCK, 1804, and to several others.

In any case, the *Rastellum* with a capital initial letter in the index (7, p. 262), accompanied by a definition, is sufficient to fulfill all obligatory conditions of the *Code* under Articles 11(c)(ii) and 12.

DATE OF PUBLICATION

The publication date of FAUJAS' work is given on the title page as the seventh year of the French Republic, which corresponds approximately to the time from September 22, 1798, to September 22, 1799. However, internal evidence in the text clearly shows that the book was published at a later date.

FIG. J153. *(Continued from facing page.)*

1-3. Planospirites ostracina LAMARCK, 1801 (Teyler's Museum Coll., no. 5137), type specimen of FAUJAS, all ×1; *1,* ant. side, umbo at bottom, attachment surface at left; *2,* LV int.; *3,* shell attachment surface, spiral umbo at top right next to imprint of echinoid ambulacral zone, same specimen as Fig. J151,*1.*

4-5. Rastellum macropterum (J. DE C. SOWERBY, 1824) (Teyler's Museum Coll., no. 11046); type specimen of FAUJAS, LV int., ×1; *4,* oblique view, and *5,* vert. view, both shown resting on matrix of "tuffeau de Maestricht," same specimen as Fig. J151,*2.*

The title page probably should be interpreted as indicating the date on which FAUJAS started to write the text or possibly that on which the printer set type for the title page. It is worthy of note that FAUJAS (7, p. 183) referred to an article by G. A. DELUC of Geneva, citing its date as the year 8 of the French Republic. On the same page FAUJAS referred to an article by FORTIS published in 1801 (compare SILVESTRI, 1929, p. 339, footnote 1) and elsewhere (7, p. 173) cited an article by the then well-known fossil collector A. G. CAMPER, which appeared in 1800 (compare VAN REGTEREN ALTENA, 1957, p. 112, footnote 1).

LAMARCK's *"Systême des Animaux sans Vertèbres"* is quoted again and again, and in a footnote (p. 136) FAUJAS stated that he saw and used extensively the proof sheets while writing his text and that LAMARCK's work was to appear in the immediate future (actually first published in January 1801). The monograph by FAUJAS may have appeared shortly afterwards. The date 1801 seems to be the earliest possible for publication of FAUJAS' work, as far as I am able to ascertain from the work itself.

However, Dr. C. O. VAN REGTEREN ALTENA (compare 30, p. 111-112) wrote me that additional information is available in the translation of FAUJAS' work into Nederduitsch made and published by J. D. PASTEUR (1802). In his introduction, dated November 15, 1802, PASTEUR (26, p. V-VIII) stated that he had received only the first five "cahiers" [=parts] of FAUJAS' work and that the remainder then had not been published. Because PASTEUR was anxious to publish his translation, he decided not to wait for the unpublished sections. Thus, PASTEUR's volume 1 (26, p. 1-136) contains the translation of only the first five cahiers of FAUJAS' work. It appears that the cited first 136 pages of FAUJAS' monograph probably were published shortly before November 15, 1802, and the remainder at a later date, possibly in 1803 or 1804.

WORK BY SCHROETER

The name *Rastellum* was used by pre-Linnean authors and taken from them by SCHROETER (1782, p. 74, 382, 390) (in ref. 39). His five-volume work (1779-88) is listed as nonbinominal by SHERBORN (1902,

p. xlviii). Pertinent here is a later publication by SCHROETER (40) in which a section (v. 3, p. 450, 1786) contains the heading "Genera and Modifications which are missing in Linné." In this list one finds: "2. Martini, General History of Nature, pt. 4, p. 158, fig. 2. The rake-like bivalved shell, or the original to the *Rastellis* [Latin, dative case of the plural of *Rastellum*] of Lister, a fossil. . . ." [All quotes translated from the German.]

This reference in no way indicates that in 1786 SCHROETER approved *Rastellum* as an acceptable generic name. On the contrary, all such names approved by him in this publication were introduced with their own proper headings, whereas *Rastellum* was not. Besides, this work is listed as nonbinominal in SHERBORN.

A search of the literature so far has revealed no nomenclaturally available use of *Rastellum* antedating the work by FAUJAS. Therefore, *Rastellum* is dated from FAUJAS, 1799 [?1802].

TYPE SPECIES

In connection with *Rastellum* FAUJAS illustrated two species (7, pl. 28, fig. 5 and fig. 7). Unfortunately, he failed to give specific names to these, and no specific names are connected with *Rastellum* anywhere in the text.

The articles of the *Code* concerning the type species of a genus are somewhat contradictory, when one attempts to apply them to the extraordinary case of *Rastellum*.

1) Article 67(f) is explicit in specifying that, when a new genus is being proposed, "only the statements or other actions of the original author are relevant in deciding" which species were originally included in the genus. The two figures (7, pl. 28, fig. 5 and 7) may be viewed as definitive actions, other than statements, of the author in the sense of Article 67(f). This action would restrict selection of a type species to the species depicted by the cited two figures, whether they are named or not. Any other species subsequently referred to *Rastellum* FAUJAS could not serve as the type species. Any author who subsequently applied a correct specific name to one of the two pertinent figures in FAUJAS *ipso facto* designated the type species of *Rastellum*. It is required simply to find the appropriate

published name that is nomenclaturally available for the species and also valid. This tedious search has been made, and direct and unquestioned references to one of the two pertinent figures of pl. 28 have been found. The name is discussed below under method 1 of type fixation.

2) Article 69(a)(i) specifies that the originally included species of a genus "comprise only those actually cited by name in the newly established nominal genus. . . ." Therefore, one might argue that *Rastellum* was introduced as a generic name with no original nominal species included in the sense of Article 69(a)(i) and (ii), and the first species (one or more) later expressly referred to *Rastellum* become(s) its original species. The consequences of these assumptions are given under method 2 of type fixation.

METHOD 1 OF TYPE FIXATION

Although VON SCHLOTHEIM (1813, p. 109-113) provided new species names for many of the figures in FAUJAS, none of these refer to his plate 28. In a later work (VON SCHLOTHEIM, 1820, p. 242), however, he referred under the newly proposed *"Ostracites crist. complicatus"* to illustrations in publications by three different authors, the first-cited of which was FAUJAS (7, pl. 24, fig. 1; pl. 28, fig. 7), evidently regarding the two figures as representing the same species.

The *"crist."* in this name stands for the subgenus *Cristacites* VON SCHLOTHEIM, 1820, which is discussed below in a separate section of this chapter. Hence, the name of the species needs to be written as *Ostracites (Cristacites) complicatus* VON SCHLOTHEIM, 1820. However, VON SCHLOTHEIM referred to these figures and the others cited by him with some caution by using *"Conf."* [=confer.], thereby indicating that the identifications were provisional at best. Until VON SCHLOTHEIM's types, which came from the vicinity of Hildesheim, Germany, are studied, *Ostracites complicatus* VON SCHLOTHEIM, 1820, remains uncertain and cannot be used for our purposes here.

The first author to supply names to the two pertinent figures in FAUJAS (7) and thereby to furnish a list of species names for *Rastellum* FAUJAS was WINKLER (1863-67, p. 251, 253). WINKLER was curator of paleontological collections at the Teyler's Museum in Haarlem, where the type specimens of the two pertinent figures in FAUJAS are on deposit. His publication which is an annotated detailed catalog of the collection, listed under no. 11046 the following: *"Ostrea macroptera* Sow./ *Rastellum* Faujas/Voyez: Sowerby, Min. Conch., T. V, p. 105, pl. CDLXVIII,/fig. 2, 3./Faujas St. Fond, Hist. mont. St. Pierre, p. 119,/pl. XXVIII.

fig. 7./de Maestricht . . . A 28./*Échantillon original de Faujas St. Fond." (Fig. J138). Under no. 5137 he listed *"Ostrea plicata?* Goldf. *sp./ Rastellum* Faujas/*Gryphaea carinata* Lamk./*Gryphaea plicata* Lamk./*Exogyra plicata* Goldf./ Voyez:/Goldfuss, *Petr. Germ.,* T. II, p. 37, pl. LXXXVII,/fig. 5/Faujas St. Fond, *Hist. mont. St. Pierre,* p. 118,/pl. XXVIII, fig. 5/Lamarck, *Anim. sans vert.,* T. VI, p. 119,/Bosquet<Staring, *Bodem v. Nederl.,* T. II, p. 386./Échantillon original de Faujas St. Fond./de Maestricht . . . A. 28." (Fig. J96).

The second one of the two specimens is identified only questionably. For that reason *Ostrea plicata?* GOLDFUSS must be excluded and cannot be designated the type species of *Rastellum.* The other species, *Ostrea macroptera* J. DE C. SOWERBY, 1824, becomes the only species eligible. Thus, this species, *sensu* WINKLER, 1863-67, is the type species of *Rastellum* FAUJAS, 1799 [?1802] by monotypy.

Rastellum macropterum (SOWERBY) is one of many similar species in the Cretaceous beds and forms with them a distinctive group of oysters (subfamily Lophinae) (Fig. J151,2; J153,4-5). This group has narrow, crescentically curved shells and a zigzag commissure with many acute-angled points, for which the name *Rastellum,* the small rake, is appropriate. The winglike posterior auricles shown by the specimens of FAUJAS and SOWERBY are probably a variable feature and not diagnostic.

The many species of this group have received too many ill-founded formal names. Most have been described from insufficient material, and their type localities are unknown or poorly known. Thus it is now an almost impossible task to untangle them and to do justice to the nomenclatural priorities of various species names.

WOODS (1913, p. 342-347) has united them all under one name, *Ostrea diluviana* LINNÉ (1767, p. 1148). If so defined, the species would encompass a very long stratigraphic span, from Aptian to Maastrichtian. This appears to be excessive.

In summary, if interpretation method 1 is accepted, only two species, namely those listed below, are eligible for selection of the type species of *Rastellum* FAUJAS, 1799 [?1802], and the second one of them becomes *ipso facto* the type species:

1) *Rastellum* species no. 1, unnamed by FAUJAS (7, pl. 28, fig. 5), =*Ostrea plicata?* GOLDF. *sp.* in WINKLER (47, p. 253).

2) *Rastellum* species no 2, unnamed by FAUJAS (7, pl. 28, fig. 7), =*Ostrea macroptera* Sow. in WINKLER (47, p. 251), =*Rastellum macropterum* (J. DE C. SOWERBY, 1824), *sensu* WINKLER, 1863-67.

For the purpose of clearing up the status of *Rastellum* FAUJAS in the event that method 1 of type fixation is not accepted, I designate the species which FAUJAS (7) figured as pl. 28, fig. 7, but did not name, as the type species and conclude that the name *Rastellum macropterum* (J. DE C. SOWERBY, 1824) is its name.

METHOD 2 OF TYPE FIXATION

In the event that method 1 of type fixation of *Rastellum* FAUJAS is found entirely unacceptable, this form must be regarded as a genus without included nominal species in the sense of *Code* Article 69(a)(ii). In such case, the first work to consider is MÖRCH (1850, p. 26), who treated *Rastellum* as a subgenus of *Ostrea* without indicating in any way the author of this subgeneric name. Also, he did not add "*n. gen.*" or "*n. subgen.*" after his *Rastellum,* nor give other indication that might be construed to mean that he was introducing a new taxon. It must be assumed that he was merely using a name familiar to him from the literature and that he was fully cognizant of earlier uses. In using *Rastellum* as a subgenus of *Ostrea* he listed three specific names with it, two of which were followed by question marks, and only *Ostrea (Rastellum) plicata Ch.*" was assigned without doubt. IREDALE (1939, p. 401) interpreted this species as the haplotype species of *Rastellum* MÖRCH, 1850. According to Article 67(g) this species becomes automatically the type of *Rastellum* FAUJAS.

IREDALE regarded *Ostrea plicata* CHEMNITZ as equivalent to *O. plicatula* GMELIN. This is not certain, however, since *O. plicata* may represent several species. It is significant that the publication by MÖRCH (1850) was a catalog expressly prepared for a public auction, as indicated by its title, and was not prepared and published for the purpose of scientific, public, permanent record as required by Article 8(2). Presumably MÖRCH (1850) is not acceptable in zoological nomenclature, although similar sales catalogs have been officially sanctioned by I.C.Z.N. in recent years.

The next authors to mention *Rastellum* FAUJAS were WINKLER (1863-67), discussed above under method 1, and PERVINQUIÈRE (1910, p. 119). The latter work is discussed below under "Subsequent Usage."

I believe that method 2 of type fixation cannot be applied, because it ignores "other actions of the original author" specified as decisive in *Code* Article 67(f).

SUBSEQUENT USAGE

Since 1802, *Rastellum* has been approved and used by MÖRCH (1850), DOUVILLÉ (1911), ROLLIER (1911, p. 268, 274-278; 1917, p. 543-547), MAIRE (1941, p. 271), CHARLES & MAUBEUGE (1951, p. 109-118), KAUFFMAN (1965, p. 30), and possibly others. It is evident that MÖRCH (1850) knew of *Rastellum* and approved of it.

FISCHER (1880-87, p. 926) listed *Rastellum* SCHROETER, 1782, as a rejected name under *Alectryonia* FISCHER DE WALDHEIM, 1807. Later French-speaking authors evidently relied on FISCHER (1880-87) for their information on authorship.

Simultaneously with DOUVILLÉ (1911), PERVINQUIÈRE (1910b, p. 119; 1911, p. 646) stated: "*Rostellum* [*sic*] has been used since long ago as a common name (Lister, Knorr, d'Argeville, Faujas de St. Fond, etc.), but it has never been delimited as a genus; besides one has applied it also to forms of the group of *O. crista galli* and of *O. hyotis;* therefore, this too is a synonym of the two preceding names [*Lopha and Alectryonia*]" [translated from the French]. The spelling *Rostellum* occurs in both publications of PERVINQUIÈRE, which are identical word for word; it is presumably a misprint. The statement does not indicate in any way that he regarded *Rastellum* as not available for nomenclatural purposes, but merely that he did not approve of it because there were two better defined synonyms available. Needless to say, PERVINQUIÈRE's reasons for declining to accord nomenclatural status to *Rastellum* FAUJAS are not well grounded.

DOUVILLÉ (1911, p. 634, footnote 2) traced the name *Rastellum* and the taxon it represents back to LISTER and made his approval of it unmistakably clear:

"*Rastellum* LISTER 1648 (pl. 486) [error for 1678] has been created for a fossil oyster from England which is *O.* [*Ostrea*] *carinata* or a near-related form. This genus has been accepted by most ancient conchologists up to MARTINI & CHEMNITZ (1778) and SCHROETER (1782); but having been omitted by LINNÉ and later by LAMARCK, it has been wrongly [*à tort*] declared null and rejected" [translated from the French].

Among authors mentioned by DOUVILLÉ, LISTER antedates the starting point (1758) of modern nomenclature, and the publication by MARTINI & CHEMNITZ (1769-95) has been officially rejected for nomenclatural purposes (see Official Index of Rejected and Invalid Works in Zoological Nomenclature, 1958, p. 5, title no. 21).

ROLLIER and CHARLES & MAUBEUGE ascribed the genus to SCHROETER (1782), apparently unaware that this author was non-binominal. The fact that these authors did not list FAUJAS-SAINT-FOND as the original author of *Rastellum* does not prove that he was unknown to them, nor does it prove that they wished to distinguish between *Rastellum* SCHROETER and *Rastellum* FAUJAS. The simplest explanation is that they did not realize that SCHROETER's names are not available according to the *International Code of Zoological Nomenclature*, but merely followed FISCHER's lead.

These publications prove that *Rastellum* is not a *nomen oblitum*.

OTHER GENERIC NAMES CONCERNED WITH RASTELLUM

If *Rastellum* FAUJAS 1799 [?1802] is accepted and restored to general use, some generic names established by various authors at later dates will become affected by it.

Arctostrea was established by PERVINQUIÈRE (1910a), with *Ostrea carinata* LAMARCK, 1806 (p. 166), from Cenomanian beds in the vicinity of Cany, Département Seine-Inférieure, northwestern France (PERVINQUIÈRE, 1910a) cited as its type species. PERVINQUIÈRE regarded *Arctostrea* as a subgenus of *Lopha* RÖDING, 1798. *Arctostrea* PERVINQUIÈRE is a junior subjective synonym of *Rastellum,* possibly useful as a subgenus of *Rastellum.*

Arcostrea CHARLES & MAUBEUGE, 1951 (p. 114-115) appears to be a *lapsus calami,* although the same spelling is consistently used in four places. These authors did not indicate that they were proposing a new name or making an emendation.

Arctaostrea HAAS, 1938 (p. 294) is a misprint, because it is also correctly spelled on the same page.

Arctostraea JOURDY, 1924 (p. 17) must be a *lapsus calami,* because elsewhere (p. 101) the spelling is correct.

PART 2—CRISTACITES

NOMENCLATURAL INQUIRY

A subgeneric name was introduced by E. F. VON SCHLOTHEIM (1820, p. 240-245) in so obscure and haphazard a fashion that it is exceedingly difficult to analyze his intentions; it is not at all certain whether the name is available according to the *International Code of Zoological Nomenclature.* For this reason somewhat detailed discussion is in order.

Throughout his publications, VON SCHLOT-

HEIM used generic names ending in *-ites* for *fossils* (e.g., *Ostracites* as generic name for fossil species which he would have assigned to the genus *Ostrea* had they been living species). He named and described many species of *Ostracites* and then followed these by a vernacular center heading "*D. Cristaciten (Hahnenkämme)*" (p. 240). Under this heading he described ten more species of *Ostracites,* listed as *Ostrac. crista galli, Ostrac. crist. planulatus, Ostrac. crist. complicatus, Ostrac. crist. ungulatus, Ostrac. crist. urogalli, Ostrac. crist. vaginatus, Ostrac. crist. hastellatus, Ostrac. crist. parasiticus, Ostrac. crist. cornucopiaeformis,* and *Ostrac. crist. difformis.* All ten were abbreviated as shown here, each of the nine "*crist.*" is printed in italics like the other parts of the ten names, and each of the nine "*crist.*" starts with a lower-case letter *c.* Because of the italics it is obvious that "*crist.*" was not used as a vernacular word. An enigma is presented by what the "*crist.*" in these names stands for and how it should be handled in nomenclature. This is the crux of the problem before us.

VON SCHLOTHEIM (1820) gave no explanation of his usage. It is not surprising that such skillful authorities as DIENER (1923, p. 128), SHERBORN (1925, pt. 7, p. 1636), and KUTASSY (1931, p. 340) arrived at an interpretation that cannot be upheld. Because of the weight of SHERBORN's authority it is now necessary to analyze the problem.

"CRIST."=CRISTA

The first of the ten names, *Ostracites crista galli,* was not VON SCHLOTHEIM's creation but contains a specific name given by LINNÉ (1758, p. 704), who established this very same species as *Mytilus crista galli.* In VON SCHLOTHEIM's time it had already been demonstrated that the living species "*Mytilus crista galli*" of LINNÉ was better placed in the genus *Ostrea.* VON SCHLOTHEIM's *Ostracites crista galli* was the name given by him to fossils that were supposed to be practically indistinguishable from specimens of the living species.

Perhaps then, *crist.* is simply part of the specific names and an abbreviation of *crista* (=crest in Latin), because the first one of the ten names is spelled out as *crista galli.* SHERBORN thought so, and every author who

commented on these species judged similarly. Opposed to this, several indications in
the text of VON SCHLOTHEIM (1820) show
that this assumption does not fit most of the
names involved.

1) No saving of space or work included between
the abbreviation *crist.* and the full word *crista* is
involved, because each requires six printer's type
blocks, which had to be picked up by hand and
assembled.

2) Specific names ending in *-us* would be grammatically wrong, because *crista* is a feminine noun
and the adjectival ending *-us* is masculine. Two of
the nine names end in *-is,* and are uncertain as to
whether the ending is masculine or feminine. This
leaves six out of ten names definitely wrong, two
indeterminate, and two grammatically correct. All
grammatical errors would disappear as soon as one
could assume that the adjectives ending in *-us* and
in *-is* refer to a noun of masculine gender. VON
SCHLOTHEIM was a well-educated man who would
not have made such simple grammatical errors.

3) Even if one were to assume that VON SCHLOT-
HEIM made six grammatical errors and if one were
to change the masculine endings to the feminine
adjectival endings *-a,* such words as *crista parasitica* would still make no sense (for what is a
parasitic crest?). On the other hand, LINNÉ's specific name *crista galli* (crest of the cock=coxcomb)
makes sense, as does *crista urogalli* (crest of the
capercaillie cock, *Tetrao urogallus* LINNÉ, 1758, a
European grouse). However, the latter was not a
new specific name in 1820 but had been established
by VON SCHLOTHEIM (1813, p. 112) previously,
spelled out in full at that time. A few of the
other names might be defended if the adjectival
gender were feminine, but most would make no
sense in connection with *crista.*

4) According to LINNÉ (1758) and even down
to the present day (see *Code* Article 26a), such
compound names as *crista galli* are acceptable in
strictly binomial nomenclature, because both
words are needed to convey one idea and because
both words together are really a unit. However,
such words as *crist. complicatus* (or *crista compli-
cata*) do not qualify as acceptable compounds under the *Code.* They simply are not compounds at
all, but two separate words. Only one of the nine
names of VON SCHLOTHEIM qualifies as an acceptable compound; it is *crista urogalli,* from now on to
be written as one word, see Article 26(a). Thanks
to LINNÉ's leadership the rules concerning compound names were well known and widely accepted by the time VON SCHLOTHEIM wrote his book.

5) The *Ostracites crist. ungulatus* VON SCHLOT-
HEIM, 1820, had already been named and established by the same author, and its original name
was *Ostracites ungulatus* VON SCHLOTHEIM (1813, p.
112). If one assumes that *"crist."* is part of the
specific name *"crist. ungulatus,"* then this would

have been a deliberate name change by VON
SCHLOTHEIM.

6) Under the description of *Ostracites crist. difformis,* VON SCHLOTHEIM compared several species,
one of which he called *"Ostrac. hastellatus"* (*37,*
p. 245) instead of *"Ostracites crist. hastellatus"*
(*37,* p. 243). This might be merely a *lapsus
calami,* or it might prove, just as the other items
enumerated above prove, that VON SCHLOTHEIM
himself did not believe in the indispensability of
the *"crist."* in these names and that he did not
regard these specific names as functional indis-
soluble compounds, except for LINNÉ's *crista galli*
and perhaps his own *crista urogalli.*

If one persists both in regarding *"crist."*
as an abbreviation of *crista* and in regarding
crista as an integral part of the specific
name, one must also recognize that such
names as *Ostracites crista planulatus* and
seven others are not binominal. One is
forced to conclude that VON SCHLOTHEIM
(1820) was not a binominal author as concerns eight names, although he was strictly
binominal in the remainder of this book.
If so, his work of 1820 does not satisfy Article 5, and the work as a whole must be rejected according to *Code* Article 11(c).

"CRIST."=CRISTACITES

It can be shown that *crist.* stands for
cristacites, and the arguments in favor of
that interpretation, listed below, are decisive.

1) In 1823 VON SCHLOTHEIM (*38*) showed clearly
what he had in mind, for he listed (*38,* p. 75) two
species: *"Cristacites complanatus* and *difformis."*
Of these, the second seems to be the same as
Ostrac. crist. difformis (VON SCHLOTHEIM, 1820, p.
245). He also listed (*38,* p. 82) the two again, as
follows: *"Ostracites cristacit. complanatus"* and
"Ostracites cristacit. difformis."

2) The abbreviation *crist.,* derived from *cristacites,* is really a saving as to space or work in
writing and printing.

3) All adjectival specific names would have to
end in *-us* or in *-is,* because they would have to
conform with the gender of *cristacites,* a noun of
masculine gender. All the specific names do this,
and it is clear that VON SCHLOTHEIM made no grammatical errors.

These arguments prove conclusively that
crist. stands for *cristacites.* It is not clear,
however, whether *cristacites* is part of the
specific names or is a subgeneric name for
some unknown reason spelled in lower-case
letters. If one assumes that *cristacites* is a

part of the specific names, the following difficulties appear:

1) Such names as *cristacites parasiticus* as specific names would still make little sense, although one might defend them.

2) These two-word specific names would remain two separate words and could not be interpreted as acceptable compounds according to *Code* Article 26(a), excepting *crista galli* and *crist. urogalli* or *cristacites urogalli*. Of the ten names, eight would remain in unsatisfactory condition.

3) The cases of *Ostracites crist. ungulatus* VON SCHLOTHEIM, 1820, versus *O. ungulatus* VON SCHLOTHEIM, 1813, and of *O. crist. hastellatus* VON SCHLOTHEIM (1820, p. 243) versus *O. hastellatus* VON SCHLOTHEIM (1820, p. 245) would still remain unexplainable. They would continue to militate against interpreting the two-word specific names as acceptable compounds.

4) Above all, eight of the ten names involved would still consist of a generic name and two specific names. These eight would still be trinominal names and would force one to reject VON SCHLOTHEIM (1820) as not consistently binominal, although it is obvious from all the other parts of his various works that he was strictly binominal.

Did VON SCHLOTHEIM lapse from binominal nomenclature in these cases, or is there another, better explanation?

CRISTACITES AS SUBGENERIC NAME

The following considerations make it highly likely that VON SCHLOTHEIM (1820) regarded *cristacites* and its plural vernacular form *Cristaciten* as a sort of subgenus:

1) VON SCHLOTHEIM (1820, p. 245) spoke of the *"Familie der Cristaciten."* No matter how one interprets his concept of a family and of this family in particular, it must have been construed as higher in rank than species.

2) He placed the center heading *"Cristaciten"* in the text so that it indicated a supraspecific taxon embracing the ten species and placed this supraspecific taxon under the genus *Ostracites.*

3) In all his works he reserved nouns ending in -*ites* for genus-group names (e.g., *Brachyurites, Bucardites, Gryphites, Mytulites*).

4) If VON SCHLOTHEIM regarded *cristacites* as a subgeneric name, he could have used *Ostracites crist. ungulatus* and *Ostracites ungulatus* interchangeably. The same is true of *O. crist. hastellatus* and *O. hastellatus.*

5) If *cristacites* is considered a subgeneric name, then all nine names involved are composed of the generic name *Ostracites* plus the subgeneric name *cristacites* plus a one-word specific name. In other words, they would be binominals as required by the *Code* (Article 6).

6) The adjectives serving as specific names would refer to the generic name *Ostracites.* They

would be grammatically correct and would be well-chosen descriptive terms that make good sense.

7) In a later work, VON SCHLOTHEIM (1823, p. 75) used *Cristacites* as a generic name and capitalized its beginning letter.

In summary, nearly all difficulties encountered in interpreting the enigmatic abbreviation *"crist."* in VON SCHLOTHEIM (1820) are resolved if the abbreviation is interpreted as standing for *cristacites* and if *cristacites* is regarded as a subgeneric name. Of the various items discussed in preceding pages the one that remains unresolved is the change from *Ostracites crista urogalli* VON SCHLOTHEIM (1813, p. 112) to *Ostracites crist. urogalli* VON SCHLOTHEIM (1820, p. 242). This change militates against interpreting the *crist.* as an abbreviation of *cristacites* and thereby makes the assumption that *crist.=cristacites* can be regarded as a subgenus an unlikely one, unless VON SCHLOTHEIM committed a *lapsus calami* here or made a deliberate unexplained change. One has to weigh this one unresolved difficulty against the numerous items resolved by these new interpretations.

QUESTIONS OF AVAILABILITY AND VALIDITY

The manner in which VON SCHLOTHEIM (1820) proposed *cristacites* as a subgenus of sorts is highly unusual from the point of view of present-day nomenclature. The name is spelled with a lower-case initial letter and is abbreviated as *crist.* in every instance. This raises many questions.

Can a subgeneric name be established acceptably if in the original publication it is abbreviated in every instance and if this abbreviation is not explained? That the name starts with a lower-case letter is perhaps not so serious an objection, because the newest *International Code of Zoological Nomenclature* nowhere states that a generic or subgeneric name must have an initial capital letter to be available when first published. It merely requires that the name once published must be corrected to start with a capital letter (*Code* Articles 17(6) and 28).

That the name was abbreviated in every case is possibly explainable. VON SCHLOTHEIM placed the abbreviations in the text under and following a clear center heading

"D. Cristaciten. (Hahnenkämme.)." Is not this sufficient explanation of the abbreviations? All a reader has to do is refer to the center heading. What was perfectly simple and obvious to the author when writing his book is not necessarily understandable to readers many years later.

The concept of a subgenus and the nomenclatural niceties needed to establish one were very uncertain in 1820. LINNÉ (1758) had some difficulties with them. These considerations probably explain the puzzling features of VON SCHLOTHEIM's *cristacites*.

Summarizing this investigation, I express the opinion that VON SCHLOTHEIM (1820, p. 240-245) had a subgeneric name of sorts in mind when he used *crist.* and *Cristaciten* (vernacular) in his work. However, the subgeneric name *cristacites* was introduced by VON SCHLOTHEIM in 1820 in such a dubious manner that whether it is available in nomenclature is uncertain.

Regardless of this uncertainty, the *"crist."* of VON SCHLOTHEIM, 1820, is herewith corrected to *Cristacites* VON SCHLOTHEIM, 1820, in accordance with *Code* Articles 28, 32, and 33. This correction is a justified emendation in the sense of Article 33(a)(i), irrespective of the availability of the name. Any and all statements made here are not to be construed as proposing or establishing a generic or subgeneric name.

Because of the obscure fashion in which VON SCHLOTHEIM introduced this subgeneric name it has remained unnoticed, unrecognized, and unused since 1823. *Cristacites* is listed neither by SHERBORN (1922-33) nor by NEAVE (1939-50). It is truly a *nomen oblitum* and to be rejected in accordance with *Code* Article 23(b).

CRISTACITES VON SCHLOTHEIM (1823)

The preceding discussions and conclusions concern only the work done by VON SCHLOTHEIM (1820) published in 1820. His later publication (VON SCHLOTHEIM, 1823) is a different matter and must be judged on its own merits.

In the later work the name *Cristacites* appears three times: 1) *Cristacites complanatus* and *difformis* (p. 75); 2) *Ostracites cristacit. complanatus* and —— ——

difformis (p. 82); 3) *Ostracites crist. difformis* (p. 111, in the explanation to pl. 36, fig. 2).

In the first of these places, *Cristacites* is used as a generic name and begins with a capital letter. In the other two it is abbreviated, does not begin with a capital letter, and is used as a subgeneric name of sorts.

According to *Code* Article 16(a)(v) the citation of one or more available specific names in combination with a new generic name constitutes an indication. Such an indication suffices to establish a generic name published before 1931 (Art. 12). All in all *Cristacites* VON SCHLOTHEIM (1823, p. 75) satisfies all requirements to become an available name. However, no one seems to have used *Cristacites* subsequently. After 143 years it remains truly an undetected name, a *nomen oblitum,* and might just as well stay that way.

REFERENCES

Bruguière, J. C.
(1) 1789-92, *Encyclopédie Méthodique; histoire naturelle des vers:* v. 1, pt. 1, p. 1-344, "1792" [1789]; pt. 2, p. 345-758 (1792) Panckoucke (Paris).
(2) 1791-1816, *Tableau encyclopédique et méthodique des trois règnes de la nature; vers testacées à coquilles bivalves:* v. 1, p. i-viii +1-83, pl. 1-95 (1791); p. 85-132, pl. 96-189 [1792]; pl. 190-286 (1797); pl. 287-390 [1798]; pl. 391-588 (1816), Panckoucke (Paris). The pl. 190-588 are attributed to LAMARCK.

Camper, A. G.
(3) 1800, *Lettre de A. G. Camper à G. Cuvier, sur les ossements fossiles de la montagne de St. Pierre, à Maestricht:* Jour. Physique, Chimie, Histoire Nat. Élément., Arts, v. 51, p. 278-291, 2 pl. [Not seen.]

Charles, R.-P., & Maubeuge, P.-L.
(4) 1951, *Les huîtres plissées jurassiques de l'est du Bassin Parisien:* Musée Histoire Nat. Marseille, Bull., v. 11 (1951), p. 101-119, 2 text fig., 3 pl.

Diener, Carl
(5) 1923, *Lamellibranchiata triadica:* Fossilium Catalogus 1: Animalia, pt. 19, 257+2 p. (Dec. 12).

Douvillé, Henri
(6) 1911, *Observations sur les ostréidés, origine et classification:* Soc. Géol. France, Bull., ser. 4, v. 10 (1910), pt. 7, p. 634-645, pl. 10-11 (May 2).

Faujas-Saint-Fond, Barthélemí
(7) 1799 [?1802-04], *Histoire naturelle de la montagne de Saint-Pierre de Maestricht:* 263 p., 54 pl., H. J. Jansen (Paris).

Fischer, Paul
(8) 1880-87, *Manuel de conchyliologie et de paléontologie conchyliologique ou histoire naturelle des mollusques vivants et fossiles suivi d'un appendice sur les brachiopodes par D. P. Oehlert:* xxiv+1369 p., 1138 text fig., frontispiece+23 pl. (September 21, 1880-June 15, 1887), F. Savy (Paris). [Fasc. 10, p. 897-1008, published April 30, 1886.]

Fischer de Waldheim, Gotthelf
(9) 1807, *Museum Demidoff. Mis en ordre systématique et descrit par G. Fischer, v. 3, Végétaux et Animaux:* ix+330 p., 6 pl., Imprimerie de l'Univ. Imp. (Moscou).

Gmelin, J. F.
(10) [1791], *Caroli a Linné Systema naturae per regna tria naturae:* edit. 13, v. 1, pt. 6, p. 3021-3910, G. E. Beer (Leipzig).

Goldfuss, G. A.
(11) 1826-44, *Petrefacta Germaniae tam ea, etc. . . . Abbildungen und Beschreibungen der Petrefacten Deutschlands und der angränzenden Länder unter Mitwirkung des Herrn Grafen zu Münster:* 3 pts., Arnz & Co. (Düsseldorf).

Haas, Fritz
(12) 1938, *Bivalvia, Teil II, 2. Lieferung:* H. G. Bronn's Klassen und Ordnungen des Tierreichs, v. 3, Mollusca, pt. 3, p. 209-466, text fig. 151-165; Akad. Verlag. (Leipzig).

I.C.Z.N.
(13) 1961, International code of zoological nomenclature adopted by the XV International Congress of Zoology, N. R. Stoll & others [edit. comm.]: Internatl. Comm. Zool. Nomenclature, 1961, xviii+176 p.; 2nd edit., 1964 (pagination unchanged) (London).

Iredale, Tom
(14) 1939, *Mollusca, Part I:* British Museum (Nat. History), Great Barrier Reef Exped. 1928-29, Sci. Rept., v. 5, no. 6, p. 209-425, 1 text fig., 7 pl. (Feb. 25).

Jourdy, le Général E.
(15) 1924, *Histoire naturelle des Exogyres:* Ann. Paléontologie (Marcellin Boule, director), v. 13, 104 p., 8 text fig., 11 pl.

Kauffman, E. G.
(16) 1965, *Middle and Late Turonian oysters of the Lopha lugubris group:* Smithsonian Misc. Coll., v. 148, no. 6, pub. 4602, 92 p., 18 text fig., 8 pl. (Oct. 6).

Kutassy, A.
(17) 1931, *Lamellibranchiata triadica II:* Fossilium Catalogus I: Animalia, pt. 51, 477 p. (Nov. 16).

Lamarck, J. B. A. P. M. de
(18) 1799, *Prodrome d'une nouvelle classification des coquilles, etc.:* Soc. Histoire Nat. Paris, Mém., v. 1, p. 63-91.
(19) 1801, *Système des animaux sans vertèbres, ou tableau général des classes, des ordres et des genres de ces animaux:* [edit. 1], viii+432 p., Deterville (Paris) (Jan.).
(20) 1806, *Suite des mémoires sur les fossiles des environs de Paris:* Annales du Muséum, v. 8, p. 166. [Not seen.]

Linné, Carl
(21) 1758, *Systema Naturae per regna tria naturae . . .:* edit. 10, v. 1, iv+824 p. and errata page, Laurentius Salvius (Stockholm).
(22) 1766-68, *Systema Naturae per regna tria naturae . . .:* edit. 12, 3 v. in 4°. [Not seen.]

Maire, Victor
(23) 1941, *Contribution à la connaissance de la faune de l'Oolithe Ferrugineuse Oxfordienne de Talant (Cote d'Or):* Soc. Géol. France, Bull., ser. 5, v. 10 (1940), pt. 7-9, p. 263-272 (Dec. 1941).

Mörch, O. A. L.
(24) 1850, *Catalogus conchyliorum quae reliquit C. P. Kierulf, MD, DR. Nunc publica auctione X Decembris MDCCCL Hafniae dividenda,* 33 p., 2 pl., Trier (Copenhagen).

Neave, S. A.
(25) 1939-50, *Nomenclator zoologicus:* 5 vols. separately paginated, Zool. Soc. London (London).

Pasteur, J. D.
(26) 1802, *Natuurlijke historie van den St. Pietersberg bij Maastricht door B. Faujas Saint Fond uit het Frensch:* v. 1, xii+185 p., 2+19 pl.; v. 2 [not seen], Amsterdam (Johannes Allart).

Pervinquière, Léon
(27) 1910a, *Ostrea carinata Lamarck, 1806:* Paleontologia Universalis, ser. 3, no. 2, fiche 197 (July 26).
(28) 1910b, *Quelques observations sur la nomenclature des ostracés, à propos de la classification phylogénétique exposée par M. H. Douvillé:* Soc. Géol. France, Comptes Rendus Sommaire des Séances, 1910, no. 13-14 (June 20, 1910), p. 119-120.
(29) 1911, *Quelques observations sur la nomenclature des ostracés, à propos de la classification phylogénétique exposée par M. H. Douvillé:* Same, Bull., ser. 4, v. 10 (1910), pt. 7, p. 645-646 (May 2).

Regteren Altena, C. O. van
(30) 1957, *Achttiende-eeuwse verzamelaars van fossielen te Maastricht en het lot hunner*

collecties (with English summary): Natuurhist. Genoot. Limburg, Publ. no. 9 (1956),
p. 83-112, 7 text fig. (Feb.).
(31) 1963, *Nieuwe gegevens over achttiende-
eeuwse verzamelaars van fossielen te Maastricht:* Natuurhist. Maandblad, v. 52, no.
2, p. 28-32, 1 text fig. (Feb. 28).

Röding, P. F.
(32) 1798, *Museum Boltenianum sive catalogus
cimeliorum e tribus regnis naturae, quae
olim collegerat Joa. Fried. Bolten:* v. 2,
viii+199 p., Typus Johan Christi Trapii
(Hamburg).

Rollier, Louis
(33) 1911, *Les faciès du Dogger ou Oolithique
dans le Jura et les régions voisines:* Fondation Schnyder von Wartensee à Zurich,
Mém. 18, v+352 p., 56 text fig., Georg &
Cie (Geneva, Basel).
(34) 1917, *Fossiles nouveaux ou peu connus des
terrains secondaires (Mesozoïques) du Jura
et des contrées environnantes, v. 1, pt. 6:*
Soc. Paléont. Suisse, Mém., v. 42 (1917),
p. 501-634+errata page, pl. 33-40.

Rumphius [Rumpf], G. E.
(35) 1705, *D'Amboinsche Rariteitkammer . . .:*
xxx+340+liii p., frontispiece, portrait, 60
pl., F. Halma (Amsterdam).

Schlottheim [Schlotheim], E. F. von
(36) 1813, *Beiträge zur Naturgeschichte der Versteinerungen in geognostischer Hinsicht:* in
Taschenbuch für die gesammte Mineralogie.,
etc., C. C. Leonhard (ed.), v. 7, I. Abh.,
p. 3-134, pl. 1-4. [Although the author
later spelled his name with one t, it is here
SCHLOTTHEIM.]
(37) 1820, *Die Petrefactenkunde auf ihrem jetzigen Standpunkte durch die Beschreibung
seiner Sammlung versteinerter und fossiler
Überreste des Thier- und Pflanzenreichs der
Vorwelt erläutert:* lxii+437 p., text vol.
only, Becker (Gotha).
(38) 1823, *Nachträge zur Petrefactenkunde,
Zweyte Abtheilung:* i+114 p., text; pl.
22-37, atlas, Becker (Gotha).

Schroeter, J. S.
(39) 1782, *Lithologisches Real- und Verballexi-
kon:* in 8 vol.; v. 5, p. 74, 382, 390 (1782),
Varrentrapp Sohn & Wenner (Frankfurt am
Main).
(40) 1783-86, *Einleitung in die Conchylienkennt-
niss nach Linné:* 3 vol.; v. 3 (1786), xvi+
596 p., pl. 8-9, J. J. Gebauer (Halle).

Sherborn, C. D.
(41) 1902, *Index Animalium . . .,* sec. 1: lix+
1195 p., Cambridge Univ. Press (Cambridge).
(42) 1922-23, *Index Animalium . . .,* sec. 2: 33
pts., British Museum (Natural History)
(London).

Silvestri, Alfredo
(43) 1929, *Protozoi Cretacei ricordate e figurati
da B. Faujas de Saint-Fond:* Accad. Pontif.
Nuovi Lincei, Atti, v. 82, Fasc. suppl., p.
327-343, 9 text fig., 1 pl.

Sowerby, James, & Sowerby, J. de C.
(44) 1812-46, *The mineral conchology of Great
Britain:* 7 vol., the author (London).

Stenzel, H. B.
(45) 1947, *Nomenclatural synopsis of supraspecific groups of the family Ostreidae (Pelecypoda, Mollusca):* Jour. Paleontology, v. 21,
no. 2, p. 165-185 (April 21).
(46) 1959, *Cretaceous oysters of southwestern
North America:* Internatl. Geol. Congress,
20th Sess., Mexico City (1956), Symposium
del Cretacico, p. 15-38, 19 text fig.

Winkler, T. C.
(47) 1863-67, *Catalogue systématique de la collection paléontologique:* viii+697 p., livr.
1 (1863), livr. 2 (1864), livr. 3 (1865),
livr. 4 (1865), livr. 5 (1866), livr. 6 (1867),
Musée Teyler, Teyler's Stichting (Haarlem).

Woods, Henry
(48) 1913, *A monograph of the Cretaceous Lamellibranchia of England, v. 2, pt. 9:* Palaeontograph. Soc. (1912), v. 66, p. 341-
473, pl. 55-62 (Feb.).

PART N ERRATA AND REVISIONS

p. N13. In caption for Fig. 10, for *Venus cam-
pichiensis mortoni* (CONRAD), read: *Mercenaria
campechiensis mortoni* (CONRAD). [H. B.
STENZEL]

p. N70. In caption for Fig. 58, for *Exogyra columba* (LAMARCK), read: *Rhynchostreon suborbiculatum* (LAMARCK, 1801). [H. B. STEN-
ZEL]

p. N93. In caption for Fig. 77, for *Ostrea virgin-*

ica, read: *Crassostrea virginica.* [H. B. STEN-
ZEL]

p. N94. In caption for Fig. 78, for *Ostrea virginica,* read: *Crassostrea virginica.* [H. B. STEN-
ZEL]

p. N235. For **Saturnia** SEGUENZA, 1877, read:
Neilonella DALL, 1881, p. 126 [*Leda (Neilonella) corpulenta;* OD] [=*Saturnia* SEGUENZA,
1877, p. 1178 (type, *Nucula pusio* PHILIPPI,

1844; M) (*non Saturnia* SCHRANK, 1802, Lepidoptera); *Austrotindaria* FLEMING, 1948 . . . .].
For **S.** (**Saturnia**), read: **N.** (**Neilonella**) . . . FIG. A5,*1*. *N.* (*N.*) *corpulenta* DALL, . . .
For **S.** (**Spinula**), read: **N.** (**Spinula**), . . .
For **S.** (**Tindariopsis**), read: **N.** (**Tindariopsis**), . . . [LEE MCALESTER]

p. *N*239. For FIG. A8,*2*, read: *Phestia*. For FIG. A8,*3*, read: *Paleyoldia*. Corresponding corrections belong to captions given with systematic text (p. *N*237, p. *N*239). [LEE MCALESTER]

p. *N*267, col. 1, line 11. For WARMKE & ABBOTT, 1961, read: STENZEL, KRAUSE, & TWINING, 1957. [H. B. STENZEL]

p. *N*289, col. 2. Under **Atomodesma,** for VON BEYRICH, 1864, read: VON BEYRICH, 1865. [CURT TEICHERT]

p. *N*292. In caption of Fig. C29, for Mayalinidae, read: Myalinidae. [JOHN WEIR]

p. *N*295, col. 1. Under ?**Dictys,** read: KHALFIN, not KHAFLIN. [JOHN WEIR]

p. *N*306, col. 1. For ?**Stefaninia** VENZO, 1934, p. 165 [*Gervilleia? ogilviae* BITTNER, 1895, p. 88; SD Cox herein], read: ?**Stefaninia** COX, 1969 (herein) *ex* VENZO, 1934 [*Gervilleia? ogilviae* BITTNER, 1895; OD]. [Availability of the generic name *Stefaninia,* including its authorship and date, was not established according to ICZN *Code* by VENZO in 1934 because he failed to designate a type species (Art. 13,b) even though he did provide statement of characters presumed to distinguish the genus (Art. 13,a,i). Cox (1969) was first to comply with stipulations of the *Code* for post-1930 generic names and thus is to be cited as the author of *Stefaninia.*] [R. M. JEFFORDS]

p. *N*312, col. 1. For **Hoernesiella** GUGENBERGER, 1934, p. 46 [*H. carinthiaca;* SD Cox herein], read: *Hoernesiella* COX, 1969 (herein) *ex* GUGENBERGER, 1934 [*H. carinthiaca* GUGENBERGER, 1934, p. 46; OD]. [Correction for same reasons as applicable to *Stefaninia* (p. *N*306).] [R. M. JEFFORDS]

p. *N*382, col. 2, line 9. Delete entire line. [MYRA KEEN]

p. *N*385, col. 2. Under Family Limidae, lines 2 and 3 should read: valve in one subgenus), small and moderately thin to large and thick-shelled (*Ctenostreon*), ovate, orbicular or sub-. [L. G. HERTLEIN]

p. *N*389, col. 1. Under **L.** (**Limaria**), for LAMY, 1833, read: LAMY, 1930. [MYRA KEEN]

p. *N*405. In caption of Fig. D10, delete *N*407. [JOHN WEIR]

p. *N*407, col. 2. Under ?**Palaeanodonta,** for EICHWALD, 1895, read: EICHWALD, 1859 (*fide* F. A. BATHER; 1861, *fide* L. R. COX). [JOHN WEIR]

p. *N*409, col. 1. For **Abiella** RAGOZIN, 1933, read: **Abiella** RAGOZIN, 1955 [*Posidonomya concinna* JONES, 1901; OD]. [Correction required for same reasons cited as applicable to *Stefaninia,* p. *N*306.] [R. M. JEFFORDS]

p. *N*409, col. 2. Under **Palaeomutela,** for *P. verneuilli,* read: *P. verneuili.* [JOHN WEIR]

p. *N*410, col. 2. For **Ferganoconcha** CHERNYSHEV, 1937, p. 18, read: **Ferganoconcha** LUMKEVICH *et al.,* 1960, p. 99 [*F. sibirica* CHERNYSHEV, 1937; OD]. [Correction required for same reasons cited as applicable to *Stefaninia,* p. *N*306.] [R. M. JEFFORDS]

p. *N*411, col. 2. For ?**Tutuella** RAGOZIN, 1938, p. 106, read: ?**Tutuella** LUMKEVICH *et al.,* 1960, p. 99 [*T. chachlovi* RAGOZIN, 1938; OD]. [Correction required for same reasons cited as applicable to *Stefaninia,* p. *N*306.] [R. M. JEFFORDS]

p. *N*411, col. 2. For ?**Utschamiella** RAGOZIN, 1938, p. 138, read: ?**Utschamiella** LUMKEVICH *et al.,* 1960, p. 99 [*U. tungussica* RAGOZIN, 1938; OD]. [Correction required for same reasons cited as applicable to *Stefaninia,* p. *N*306.] [R. M. JEFFORDS]

p. *N*489, col. 2. For **Sainschandia** MARTINSON, 1957, p. 287, read: **Sainschandia** MARTINSON, 1961, p. 209 [*S. turensis* MARTINSON, 1957; OD]. [Correction required for same reasons cited as applicable to *Stefaninia,* p. *N*306.] [R. M. JEFFORDS]

p. *N*492. **Lucina** BRUGUIÈRE, 1797, for type species read: *Venus edentula* LINNÉ, 1758; SM LAMARCK, 1799 (=*Anodontia alba* LINK, 1807, type-species of *Anodontia,* OD). Application to ICZN submitted by MYRA KEEN and R. TUCKER ABBOTT for conservation of *Lucina* in sense accepted by CHAVAN. At bottom of col. 2 substitute footnote statement reading: Generic names published with figures but no descriptions by DESHAYES, 1857, are available, for the zoological *Code* (Art. 16,a,i) stipulates that pre-1931 names accompanied by illustrations but no descriptions are validly "indicated." [MYRA KEEN]

p. *N*494. For type species of **Callucina** DALL, 1901, read: [*Callilucina keenae* CHAVAN, herein (*pro Lucina radians* CONRAD, 1841, *non* BORY DE ST. VINCENT, 1824)]. [ANDRÉ CHAVAN]

p. *N*494, col. 2. Under **Ctena,** add in synonymy with *Lucina pectinata* CARPENTER, 1857: (= *Codakia mexicana* DALL, 1901). [MYRA KEEN]

p. N499, col. 2. Under **Myrtea,** for *Cyrachaea* LEACH, 1819, read: *Cyrachaea* LEACH in GRAY, 1847. [ANDRÉ CHAVAN]

p. N500, col. 2. Under **Lucinoma,** line 4, add after 1846): (type, *Venus borealis* LINNÉ, 1766; M). [ANDRÉ CHAVAN]

p. N502, col. 1. Under **Miltha,** for [*Lucina childreni* GRAY, 1825; OD], read: [*Lucina childrenae* GRAY, 1825 (=*L. childrinae* GRAY, 1824 (misspelling); *L. childreni* GRAY, 1825); M]. In third line, for *neozelandica,* read: *neozelanica.* [ANDRÉ CHAVAN]

p. N508, col. 2. Under **Thyasira,** for *Bequania* LEACH in BROWN, 1827, read: *Bequania* LEACH in BROWN, 1844. After *Ptychina* PHILIPPI, 1836, add: (type, *P. biplicata;* M). [ANDRÉ CHAVAN]

p. N518. For Superfamily Leptonacea Gray, 1847, read: Superfamily Galeommatacea Gray, 1840 [*nom. transl.* VOKES, 1967 (*ex* Galeommatidae GRAY, 1840, *nom. correct.* DALL, 1899, *pro* Galeommidae GRAY, 1847, =Galeommidii GRAY, 1840)] [=Leptonacea GRAY, 1847; Erycinacea FISCHER, 1887]. [MYRA KEEN]

p. N523, col. 1. Under **B. (Byssobornia),** add to fourth line: Japan. [ANDRÉ CHAVAN]

p. N525, col. 2. Under **Pseudopythina,** for P. FISCHER in DI MONTEROSATO, 1884, read: P. FISCHER, 1878. Separate into subgenera **P. (Pseudopythina)** and **P. (Borniopsis).** Characteristics of **P. (Borniopsis)** include: Inflated beaks, ligamental socket elongate and deep, with protruding lower margin; tooth *1* well separated, *2* thicker than in *Pseudopythina,* with its posterior end enveloped by peculiar prolongation of the lunular margin; *4b* more distinct and directed somewhat backward. [ANDRÉ CHAVAN]

p. N531, col. 1. To description of **Montacutona,** add: Fine radials sometimes apparent, pallial line well marked. [ANDRÉ CHAVAN]

p. N533, col. 1. For **Galeomma** SOWERBY in TURTON, 1825 [**G. turtoni* SOWERBY, 1825, read: **Galeomma** TURTON, 1825 [**G. turtoni* SOWERBY *et al.* in TURTON, 1825. [TURTON proposed the new genus *Galeomma* with mention that the species also was new, but without giving it a specific name. The editors printed his article as submitted but in a footnote proposed the specific name *G. turtoni.* No editorial names were cited; later authors have inferred that it was SOWERBY, but on the title page of the journal his is one of several listed. I therefore suggest that authorship of the species be cited as "SOWERBY *et al.* in TURTON." The

generic name is definitely TURTON's alone. It could be interpreted as a genus without named species, in which case the first specific name would be by "The Editors, Zoological Journal."] [MYRA KEEN]

p. N535, col. 2. Under **Lactemiles,** add: Prosogyrous beaks. [ANDRÉ CHAVAN]

p. N546, col. 2. Under **?Redonia,** line 5, delete: and transversely striated. [ANDRÉ CHAVAN]

p. N553. Fig. E53,2a,b, should read *Carditella,* not *Coripia.* [ANDRÉ CHAVAN]

p. N554, col. 1. Under **P. (Coripia),** line 4, for ligament partly internal, read: *3b* elongate (more than in *Pteromeris*). [ANDRÉ CHAVAN]

p. N561, col. 2. Under **?Aenigmoconcha,** last line, for RAGOZIN, 1955, read: BENEDICTOVA, 1955. Under **?Yavorskiella,** last line, for RAGOZIN, 1955, read: BENEDICTOVA, 1955. [JOHN WEIR]

p. N592, col. 1. Under **Eoprosodacna** delete [= *Limnopappia* SCHLICKUM, 1962 (type, *L. schuetti;* OD)]. Same, under **E. (Succuridacna)** KOROBKOV, 1954 delete [=*Limnopagetia* SCHLICKUM, 1963 (type, *Cardium friabile* KRAUSS, 1852; OD)]. Same, next after **Limnodacna** EBERSIN, 1936, add following new paragraphs:

Limnopagetia SCHLICKUM, 1963 [**Cardium friabile* KRAUSS, 1852; OD]. *U.Mio.,* Eu. (Ger.).

Limnopappia SCHLICKUM, 1962 [**L. schuetti;* OD]. *U.Mio.,* Eu.(Ger.). [SCHLICKUM & ČTYROKÝ (1965) recommended that *Eoprosodacna, Succuridacna, Limnopagetia,* and *Limnopappia* should be regarded as distinct genera grouped in a subfamily named Limnopappiinae SCHLICKUM, 1963.] [MYRA KEEN]

p. N593. In caption for Fig. E89 delete Lahillidae, substituting for it: Lymnocardiidae. [MYRA KEEN]

p. N608, col. 2. For MESODESMATIDAE Gray, 1839, line 7, read: MESODESMATIDAE Gray, 1840; line 8, for GRAY, 1839, read: GRAY, 1840. [MYRA KEEN]

p. N638. Delete *Solecurtellus* GHOSH, 1920, as synonym of *Solecurtus* DE BLAINVILLE, 1824. [MYRA KEEN]

p. N639. Add *Solecurtellus* GHOSH, 1920, as synonym of *Tagelus* GRAY, 1847. [MYRA KEEN]

p. N668. For DESHAYES, 1858 (col. 2, line 4) read: DESHAYES, 1855. [MYRA KEEN]

p. N675. **Pitar (Omnivenus)** and **P. (Rhabdopitaria)** were studied by STENZEL (in STENZEL, KRAUSE, & TWINING, 1957, p. 151-154) who proved that the two are synonyms and related closer to *Mercenaria* than to *Pitar*. *Rhabdopitaria* was selected as the name to use and regarded as a genus related to *Mercenaria*. Contrary to statements made by PALMER and KEEN, the nymphs of both, *Rhabdopitaria* and *Omnivenus*, are rugose. The types of the type species of *Rhabdopitaria* have rugose nymphs as STENZEL ascertained through personal inspection. [H. B. STENZEL]

p. *N688*. Insert *Rhabdopitaria* as follows:

Rhabdopitaria PALMER, 1927 [**Callocardia astartoides* GARDNER, 1923; OD] [=*Omnivenus* PALMER; 1927 (type, *Cytherea discoidalis* CONRAD, 1833; OD)]. Shell smooth but middle layer radically ribbed, because it represents growth tracks of marginal crenulations; inside valve margins crenulated along entire periphery; nymphs rugose as in *Mercenaria (711;* Stenzel, Krause, & Twining, 1957). *Eoc.,* E.N.Am. [H. B. STENZEL]

p. *N779*, col. 1. For **Epidiceras** DOUVILLÉ, 1936, p. 332, read: **Epidiceras** DECHASEAUX, 1952, p. 326 [**Diceras sinistrum* DESHAYES, 1824, p. 466; OD]. [Correction required for same reasons cited as applicable to *Stefaninia*, p. *N306*.] [R. M. JEFFORDS]

p. *N803*, col. 2. Delete **Pseudobarretia** MÜLLERRIED, 1931, p. 255 [**P. chiapasensis (nom. nud.);* OD]. [Name unavailable.] [R. M. JEFFORDS]

p. *N857*, col. 2, line 6 from bottom. For 1843, read: 1843 (1844). [MYRA KEEN]

p. *N858*, col. 1, line 4. Under Clavagellidae, for 1843, read: 1843 (1844) (often erroneously cited as 1843). [MYRA KEEN]

p. *N860*, col. 2. **Avardaria** is a gastropod genus. For **Cardiarlus**, read: **Cardiarius**. **Cardiarius** DUMÉRIL, 1806, is a *nom. van.* for **Cardium** LINNÉ, 1758. **Cartissa** is a *nom. null.* for *Cardissa* MEGERLE VON MÜHLFELD, 1811 (syn. of **Corculum** RÖDING, 1798, p. *N588*). [MYRA KEEN]

p. *N861*, col. 1. **Suchumica** is a gastropod genus. [MYRA KEEN]

p. *N925*, col. 1. Following Elimata, **N389,** add *Elizia*, N633. [MYRA KEEN]

p. *N931*, col. 1. For *Kymatox*, N601, read: *Kymatox*, N606.

p. *N937*, col. 3. For Nyassa, **N407,** read: Nyassa, **N411.** [JOHN WEIR]

p. *N937*, col. 3. Following *Odontocineta*, N850, add Odontogena, **N523.** [MYRA KEEN]

ADDITIONAL REFERENCES

p. *N870*:

Benediktova [Benedictova], R. N.

(35a) 1955, *Plastinchatozhabernye Gorlovskovo Basseina:* in L. L. Khalfin, Atlas rukovodyashchykh form iskopaemykh faunu i floru zapadnoi sibiri [Atlas of leading forms of the fossil fauna of western Siberia], v. 2, p. 39-42, pl. 7, Zapadno-Sibir.Geol. Uprav.-Tomsk Politekh. Inst., Gosudar. Nauch.-Tekh. Izd. Lit. Geol. i Okhrane Nedr (Moskva). [*Lamellibranchs of the Gorlovsk Basin.*] [JOHN WEIR]

p. *N897*:

Ragozin, L. A.

(771a) 1955, *Plastinchatozhabernye Kuznetskovo Basseina:* in L. L. Khalfin, Atlas rukovodyashchykh form iskopaemykh faunu i floru zapadnoi sibiri [Atlas of leading forms of the fossil fauna and flora of western Siberia], v. 2, p. 8-38, text fig. 1-45, pl. 1-6, Zapadno-Sibir. Geol. Uprav.-Tomsk. Politekh. Inst., Gosudar. Nauch.-Tekh. Izd. Lit. Geol. i Okhrane Nedr (Moskva). [*Lamellibranchs of the Kuznetsk Basin.*] [JOHN WEIR]

p. *N902*:

Stenzel, H. B., Krause, E. K., & Twining, J. T.

(885a) 1957, *Pelecypoda from the type locality of the Stone City Beds (Middle Eocene) of Texas:* Univ. Texas Publ. 5704, 237 p., 31 text fig., 22 pl. (Feb. 15). [H. B. STENZEL]

INDEX

Italicized names in the following index are considered to be invalid; those printed in roman type, including morphological terms, are accepted as valid. Suprafamilial names are distinguished by the use of full capitals and author's names are set in small capitals with an initial large capital. Page references having chief importance are in boldface type (as **N327**). Some divergences in classification reflect differences of authors concerning validity of nomenclature.

POSTSCRIPT

By H. B. Stenzel

(Louisiana State University, Baton Rouge)

I am grateful to the Editor for allowance to place these few lines at the end of *Treatise* Part N, Volume 3, even after its index had been completed. Their purpose is to take account of an interesting article by N. D. Newell and D. W. Boyd entitled *Oyster-like Permian Bivalvia* (*American Museum Natural History,* Bulletin, vol. 143, art. 4, December, 1970). Issued too late for my attention in preparing *Treatise* materials on fossil oysters, its descriptions and discussions of numerous taxonomic units, accompanied by exceptionally fine illustrations, are worthy of close study not only by workers on oysters but by paleontologists generally.

Commendation of the Newell and Boyd contribution needs qualifications to the extent of objection to its use of some new morphological terms for oysters and their shells without any indication that they come from my own work on oysters for the *Treatise.* Also, a new genus of mine (*Hyotissa*), intended for first publication in the *Treatise* (p. N1107) was mentioned by Newell and Boyd (p. 226). I can only infer that the unfortunate situation came about as a result of oversight by Newell of observations on my *Treatise* typescript and illustrations referred to him in 1966-67 by Moore for editorial assistance.

I am gratified to notice that Newell and Boyd agree with several of the major results of my own work in volume N-3, for instance, in that the oysters, as commonly understood by various authors, are not a monophyletic family, but consist of two families (Ostreidae *sensu stricto* and Gryphaeidae *nom. transl.* Stenzel, herein) and that *Gryphaea* need not be derived from *Liostrea* or an *Ostrea*-like ancestor, but may be descended without an intermediary genus directly from a genus of the Pseudomonotidae.

Newell writes (letter to me of March 30, 1971): "I am sure that I learned of these things in our several conversations. Evidently, I absorbed much knowledge from you without distinguishing your original contribution from the general store of common knowledge."

April 2, 1971